Managed Pressure Drilling
Modeling, Strategy and Planning

Wilson C. Chin, Ph.D.
Stratamagnetic Software, LLC
Houston, Texas

ELSEVIER

AMSTERDAM • BOSTON • HEIDELBERG • LONDON
NEW YORK • OXFORD • PARIS • SAN DIEGO
SAN FRANCISCO • SINGAPORE • SYDNEY • TOKYO

Gulf Professional Publishing is an imprint of Elsevier

Gulf Professional Publishing is an imprint of Elsevier
225 Wyman Street, Waltham, MA 02451, USA
The Boulevard, Langford Lane, Kidlington, Oxford, OX5 1GB, UK

Notices

Knowledge and best practice in this field are constantly changing. As new research and experience broaden our understanding, changes in research methods, professional practices, or medical treatment may become necessary.

Practitioners and researchers must always rely on their own experience and knowledge in evaluating and using any information, methods, compounds, or experiments described herein. In using such information or methods they should be mindful of their own safety and the safety of others, including parties for whom they have a professional responsibility.

To the fullest extent of the law, neither the Publisher nor the authors, contributors, or editors, assume any liability for any injury and/or damage to persons or property as a matter of products liability, negligence or otherwise, or from any use or operation of any methods, products, instructions, or ideas contained in the material herein.

Library of Congress Cataloging-in-Publication Data
Chin, Wilson C.
  Managed pressure drilling: modeling, strategy and planning/Wilson C. Chin.
    p. cm.
  Includes bibliographical references and index.
  ISBN 978-0-12-385124-6
  1. Managed pressure drilling (Petroleum engineering) 2. Oil well drilling.
3. Wells—Fluid dynamics—Mathematical models. I. Title.
TN871.26.C45 2012
622′.3381–dc23   2011035451

British Library Cataloguing-in-Publication Data
A catalogue record for this book is available from the British Library.

For information on all Gulf Professional Publishing publications,
visit our website at *www.elsevierdirect.com*

Contents

Preface

My first exposure to the importance of good hole cleaning and pressure analysis occurred in 1981 when I was initiated into the petroleum industry, having left the aerospace industry, for which I had trained diligently. The new subject matter was not glamorous, to say the least, but years later I would come to understand its significance in both drilling and cementing. The advent of deviated and horizontal wells elevated the role of annular flow in oilfield operations.

A decade later, I published my first book on borehole flow modeling, introducing the use of curvilinear grid systems to accurately capture the physics. Over the years, this effort was self-funded and undertaken as a labor of love. However, another decade later I launched my consulting company, Stratamagnetic Software, LLC, supported by the U.S. Department of Energy through its Small Business Innovation Research Program, under Grant DE-FG03-99ER82895, to improve grid generation techniques for the oil industry. Related work in this area with several clients continued over the years in different and varied applications.

In 2009, the Department of Energy awarded a contract to support my technical proposal "Advanced Steady-State and Transient, Three-Dimensional, Single and Multiphase, Non-Newtonian Simulation System for Managed Pressure Drilling." This comprehensive effort was administered by the Research Partnership to Secure Energy for America (RPSEA) through its Ultra-Deepwater Program under Subcontract No. 08121-2502-01. This award enabled my colleagues and I to "tie up loose ends" and integrate numerous models developed over two decades. More important, it provided us the opportunity to significantly extend our models in numerous directions—rotating flow, fully transient effects, three-dimensionality, multiphase, and so on—and to perform research and develop software models that we felt would have a lasting influence on the petroleum industry.

We are very fortunate that many in the industry have recognized our efforts. Aside from those who have provided us this source of important funding, anonymous reviewers have made it possible for us to publish five recent papers: four for the American Association of Drilling Engineers (AADE) National Technical Conference and Exhibition, during April 2011 in Houston and one for the Offshore Technology Conference during May 2011, also in Houston. We are of course gratified that Gulf Professional Publishing/Elsevier has agreed to publish this book, *Managed Pressure Drilling: Modeling, Strategy and Planning*, which will no doubt achieve wide dissemination of our ideas.

Consistent with my belief that scientific research should be openly shared by industry, this book and the papers my colleagues have presented disclose all elements of the new annular flow models: mathematical theory, numerical implementation, source code examples, and computational validations, with comparisons to laboratory and field data and results whenever possible. Because of our research focus, and because our ideas are always evolving, the methods developed here and implemented in software are provided "as is" and no claim is made that they address all potential technical issues.

It is hoped, however, that others will study the models and help to improve them through use and research. Over the next several months, the plan is to widely disseminate the software, on which great effort has been expended in order to optimize the user's experience through a versatile and intuitive interface so we can obtain the feedback needed to support continued product development. Access to the fully functional software system flow simulation modules executable over the Internet are available from the book's website at *gulfpp.com/9780123851246*

I am deeply appreciative of the U.S. Department of Energy and the Research Partnership to Secure Energy for America for the opportunity they have provided me to work in this exciting technology area, and I look forward to a long collaboration with them and all interested parties.

Acknowledgments

My colleagues and I gratefully acknowledge the U.S. Department of Energy for its support of our technical proposal "Advanced Steady-State and Transient, Three-Dimensional, Single and Multiphase, Non-Newtonian Simulation System for Managed Pressure Drilling" during the period 2009 through 2011. This support was administered and directed by the Research Partnership to Secure Energy for America (RPSEA) through its Ultra-Deepwater Program under Subcontract No. 08121-2502-01. Our curvilinear grid generation research was also supported by the U.S. Department of Energy, under Small Business Innovation Research Grant DE-FG03-99ER82895 from 1999 through 2000.

We thank all of the industry partners we have been privileged to work with for motivating many of the problems and methods addressed in this research. We are indebted to Art Schroeder of Energy Valley, to Jim Chitwood of Chevron, and to James Pappas of RPSEA for their encouragement and advice. We are especially grateful to John Lofton of Chevron for his engineering insights and guidance related to several areas of our modeling of rotating pipe flow effects. James Pappas, in particular, contributed significantly to the manuscript through his meticulous reading and valued comments.

Finally, we thank Ken McCombs, senior acquisitions editor at Elsevier, for his interest in the book and for his support and encouragement throughout the research and writing. Marilyn Rash, with assistance from Dianne Wood and Deborah Prato; Jill Leonard; and other Elsevier staff contributed greatly to the editorial and production efforts and their labors are deeply appreciated.

The views expressed here, of course, are my own and are not necessarily the opinions of any program sponsors or individuals.

About the Author

Wilson C. Chin earned his Ph.D. from the Massachusetts Institute of Technology and his M.Sc. from the California Institute of Technology. His early interests focused on applied mathematics, fluid mechanics, and electrodynamics. Prior to founding Stratamagnetic Software, LLC, in Houston in 1999, he was affiliated with Boeing Aerospace, United Technologies, Schlumberger Anadrill, BP Exploration, and Halliburton.

Mr. Chin has authored more than 80 technical papers, received almost 20 U.S. patents in oilfield technology, and won 5 Department of Energy awards. In addition, he has written 8 other textbooks on advanced research in petroleum technology:

- *Borehole Flow Modeling in Horizontal, Deviated and Vertical Wells* (Gulf Publishing, 1992)
- *Modern Reservoir Flow and Well Transient Analysis* (Gulf Publishing, 1993)
- *Wave Propagation in Petroleum Engineering, with Applications to Drillstring Vibrations, Measurement-While-Drilling, Swab-Surge, and Geophysics* (Gulf Publishing, 1994)
- *Formation Invasion, with Applications to Measurement-While-Drilling, Time Lapse Analysis and Formation Damage* (Gulf Publishing, 1995)
- *Computational Rheology for Pipeline and Annular Flow* (Elsevier, 2001)
- *Quantitative Methods in Reservoir Engineering* (Elsevier, 2002)
- *Formation Testing Pressure Transient and Contamination Analysis* (E&P Press, 2008)
- *MWD Signal Analysis, Optimization and Design* (E&P Press, 2011)

Mr. Chin may be contacted by email at wilsonchin@aol.com or by telephone at (832) 483-6899.

Fluid Mechanics Challenges and Technology Overview

The author's earlier book on annular flow, entitled *Borehole Flow Modeling in Horizontal, Deviated and Vertical Wells* (Gulf Publishing, 1992), was the first to use boundary-conforming, curvilinear grid systems to host highly eccentric annular cross sections that contained cuttings beds, washouts, and local fractures. That work also addressed related problems—for example, helical flow and steady rotation of Power law fluids in concentric annuli, as well as the role of barite sag in promoting local recirculation zones that dangerously block oncoming mud.

Ten years later, the second edition, renamed *Computational Rheology for Pipeline and Annular Flow* (Elsevier, 2001), expanded the initial scope to model effects such as borehole axis curvature, flows in noncircular pipe (nonannular) ducts, and half-clogged annular domains. These two works focused on steady, two-dimensional flows without rotation and then single-phase non-Newtonian rheologies without yield stress. Many of the algorithms have been adopted by operating and oil service companies over the past two decades.

The present book, which summarizes major improvements in accuracy, speed, and engineering focus, represents a significant contribution that renders the prior works almost obsolete. Even so, the curvilinear grid technology employed in the early books remains state of the art and thus provides the mathematical foundation for the newer algorithms developed here. Improvements in formulation and solution accuracy are provided, but the new book substantially extends the range of problem-solving capabilities.

The present work gained significant impetus with the award of U.S. Department of Energy Contract No. 08121-2502-01 for "Advanced Steady-State and Transient, Three-Dimensional, Single and Multiphase, non-Newtonian Simulation System for Managed Pressure Drilling," administered by the Research Partnership to Secure Energy for America (RPSEA). This award provided the opportunity to integrate past work, tie up "loose ends," introduce new extensions, and provide the software platform to bring much-needed algorithms to the industry for deepwater drilling and cementing applications.

Although the prior works are by no means old, at least chronologically, the methods developed therein are often cited as "new." However, in the context of this book they are antiquated and entirely replaced by newer algorithms. The methodologies here are described in their entirety so that interested researchers can develop, improve, and extend the models from first principles. They are "open source" in this sense. To set the stage for the presentations that follow, we explain the limitations behind the previous works as well as extensions that have appeared in papers published up to the late 2000s.

The recent works in Chin and Zhuang (2011a,b,c,d), presented at the American Association of Drilling Engineers (AADE) National Technical Conference and Exhibition in Houston in April 2011; the paper by Chin and Zhuang (2011e), given at the Offshore Technology Conference in Houston in May 2011; and the work by Chin and Zhuang (2010), presented at the CPS-SPE International Oil & Gas Conference and Exhibition in Beijing in June 2010, *do* represent up-to-date contributions. However, the present book provides much more information than is available in the published summaries, in addition to special solutions that have not yet appeared in print.

CHALLENGES IN ANNULAR FLOW MODELING

The problems confronting borehole flow modeling are numerous. First, the governing partial differential equations are highly nonlinear and difficult to solve: Classical superposition methods do not apply. This is so even for flows without rotation, where a single equation for axial velocity is found. When rotation exists, azimuthal flow coupling generally leads to numerical instabilities, which have only recently been satisfactorily addressed. When multiphase effects are considered, difficulties in the solution process are compounded by the introduction of a convective–diffusive equation for species concentration. Depending on the problem, steady solutions require stable iterative algorithms, whereas transient solutions require robust algorithms admitting larger integration time steps.

Second, annular geometries are complicated. A highly eccentric geometry typical of those encountered in the drilling and cementing of modern deviated and horizontal wells is shown in part (a) of Figure 1.1. A less than ideal annulus is sketched in part (f) of the figure, which indicates a washout, although fractures and cuttings beds are also possible. The governing equations must be solved for practical geometries, and satisfactory coordinate systems provide the key to success. In engineering simulation, solutions for reservoir flow from single wells, for instance, are developed naturally with circular coordinates. Temperatures in rectangular plates, in contrast, are obtained in rectangular coordinates.

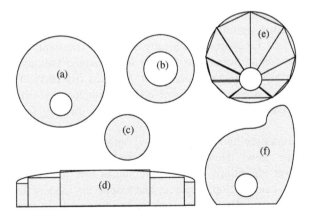

FIGURE 1.1

Real and idealized annular geometry models.

For parts (a) and (f) in Figure 1.1, neither circular nor rectangular variables apply. Therefore, researchers have introduced simplifying methods that render the equations amenable to solution. For instance, the concentric model in part (b) of Figure 1.1 can be solved for steady Power law flows; however, yield stress formulations have so far defied rigorous analysis, with solutions available only for circular pipe flows. Out of necessity, real annuli are crudely modeled by "close" concentric annular flows.

Equivalent hydraulic radii approaches model the eccentric annulus as an equivalent pipe flow, as suggested in part (c) of Figure 1.1. Such approaches are completely ad hoc and cannot be extended to other situations. Slot flow methods are suggested in part (d) of the figure, in which a narrow eccentric or almost concentric annulus is "unwrapped" and approximated by a series of parallel plate problems that can be solved. However, circumferential inertia terms cannot be properly modeled, and extensions to transient flow are impossible. Finally, the "pie slice" methods indicated in part (e) remove some limitations inherent in slot flow approaches. Here, concentric solutions are applied to different slices of the annulus, but, again, the final solution, somewhat crudely, provides only as many simulation options as those available for the concentric annuli—and these are few.

Third, yield stress effects have introduced significant difficulties in obtaining solutions that are consistent with reality. In flows with nonvanishing yield stress, plug flows are found that move as solid bodies; they are embedded in the sheared flows we are accustomed to. For flows in circular pipes, simple formulas are available for plug radius, volumetric flow rate, and so on. For eccentric annuli, plug zone size and shape are generally unknown, so solutions to this important problem cannot be obtained at all. Even if plug zone location and geometry are available, mathematical issues associated with matching regional solutions across internal boundaries are overwhelming.

Finally, we cite issues associated with utility and user friendliness. Even if all of the problems just described can be solved, they must be solved quickly and stably with minimal trial and error. Results must automatically display in three-dimensional color graphics and movies. Computational and engineering expertise should not be required to obtain practical solutions. Simple definitions of annular geometry, rheological properties, and run-time inputs such as flow rate or pressure gradient and pipe axial and rotational speed are all that should be required. Only when such conditions are fulfilled will the models find real use.

Simulation challenges met and exceeded

We are pleased to report that the difficulties just cited have been overcome through combined use of rigorous mathematics and state-of-the-art numerical analysis. In addition, careful emphasis and focus on graphical interfaces and ease-of-use issues promise to make the algorithms relevant to modern drilling, cementing, and deepwater applications requiring immediate answers. These require fast solutions operable at field offices and rig sites. The applications are demanding because they require methods that work the first time and every time. As suggested earlier, we explain in the following paragraphs the limitations behind older models (including the author's) versus the newer models to set the stage for the remainder of this book.

While *Borehole Flow Modeling*, *Computational Rheology*, and several company algorithms do model eccentric annuli using the author's curvilinear grid systems, past transformed differential equations neglected partial derivatives of the (variable) apparent viscosity because they led to numerical instabilities. This approximation has been removed.

In the author's prior models, plug zone size and shape were determined by a shock-capturing method that did not always satisfy conservation laws. The new method, using an "extended Herschel-Bulkley" constitutive relation, recognizes that real fluids vary continuously and do not solidify suddenly. Computations therefore reach into the plug zone, and plug size, shape, and interfacial gradients and details are calculated iteratively as part of the solution. The approach mirrors the author's prior aerospace approaches in gas-dynamic shock capturing. In high-speed aerodynamics, shockwaves (or pressure discontinuities) can form at wing surfaces. In the late 1970s, the author employed the so-called "viscous transonic equation" to naturally compute evolving shocks that satisfy physical conservation laws and standard thermodynamic and entropy constraints without partitioning the flow into multiple domains. A similar approach was undertaken here.

Steady rotating pipe flow modeling for eccentric problems is beset with numerical instability problems that have not been satisfactorily addressed. Although a limited number of papers do report solutions, they are lacking in numerical detail and the models do not appear to be available for general use. The present approach, which is robust and numerically stable, calculates steady rotating pipe flow solutions as the asymptotic limit of a transient problem. Many steady and transient applications are given in this book.

Steady, single-phase models with rotation have not been generally addressed in the literature—and unsteady methods hosted on curvilinear grids are even rarer. Here we provide the first such algorithms for annular borehole flow. These represent more than the obvious "$\partial/\partial t$" appendages to steady flow operators and simple time integration. Contradictory and confusing issues were addressed. Prior to 1990, mathematical solutions and field observations indicated that the effect of pipe rotation was increased flow rate (for a given pressure gradient) due to shear thinning; equivalently, when volumetric flow rate is prescribed, rotation leads to a weaker pressure gradient. Subsequent to that period, field observations were completely opposite: Pipe rotation decreases flow rate for the same applied pressure gradient. These contradictions have been cited often in industry discussions, and the fact that field observations do not represent controlled experiments does not help.

Our work has provided a simple explanation for the apparent contradictions. In the early work, concentric annuli were the main focus in vertical well applications. For such problems, the effects of rheology appear only through shear thinning, and this is responsible for the reduced resistance observed. In recent publications, the focus lies in deviated and horizontal wells where eccentricities can be large. Although shear thinning is nonetheless present, the loss of symmetry introduces certain nonlinear convective terms to the governing equations that modify the effective pressure gradient. When these terms are included in the model, post-1990s conclusions are computed naturally; they are consistent with pre-1990s observations for concentric annular flow. The calculations are operationally significant in managed pressure drilling because pipe rotation now provides additional means for pressure control at the drillbit.

Conventionally, borehole pressures are adjusted by changing dynamic friction using different pump rates (a procedure that may not be safe, since sudden pump transients are involved); by altering mud rheology and weight (a process that is slow); or simply by adjusting the surface choke. In our work, we demonstrate that drillstring rotation can affect pressures significantly without the limitations just indicated, thus providing an important tool useful in navigating narrow pressure windows often found in ultra-deepwater drilling.

Transient, three-dimensional, multiphase flow modeling is important to modern drilling and completions. The work of Savery, Darbe, and Chin (2007), with laboratory validations reported in Deawwanich et al. (2008), Nguyen et al. (2008), Savery et al. (2008), and Savery, Chin, and Babu Yerubandi (2008), describes a successful and flexible computational scheme used to solve the fully coupled equations governing axial and azimuthal velocities and species concentration.

The work models miscible mixing and predicts, for example, diffusion thicknesses as a function of cross-sectional location, time scales required for different mixing processes, and the locations of nonplanar interfaces separating multiple fluid slugs. However, the method described is extremely intensive in computation and memory. Using boundary layer approximations, equally accurate solutions are obtained in Chapters 8 and 9 of this book that are orders of magnitude faster and therefore suitable for job-planning applications.

For managed pressure drilling applications where the details of interfacial mixing are not important, pressure profiles along the borehole (and, in particular, at the drillbit) as functions of time can be obtained in minutes. For cementing applications, the ability to "zoom in" in order to examine interfacial mixing details is provided. Unlike the 2007 model of Savery et al., however, computations require only inputs associated with just two contiguous fluids, thus ensuring fast solutions that can be easily repeated for multiple "what if" analyses. Rapid access to answers ensures relevance to job-planning activities.

WHY COMPUTATIONAL RHEOLOGY?

Students accustomed to steady Newtonian fluid mechanics are familiar with "obvious" rules of thumb. But when petroleum applications are encountered, these must be abandoned without suitable or useful replacements. We offer numerous examples.

- For Newtonian flows with stationary boundaries, doubling the pressure gradient doubles the flow rate, while doubling the viscosity halves the volumetric flow. In fact, many solutions contain the simple lumped parameter "$1/\mu \, dp/dz$," where μ is viscosity and dp/dz is pressure gradient. These observations are not true for non-Newtonian flows, and analogous scaling laws do not exist.

- In Newtonian fluid flow problems, barring dependencies on pressure and temperature, viscosity is a constant throughout the entire domain. This is not true of non-Newtonian flows, where the "apparent viscosity" varies throughout the cross section and also depends on pressure gradient or volumetric flow rate. Thus, while intrinsic parameters like "n" and "K" can be inferred from viscometer readings, instrument readings for viscosity are largely irrelevant for applications in which, say, the same fluid is flowing in a real borehole annulus.

- For steady concentric Newtonian annular flows, axial and azimuthal velocity fields completely decouple despite the nonlinearity of the Navier-Stokes equations. Axial velocities depend on applied pressure gradient and not on rotational rate, while azimuthal velocities depend on rotational rate and not on pressure gradient. This is not the case for non-Newtonian flows, even in concentric applications, because the apparent viscosity function depends on both velocities. Thus, for example, laboratory and field observations obtained for Newtonian flows are completely inapplicable to non-Newtonian flow, and any expense incurred is wasted.

- While it is not obvious without studying the governing equations, the time scales associated with, for instance, flowline startup and shutdown or time to steady state in rotating pipe applications, are completely different for Newtonian versus non-Newtonian flows.

It is clear from these examples that no simple methods exist for non-Newtonian flow prediction except for well-validated computational methods. This book develops a wealth of proven algorithms that, importantly, have been integrated for convenient use within a software framework.

BROAD PRINCIPLES AND NUMERICAL CONSISTENCY

Truisms, such as "death and taxes," are often difficult to prove. Such truisms are to be found in annular flow—for instance, for the same applied pressure gradient and pipe and hole diameters, higher borehole eccentricities move greater amounts of fluid whatever the rheology. Another apparent truism is found for concentric rotating flows—rotation reduces resistance because of shear thinning and increases flow rate. And for most eccentric annular cross sections, rotation seems to decrease flow rates because axial pressure gradients are altered in subtle ways. There are most likely exceptions to such "truths" and others. We are gratified that flow simulation results seem to be consistent with them. More important, however, is that predictive means are now available to provide "numbers" and, of course, appropriate guidance when truths are not so true.

What we're not

Although this book focuses on a wide range of problems, it is important to summarize and, if necessary, briefly explain those issues not covered in addition to technical areas that the simulators do not address.

- Secondary flows in pipes and annuli are not investigated (e.g., Taylor vortices are not studied).
- The simulators do not model gas kicks.
- Pressure and temperature effects on fluids, while important operationally, are not addressed because they can only be determined from empirical measurement. We assume that rheological properties for the particular downhole environment are available and "go" with those inputs.
- Present turbulence-modeling methods are highly empirical and do not fit within the predictive framework of the software research. Thus, conventional models are not included, although an analogy to "small n" rheologies is developed in some detail.
- Swab-surge effects critically affect drilling operations. A significant portion of our research and software development focused on accurate modeling of yield stress fluids in complicated annular domains. The work addressed constant density applications. Transient compressible effects are also important in practice (i.e., "water hammer" effects), but these are reserved for future investigation.

What we can solve and more

No research or software development effort is ever complete, and ours is no exception. Since 2010, six papers have been presented by the author at various conferences, and audiences have raised common questions. We repeat these together with our remarks.

How well validated is the methodology? Early single-phase flow checks are reported in *Borehole Flow Modeling*, and pipeline validations are given in *Computational Rheology*. The transient, three-dimensional, multiphase flow references cited earlier give experimental details with photographs for non-Newtonian, rotating, annular flows obtained at a major university.

Are torque and drag calculations possible? Calculations can be performed with simulator output, but they are not yet part of any automated postprocessing scheme.

Can the gridding technology be extended to three dimensions in order to model real boreholes? Yes, definitely. In fact, a large oil service company has parameterized our model by adding a third "z" direction to the mappings. Mappings are repeated as needed, and local cross-sectional geometries are developed with the aid of caliper logs. The flow equations are written to this three-dimensional grid and solved. The resulting software is numerically intensive and demanding of memory resources.

How general are our transient pipe motion options? The "canned" options available in the "Transient 2D" and "Transient 3D multiphase" options are very flexible, although, of course, they cannot cover every possible scenario. We do emphasize that nothing in the general algorithm and formulation precludes the most arbitrary reciprocating and rotating pipe or casing motions envisioned by users. Implementation is straightforward but requires source code access.

Is it possible to model rheologies besides Newtonian, Power law, Bingham plastic, and Herschel-Bulkley (also known as Herschel-Buckley)—that is, fluids with "memory" effects? Yes, arbitrary rheologies can be modeled. We have selected the Herschel-Bulkley as our primary candidate only because this is widely used. However, other constitutive relations can be used by modifying a short apparent viscosity update module. This applies also to memory fluids. For such flows, the same update procedure is used at the end of each time step during the integration process, with fluid strains from previous time steps now entering current apparent viscosity definitions.

Can rheological flow models be used to predict movements of single cuttings chips? Although many in the profession and many software salespeople will argue yes, the answer is a definitive no. The author's background as a research aerodynamicist at Boeing and as a turbomachinery manager at Pratt & Whitney Aircraft is, in part, responsible for this negativism: It is difficult to model inviscid flow past a single fixed airfoil, let alone low Reynolds number viscous flow past an unrestrained rock chip with unknown geometry and origin.

Computational models can, of course, be used for correlative purposes. For instance, in *Borehole Flow Modeling*, the author shows that high viscous stresses at the top of cuttings beds in deviated wells are associated with good hole cleaning; obviously, mechanical friction plays a crucial role in cuttings bed erosion and removal. On the other hand, high axial velocities and viscosities are instrumental for debris removal in vertical wells, a conclusion obvious from Stokes' formula for slow-moving spheres. As another example, low apparent viscosities are associated with spotting fluid effectiveness in freeing stuck pipe. Because correlative methods offer significant potential in explaining physical phenomena and in offering solutions, the software models reported in this book also provide detailed solutions for apparent viscosity, shear rate, and viscous stress in addition to solutions for velocity. Their availability allows engineers to identify and explain new observations with greater ease and hopefully to make drilling and cementing safer and more economical.

CLOSING INTRODUCTORY REMARKS

Before delving into the heart of annular flow modeling and, in particular, managed pressure drilling and cementing applications, we offer several remarks (amply illustrated) about the mathematical approaches taken, the technical problems solved, and the overall system and objectives addressed.

Insofar as theory is concerned, two fundamental building blocks are used. The first employs "boundary-conforming, curvilinear grid" systems to represent complicated geometries. An example of an interesting annulus is shown in Figure 1.2, where the borehole wall is shaped with Texas boundaries and the "hole" is an elliptical fracture. One would write the governing equations to these coordinates and solve them with iterative methods for steady flows and time-marching integration schemes for transient problems. The mathematics is developed entirely from first principles in terms of basic concepts from calculus. The usual references to differential geometry are not necessary and not used.

The conventional constitutive relations used for Newtonian, Power law, Bingham plastic, and Herschel-Bulkley fluids are illustrated in the stress and shear rate diagrams of Figure 1.3. As will be explained, computations for fluids without yield stress are straightforward in a sense, but in the case of yield stress problems, the size, shape, and location of plug zones (which move as solid bodies within sheared flows) are unknown, rendering the computational problem intractable. Until the work of Chin and Zhuang (2010), calculations for eccentric annular flow with plug zones were not possible.

A major breakthrough offering fast practical solutions without compromising the mathematics was achieved, using an "extended Herschel-Bulkley" law proposed in Souza Mendes and Dutra (2004), which provides for continuous flow solutions reaching across and into typical plug zones. This approach, generalizable to rheologies beyond those commonly used, promises to broaden the reach of computational methods for modeling newer muds and cements now being introduced commercially to the industry.

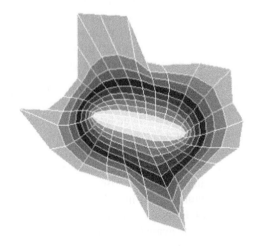

FIGURE 1.2

Annulus with Texas boundaries and elliptical hole.

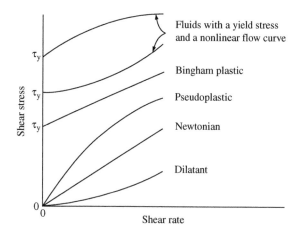

FIGURE 1.3

Constitutive relations for basic rheologies.

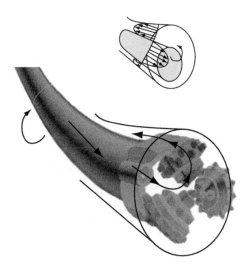

FIGURE 1.4

Eccentric annular flow model.

Taken together, our grid generation and plug zone methodologies and our advanced numerical methods for iterative and time-marching schemes (for application to systems of partial differential equations) allow solutions to the most complicated annular flow problems.

The schematic in Figure 1.4 outlines the scope of the technical problem areas that can be addressed. These include high annular eccentricity, geometric anomalies like washouts and cuttings beds, general axial reciprocation in time together with arbitrary unsteady rotational pipe motions, plug zone modeling associated with yield stress fluids, general pump schedules allowing multiple

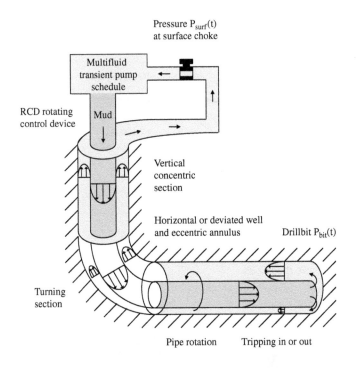

FIGURE 1.5

Managed pressure system simulation.

slugs of non-Newtonian fluids, borehole axis curvature, pressure loss at the drillbit, steady and transient flow analysis, and so on.

The foregoing capabilities have been assembled to form a user-friendly software framework with convenient color graphics to model the managed pressure drilling requirements in Figure 1.5. In effect, we consider the entire problem, from general pumping schedule to flow down the drill-pipe, through the drillbit, and finally up a highly eccentric annulus. Our objective is accurate computation of a borehole pressure profile and, particularly, pressures at the bit as functions of time; this computation is essential to job planning in drilling and cementing ultra-deepwater offshore wells. Fast computing speeds and ease of use, of course, are important to rapid decision making in environments constrained by dangerous narrow pressure windows. In the end, safety is the prime motivator.

SECTION 1.1

Managed Pressure Drilling Fluid Flow Challenges

We have satisfactorily answered "Why study rheology?" In petroleum engineering, we emphasize that *rheology* here necessarily implies *computational rheology*. Operational questions bearing

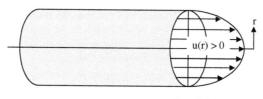

Note: du/dr < 0

FIGURE 1.6

Axisymmetric pipe flow.

important economic implications cannot be answered without dealing with actual clogged pipeline and annular borehole geometries that can only be conveniently studied using simulation methods. Before delving into our subject matter, it is useful to review several exact closed-form solutions. These are useful because they provide important validation points for calculated results, and they are instructive because they show how limiting analytical methods are. For our purposes, we will not list one-dimensional, planar solutions, which have limited petroleum industry applications, but will focus first on pipe and annular flows. Rectangular duct solutions will be treated in Chapter 5.

Newtonian pipe flow

What can be simpler than flow in a pipe? In this chapter, we will find that most "sophisticated" analytical solutions are available for pipe flows only, and then limited to just several rheological models.

Figure 1.6 illustrates straight, axisymmetric pipe flow, where the axial velocity $u(r) > 0$ depends on the radial coordinate $r > 0$. With these conventions, the "shear rate," $du/dr < 0$, is negative; that is, $u(r)$ decreases as r increases. Very often, the notation $d\gamma/dt = -du/dr > 0$ is used. If the viscous shear stress τ and the shear rate are linearly related by

$$\tau = -\mu\, du/dr > 0 \tag{1.1a}$$

where μ is a constant viscosity, then two simple relationships can be derived for pipe flow.

Let $\Delta p > 0$ be the (positive) pressure drop over a pipe of length L, and R be the inner radius of the pipe. Then the radial velocity distribution satisfies

$$u(r) = [\Delta p/(4\mu L)](R^2 - r^2) > 0 \tag{1.1b}$$

Note that u is constrained by a "no-slip" velocity condition at $r = R$. If the product of "u(r)" and the infinitesimal ring area "$2\pi r\, dr$" is integrated over (0, R), we obtain the volumetric flow rate expressed by

$$Q = \pi R^4 \Delta p/(8\mu L) > 0 \tag{1.1c}$$

Equation 1.1c is the well-known Hagen-Poiseuille formula for flow in a pipe. These solutions do not include unsteadiness or compressibility. These results are exact relationships derived from the Navier-Stokes equations, which govern viscous flows when the stress-strain relationships take the linear form in Equation 1.1a. We emphasize that the Navier-Stokes equations apply to Newtonian flows only and not to more general rheological models.

Note that viscous stress (and the wall value τ_w) can be calculated from Equation 1.1a, but the following formulas can also be used,

$$\tau(r) = r \, \Delta p/2L > 0 \tag{1.2a}$$

$$\tau_w = R \, \Delta p/2L > 0 \tag{1.2b}$$

Equations 1.2a and 1.2b apply generally to steady laminar flows in circular pipes and—importantly—whether the rheology is Newtonian or not. But they do not apply to ducts with other cross sections or to annular flows, even concentric ones, whatever the fluid.

Finally, for Newtonian flows, we show how the effects of pipe rotation are easily modeled. If we turn to the general Navier-Stokes equations in Equations 2.1 through 2.4 and set $v_r = 0$, $\partial/\partial\theta = \partial/\partial t = \partial/\partial z = 0$, and assume vanishing body forces, the continuity equation is automatically satisfied. The three main equations become $\partial p/\partial r = \rho v_\theta^2/r$, $\partial^2 v_\theta/\partial r^2 + 1/r \; \partial v_\theta/\partial r - v_\theta/r^2 = 0$, and $\partial^2 v_z/\partial r^2 + 1/r \; \partial v_z/\partial r = 1/\mu \; \partial p/\partial z$. The azimuthal momentum equation is solved by $v_\theta = \omega r$—that is, the fluid executes solid body rotation as the pipe turns at a constant speed ω and there is no influence from the axial pressure gradient. The solution for the axial momentum equation is just the u from Equation 1.1b and does not involve ω. In other words, the two velocities behave independently. The radial pressure gradient is obtained from the first equation as $\partial p/\partial r = \rho\omega^2 r$. Note that non-Newtonian pipe flows do not behave so simply.

Bingham plastic pipe flow

Bingham plastics satisfy a slightly modified constitutive relationship, usually written in the form

$$\tau = \tau_0 - \mu \, du/dr \tag{1.3a}$$

where τ_0 represents the yield stress of the fluid. In other words, fluid motion will not initiate until stresses exceed yield; in a moving fluid, a "plug flow" moving as a solid body is always found below a "plug radius" defined by

$$R_p = 2\tau_0 L/\Delta p \tag{1.3b}$$

The "if-then" nature of this model renders it nonlinear, despite the (misleading) linear appearance in Equation 1.3a. Fortunately, simple solutions are known:

$$u(r) = (1/\mu)[\{\Delta p/(4L)\}(R^2 - r^2) - \tau_0(R - r)], \quad R_p \leq r \leq R \tag{1.3c}$$

$$u(r) = (1/\mu)[\{\Delta p/(4L)\}(R^2 - R_p^2) - \tau_0(R - R_p)], \quad 0 \leq r \leq R_p \tag{1.3d}$$

$$Q/(\pi R^3) = [\tau_w/(4\mu)][1 - 4/3(\tau_0/\tau_w) + 1/3(\tau_0/\tau_w)^4] \tag{1.3e}$$

Power law fluid pipe flow

These fluids, without yield stress, satisfy the Power law model in Equation 1.4a and the rate solutions in Equations 1.4b and 1.4c.

$$\tau = K(-du/dr)^n \tag{1.4a}$$

$$u(r) = (\Delta p/2KL)^{1/n}[n/(n+1)](R^{(n+1)/n} - r^{(n+1)/n}) \tag{1.4b}$$

$$Q/(\pi R^3) = [R\Delta p/(2KL)]^{1/n}n/(3n+1) \tag{1.4c}$$

Nonlinear "Q versus Δp" graphical plots are given in Chapter 4. We emphasize that linear behavior applies to Newtonian flows exclusively.

Herschel-Bulkley pipe flow

This model combines Power law with yield stress characteristics, with the result that

$$\tau = \tau_0 + K(-du/dr)^n \tag{1.5a}$$

$$\begin{aligned} u(r) = K^{-1/n}(\Delta p/2L)^{-1}\{n/(n+1)\} \\ \times [(R\Delta p/2L - \tau_0)^{(n+1)/n} - (r\Delta p/2L - \tau_0)^{(n+1)/n}], \quad R_p \leq r \leq R \end{aligned} \tag{1.5b}$$

$$\begin{aligned} u(r) = K^{-1/n}(\Delta p/2L)^{-1}\{n/(n+1)\} \\ \times [(R\Delta p/2L - \tau_0)^{(n+1)/n} - (R_p\Delta p/2L - \tau_0)^{(n+1)/n}], \quad 0 \leq r \leq R_p \end{aligned} \tag{1.5c}$$

$$\begin{aligned} Q/(\pi R^3) = K^{-1/n}(R\Delta p/2L)^{-3}(R\Delta p/2L - \tau_0)^{(n+1)/n} \\ \times [(R\Delta p/2L - \tau_0)^2 n/(3n+1) + 2\tau_0(R\Delta p/2L - \tau_0)n/(2n+1) + \tau_0^2 n/(n+1)] \end{aligned} \tag{1.5d}$$

where the plug radius R_p is again defined by Equation 1.3b.

Ellis fluid pipe flow

Ellis fluids satisfy a more complicated constitutive relationship, with the following known results:

$$\tau = -du/dr/(A + B\,\tau^{\alpha-1}) \tag{1.6a}$$

$$u(r) = A\,\Delta p(R^2 - r^2)/(4L) + B(\Delta p/2L)^{\alpha}(R^{\alpha+1} - r^{\alpha+1})/(\alpha+1) \tag{1.6b}$$

$$\begin{aligned} Q/(\pi R^3) &= A\tau_w/4 + B\tau_w^{\alpha}/(\alpha+3) \\ &= A(R\Delta p/2L)/4 + B(R\Delta p/2L)^{\alpha}/(\alpha+3) \end{aligned} \tag{1.6c}$$

Other rheological models appear in the literature. Typical qualitative features of the main models for velocity are shown in Figure 1.7.

Annular flow solutions

The only known exact, closed-form analytical solution is a classic one describing Newtonian flow in a concentric annulus. Let R be the outer radius and κR be the inner radius, so that $0 < \kappa < 1$. Then

$$\begin{aligned} u(r) = [R^2\Delta p/(4\mu L) \\ \times [1 - (r/R)^2 + (1 - \kappa^2)\log_e(r/R)/\log_e(1/\kappa)] \end{aligned} \tag{1.7a}$$

$$Q = [\pi R^4\Delta p/(8\mu L)][1 - \kappa^4 - (1 - \kappa^2)^2/\log_e(1/\kappa)] \tag{1.7b}$$

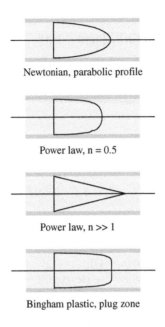

Newtonian, parabolic profile

Power law, n = 0.5

Power law, n >> 1

Bingham plastic, plug zone

FIGURE 1.7

Typical non-Newtonian velocity profiles.

noting that this solution assumes stationary walls. Here, the slope formed by Q versus $\Delta p/L$ is determined once and for all by the geometry and the value of viscosity. In fact, Q is inversely proportional to μ, varies directly with $\Delta p/L$, and depends only on the lumped quantity $1/\mu\ \Delta p/L$. The net proportionality constant just given can be determined by experiment if desired.

Note that for non-Newtonian flows, even for concentric geometries, numerical procedures are required—see Fredrickson and Bird (1958), Skelland (1967), or Bird, Stewart, and Lightfoot (2002). The limited number of exact nonrotating solutions unfortunately summarizes the state of the art, and for this reason recourse must be made to computational rheology for the great majority of practical problems. We will, however, derive an exact analytical solution for Herschel-Bulkley yield stress fluids in concentric annuli without pipe movement in Chapter 5.

SECTION 1.2

MPD Flow Simulator: Steady, Two-Dimensional, Single-Phase Flow

Our "MPD Flow Simulator" system consists of three distinctive capabilities hosted by three different but linked menu interfaces. These are "Steady 2D," "Transient 2D," and "Transient 3D multi-phase." All of the modules residing in each of the interfaces satisfy rigorous mathematical formulations that are described in detail in this book. Moreover, they are hosted by fast and numerically stable algorithms, and are tightly integrated with automated two- and three-dimensional color graphics displays that together provide detailed solutions within minutes, if not seconds.

While substantial research into fluid mechanics formulations and their computational solutions supports the totality of our efforts, only those models that operate quickly and stably are offered for general public use. We emphasize that all of our models have been validated, many with detailed field and laboratory data and, importantly, that all are consistent with each other to within 2 to 3 percent in their areas of common overlap. The work builds on the author's books *Borehole Flow Modeling in Horizontal, Deviated and Vertical Wells* (Chin, 1992) and *Computational Rheology for Pipeline and Annular Flow* (Chin, 2001), which dealt with steady, two-dimensional, single-phase flows. These models are available to petroleum organizations, as are recent extensions to transient, two- and three-dimensional, single-phase and multiphase applications, which, however, have seen only limited publication until now.

We describe our overall capabilities in this introductory chapter to give a flavor of the final product—in particular what individual software models do and how they deliver their results. In this manner, we "personalize" our partial differential methods and render them less intimidating. While great care is taken to explain our formulations, especially in the context of conventional models, the software product is designed so that no user expertise in theoretical fluid mechanics, mathematics, computer modeling, or numerical analysis is required. Aside from an appreciation of basic annular flow problems and applications, and the practical implications behind yield stress rheologies, nothing is required except the ability to "point and click," all the time attempting to understand the consequences of the calculations.

The main "Steady 2D" interface appears in Figure 1.8. The leftmost "Start" menu, expanded in Figure 1.9, provides access to introductory, self-explanatory functions. "QuikStart" provides enough information for the user to perform his first simulations the very first minute. The "Install graphics"

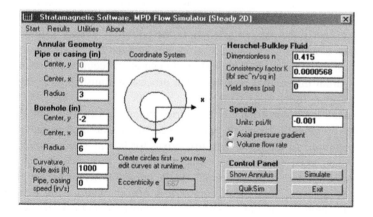

FIGURE 1.8

"Steady 2D" main menu.

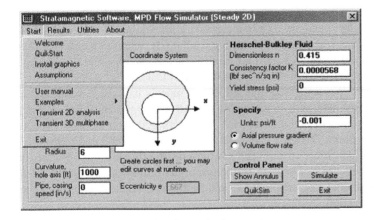

FIGURE 1.9

Introductory functions. For two circles with inner and outer radii R_i and R_o, and center separation Δ, "eccentricity" is defined as $\varepsilon = \Delta/(R_o - R_i)$. Thus, $\varepsilon = 0$ if concentric and 1 if in contact. Eccentricity is not often used in this book because annuli may be very general to include washouts and cuttings beds.

function needs to be executed only once from any of the graphics menus in this or in Section 1.3 or 1.4. "User manual" allows direct access to an integrated pdf document. "Examples" loads stored data for instructive simulations that can be run by simply clicking "Simulate." "Transient 2D analysis" links the present module to the "Transient 2D" simulator, while "Transient 3D multiphase" provides access to the more complicated transient, three-dimensional, multiphase flow simulator. All of these modules are cross-validated later in the book under challenging scenarios to demonstrate their consistency and accuracy.

The upper left portion of the interface in Figure 1.10 displays input boxes for eccentric annulus geometry definition, requiring numbers for circle centers and radii following the convention shown. Eccentricity is calculated in the passive text box at the bottom. Clicking "Show Annulus" displays the eccentric annulus assumed, together with a coarse, boundary-conforming, curvilinear grid that might be used to host the simulation. Although grid generation requires the solution of a system of coupled nonlinear partial differential equations, the process is transparent to the user and requires seconds, including display time.

Note that our annuli are not restricted to eccentric circles—at run time, inner and outer contours may be edited point by point at the user's option. In addition, fine meshes may be selected to provide still higher resolution. At the lower left, text boxes for borehole axis curvature (which models centrifugal effects) and axial drillstring or casing speed (for zero, positive, or negative constant speed movement) are available for general input. At the present time, our steady flow simulator does not support rotation, since the numerical algorithm is not unconditionally convergent. However, transient rotations *are* supported in our "Transient 2D" algorithm and may be used to compute steady-state effects. Refer to our transient flow write-ups for further discussion.

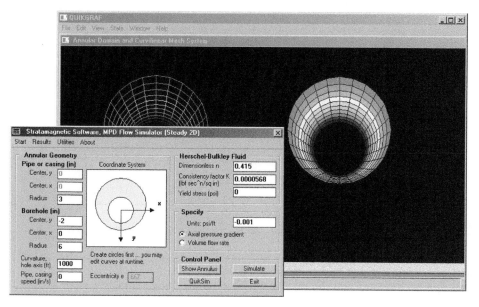

FIGURE 1.10

"Show Annulus" display and grid generation capability.

Once annular geometry is defined and Herschel-Bulkley parameters are entered at the top right, the user selects "axial pressure gradient specified" (to compute flow rate) or "volumetric flow rate specified" (to compute pressure gradient). Depending on the density of the mesh selected, the iterative calculations (used to solve the nonlinear momentum equations for the particular rheology written to the specified grid) may require up to several seconds of computation. Again, the process is completely automated.

There are two simulation modes, "QuikSim" and "Simulate." Both use the same numerical engine, solving the same complete equations hosted by the mapped equations; hydraulic radius, narrow annulus, and slot flow models are never used in our work. "QuikSim" assumes the finest mesh permissible and does not allow editing of inner and outer borehole wall shapes. Also, in order to expedite calculations, limited text output options are offered, although all possible color plots are available. "Simulate," on the other hand, offers detailed geometry-editing capabilities and detailed text output, tabulated results, and "numbers overlaid on annular geometry" capabilities, where "numbers" refers to velocity, apparent viscosity, shear rate, and viscous stress results (see Figure 1.11).

Once "QuikSim" or "Simulate" is clicked and the simulations are run to completion—a process requiring seconds for "axial pressure gradient specified" but possibly up to a minute for "volumetric flow rate specified"—all simulation results are available through the "Results" menu shown. "Text output" refers to summaries of geometric annular attributes and rheological parameters, tabulated results for velocity, apparent viscosity, shear rate and viscous stress, and detailed numbers

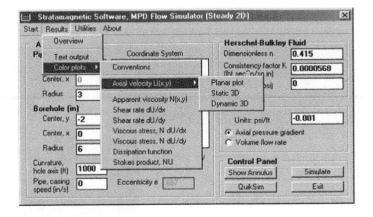

FIGURE 1.11

"Results" menu offering detailed color plots.

plots that contain more information than is available through more attractive color graphics results. For instance, this may include ASCII results, such as the following velocity display, which plots the first two significant digits of axial velocity (in units of in./sec) for an eccentric annulus with a user-supplied cuttings bed.

```
COMPUTED AXIAL VELOCITY (IN/SEC):

                              0     0     0
                        0     9    15     9     0
                        9    15    19    15     9
                  0    15    21    21    21    15          0
                        9   2121   21    19    21   2121          9
                       19    19    17    12    17    19    19
             0  9          21   12  6  6     0     6  612   21              9  0
                  1521       17     0                   0    17      2115
                     211912     0                        0   121921
         0  915          6  0                              0  6          19  9  0
                  212015  6                                   6151921

        01519201912  0                                        01219201915  9  0

                  202015  6                                   6151820
         0  915          5  0                              0  5          15  9  0
                     1810     0                              0   101718
                  1519      11     0                      0    11      1915
             0  9          15    8  5  1      1     1  2  8   1515              9  0
                       13    9  9     5       5     5     9  9   1513
                        8     5  8     4  5       4     5  4     8  5     8
                        0     0     0     0       0     0     0
```

Axial velocities are also available as "planar," "static 3D," and "dynamic 3D" plots, as shown in Figures 1.12, 1.13, and 1.14, respectively. All may be copied to the Windows clipboard, saved, and inserted into Windows documents, worksheets, and presentation software. Figure 1.13 illustrates contour-plotting capabilities, while Figure 1.14 provides the ability to rotate about any (and all) of three axes, as well as "zoom" and "move" functions.

In addition to velocities, detailed field properties for apparent viscosity, x and y component shear rate and viscous stress, dissipation function, and Stokes product, in color and ASCII text

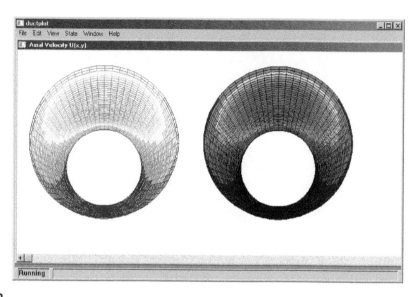

FIGURE 1.12

Planar velocity plot.

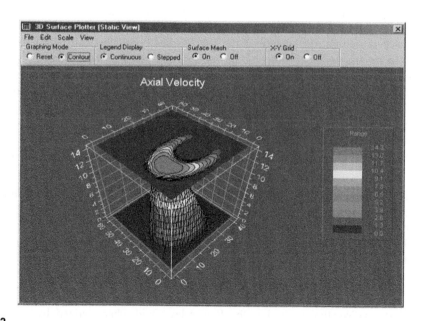

FIGURE 1.13

"Static 3D" display with contour plot generation.

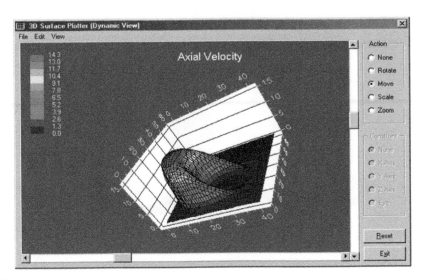

FIGURE 1.14

"Dynamic 3D" display with mouse-rotatable perspective views.

plots, are available for user viewing (Figure 1.15). These properties are obtained by postprocessing exact computed results for velocity and may be useful for different correlation applications. For instance, apparent viscosity is used in evaluating spotting fluid effectiveness for stuck pipe applications, while viscous stress at the surface of the cuttings bed correlates well with cuttings transport efficiency. Velocity and apparent viscosity are individually useful in hole-cleaning applications in vertical wells.

While the menu in Figure 1.8 represents our flagship arbitrary geometry capability under the "Steady 2D" heading, a number of simpler analysis functions are offered in the "Utilities" menu. These application programs are more restrictive; however, they can be very useful and powerful, since they are based on closed-form analytical and often exact solutions. For example, three annular programs are offered under "Concentric steady flow." Specifically, as shown in Figures 1.16 and 1.17:

- "Newtonian, nonrotating, axial pipe motion" refers to an exact, analytical solution of the Navier-Stokes equations for Newtonian fluids, assuming a nonrotating inner pipe that may be stationary axially or moving at constant speed in either direction.
- "Herschel-Bulkley, no rotation or pipe movement" refers to the first exact solution available for non-Newtonian yield stress fluids, assuming a stationary pipe without rotational or axial movement.
- "Power law, rotating, no axial pipe movement" refers to a closed-form analytical solution assuming Power law fluids for a rotating inner cylinder without axial pipe movement.

The simplest non-Newtonian fluid is the Power law fluid, characterized by a dimensionless exponent "n" and a consistency factor "K." This includes Newtonian fluids (such as air and water) when $n = 1$ and K reduces to the constant viscosity μ.

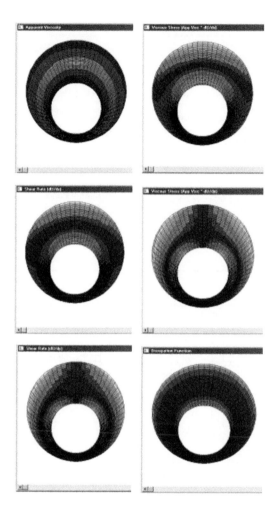

FIGURE 1.15

Detailed physical properties.

In Newtonian applications, the viscosity (aside from changes due to pressure and temperature) is a constant throughout the flow domain, which can be separately measured in a viscometer. For non-Newtonian fluids, the "apparent viscosity" varies throughout and additionally depends on pressure gradient or flow rate. For Power law fluids, however, the intrinsic properties n and K are instead constant and determined from separate viscometer measurements. To assist in their determination, two utilities, as shown in Figure 1.18, are provided. These parameters can be obtained from knowledge of two Fann dial readings or from viscosity and shear rate data. Values obtained from these programs can be used in the rheology menu of Figure 1.8.

Newtonian and non-Newtonian fluids differ significantly in their dynamical properties. We indicated how viscosities are constant for the former but variable (and problem dependent) for the latter.

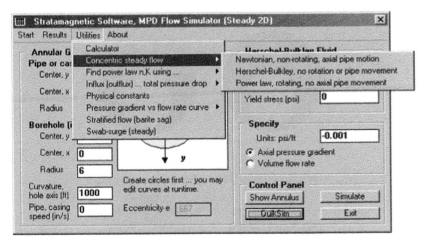

FIGURE 1.16

"Utilities" menu functions.

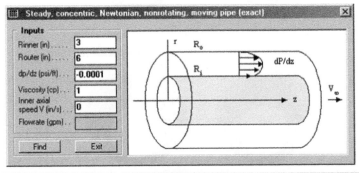

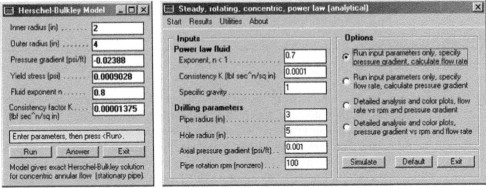

FIGURE 1.17

Concentric steady flow programs.

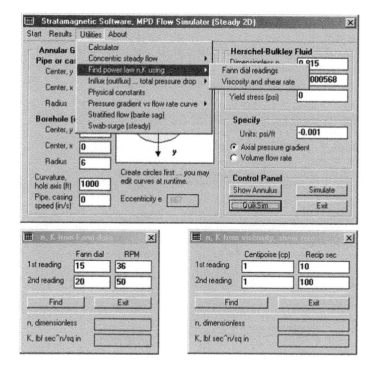

FIGURE 1.18

Finding n and K for Power law fluids.

Other important differences are found. For Newtonian flows, the relationship of pressure gradient versus flow rate is linear for a given viscosity, with a proportionality constant that depends on geometry. Once this constant is available, either analytically, computationally, or experimentally, the pressure gradient for a given flow rate is easily determined; the net pressure drop is just the product of pressure gradient and conduit length. For non-Newtonian fluids, the relationship between pressure gradient and flow rate is nonlinear. Its determination by experimental means is inconvenient and expensive, and computational methods provide an important and practical alternative.

The simulation implied by Figure 1.8 must be performed for numerous values of assumed pressure gradient to obtain the required curve. When influx or outflux is found along the flow channel, the total volumetric flow rate along the direction of flow changes, so that the local axial pressure gradient likewise changes. The total pressure drop utility in Figure 1.19 automatically determines the requisite curve and sums all pressure drops when a user-prescribed influx schedule is available. Thus, total pressure loss is known for a given influx (outflux) profile, and it is implied that, if pressure deviations from ideal values are known, net influx rates can be back-calculated.

The "pressure gradient versus flow rate curve" function in Figure 1.20 provides the curve only and represents a subcapability of Figure 1.19. In both cases, the curve is displayed in "dp/dz versus

Q" and "Q versus dp/dz" formats, where dp/dz represents the axial pressure gradient and Q denotes the volumetric flow rate.

The foregoing dp/dz and Q relationships, which, importantly, allow axial pipe movement, are very useful in managed pressure drilling applications. Although their calculation may appear straightforward (e.g., specifying a set of dp/dz's and determining the corresponding Q's), it is quite

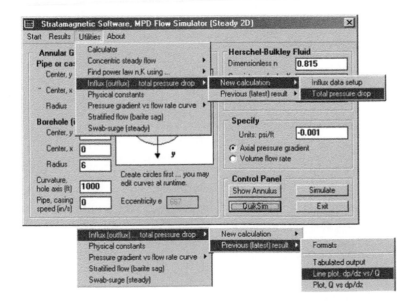

FIGURE 1.19

Influx (outflux) modeling—finding total pressure drop.

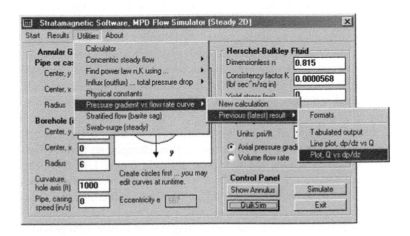

FIGURE 1.20

Pressure gradient versus flow rate relationship.

the contrary. Depending on rheology inputs, the characteristic pressure gradient for any particular problem may range from 0.00001 to 0.1 psi/ft. Simply selecting the smallest possible dp/dz and finely incrementing it over a wide range of values requires hours of calculations. To reduce computation times to one or two minutes, this strategy is employed.

A typical maximum rate of 1,500 gpm is assumed for the annular geometry and fluid rheology considered, and the "volumetric flow rate specified" option in Figure 1.8 is selected. This uses a rapid "half-step iteration" for determining the corresponding maximum gradient, one where sequential guesses are altered geometrically if they do not converge. Once the maximum gradient is available, it may be subdivided into convenient coarser intervals for dp/dz versus Q analysis, providing the required curve.

In the early 1990s, Mario Zamora, then with M-I Drilling Fluids, alerted the author to some interesting fluid-dynamical phenomena he had observed in his flow loop, one in which density stratification can be controlled along with flow rate and deviation angle. While one normally envisions flow moving "simply" from regions of high to low pressure, what the author witnessed was surprising. Under certain conditions, flow visualizations showed that recirculating vortex zones formed that, for all practical purposes, behaved like solid obstacles that blocked or impeded flow. The consequences were unimaginable, implying high risks for drilling arising from barite sag, in the deviated wells that were becoming commonplace.

These possibilities also came at a time when a refinery explosion was found to have occurred after prolonged shutdown for similar reasons. At the time, this phenomenon could not be explained by various modeling attempts and danger avoidance strategies could not be developed. It turns out that similar phenomena have been observed in geophysical and meteorological applications, in atmospheric and oceanic flow settings.

Fortunately, analytical solutions had been developed over the years for weather prediction and military use. These formulations were researched and reinterpreted for drilling application—for example, the Froude number was reintroduced with a dependence on drillstring deviation angle. The time-tested model is accessible from the menu in Figure 1.21, which provides fast solutions together with integrated color graphics output and tabulated solutions.

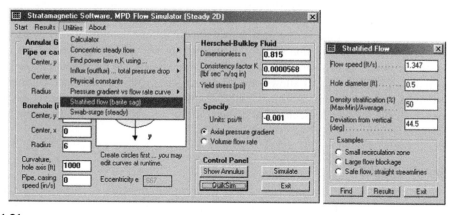

FIGURE 1.21

Stratified flow analysis.

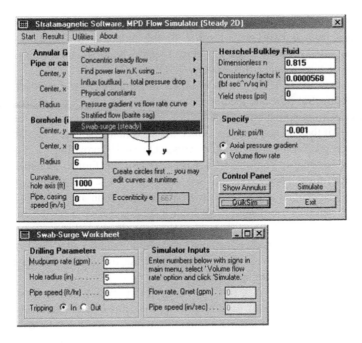

FIGURE 1.22

Swab-surge analysis.

Finally, we address swab-surge analysis in the context of our "Steady 2D" modeling capabilities. Numerous papers have addressed the subject over the years, all assuming simple concentric annular flow and usually Newtonian or Power law rheologies. Investigations have included studies for open and closed drillbits, ranging from steady flow to transient flows. For steady flow, analytical formulas are available from the open literature, while for transient analysis, the models remain proprietary.

It is important, however, to emphasize that all transient analyses reviewed by this author do not distinguish between inertia (that is, constant density "$\rho\ \partial u/\partial t$" momentum) effects and water hammer effects arising from fluid compressibility. Equations are not offered, so their merits cannot be evaluated. In our treatment of swab-surge, we will restrict ourselves to steady-state analysis via the menu in Figure 1.22. This analysis allows general eccentricity, with washouts and cuttings beds, in addition to axial pipe movement. Transient analysis with general combined reciprocation and rotation is also available, as will be seen from the next discussion. However, fluid compressibility is not considered.

Consistency Checks

To encourage confident and diligent use of the steady-state simulator for swab-surge and other applications, consistency checks that validate pressure gradient and flow rate predictions (in the presence of yield stress, pipe reciprocation, hole axis curvature, eccentricity, and so on) are important. In Figure 1.23, we have defined one example in which all input text boxes are populated with

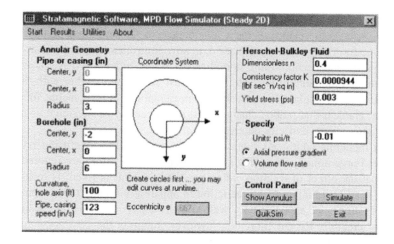

FIGURE 1.23

Parameters for comprehensive consistency check.

large numerical values so that all possible logic branches in the underlying simulator source code are executed. Some of the numbers and results may be unrealistic in practice, but our objective in this example is computational validation.

For the "pressure gradient specified" option shown, in which a pressure gradient of −0.01 psi/ft is assumed, clicking on "QuikSim" gives a computed flow rate of 2,823 gpm. We next select the "volumetric flow rate specified" option and enter "2823" in the input box. After approximately thirty seconds of iterations, we obtain the required pressure gradient, −0.009961 psi/ft, compared to an original −0.01 psi/ft.

In this example, a positive tripping speed of 123 in./sec was assumed. We now repeat the "pressure gradient specified" calculation with −123 ft/sec and −0.01 psi/ft, with the simulation now leading to a flow rate of 850.5 gpm. Next we revert to a "flow rate specified" mode and enter 850.5 in the input text box. We obtain the pressure gradient −0.009961 psi/ft, in agreement with the original −0.01 psi/ft used earlier. These calculations used data consistent with highly non-Newtonian fluids with yield stress, positive and negative drillstring speeds, rapid borehole curvature, and large annular eccentricity. All computations were stably executed. The results showed excellent consistency, whether the pipe was moving upward or downward in both "flow rate" and "pressure gradient specified" simulation modes.

SECTION 1.3

MPD Flow Simulator: Transient, Two-Dimensional, Single-Phase Flow

For steady non-Newtonian flows in arbitrary eccentric annuli without rotation, the nonlinear convective (that is, acceleration) terms vanish identically and the axial momentum equation (together with mass conservation) alone governs the flow. When transient effects are to be considered, the density-dependent inertia terms "$\rho \ \partial u/\partial t + \cdots$" are important and must be included in the analysis. This completely changes the nature of the mathematics, so the nonlinear partial differential equation, originally one for a single elliptic equation, is now controlled by parabolic or diffusive effects. The parabolic system can become quite complicated. When rotation exists, an additional coupled equation for azimuthal momentum appears, which must be solved together with the axial equation. This coupling, as will be explained in the theory presentation, is responsible for modifications to effective pressure gradient that induce changes to flow rate and cross-sectional velocity distribution, with these being strongly eccentricity dependent.

Our transient modeling addressed subtle questions motivated by confusing issues raised in the drilling literature. For simplicity, the effects of borehole axis curvature are ignored. The older literature suggests that the effect of rotation, when pressure gradient is fixed, is to increase flow rate. The explanation is shear thinning or the reduction in apparent viscosity that accompanies pipe rotation. Field observations are supported by well-known analytical modeling results, and the conclusions are very credible. Recent literature, however, suggests the contrary—that is, for a fixed pressure gradient, the effect of rotation is to decrease flow rate. Again, field observations are cited, understanding, of course, that field results in either case are not well controlled—washout, cuttings bed, pump transient, and rheological uncertainties are all likely.

From this perspective, the transient flow research required more than simply introducing "$\rho \ \partial u/\partial t + \cdots$" into the governing equations and integrating. Numerous questions arose. Why do two conflicting observations exist and how do we reconcile them? And if we can, how can we devise a predictive scheme that helps with job planning? The required explanations were simple enough. The older literature dealt with concentric annuli (for vertical wells) where all of the convective terms vanish identically—the only manner in which rheology enters the physics is through viscosity reduction via shear thinning.

The more recent literature, drawn from deviated and horizontal well applications, usually applies to highly eccentric annuli for which the convective terms are always present. To assess their importance, our initial research focused on Newtonian fluids where viscosity is constant, so that effects related to shear thinning were isolated. The rotational convective terms were shown to decrease flow rate when the same pressure gradient is assumed by changing the effective applied pressure gradient, thus explaining a long-standing drilling paradox.

In general, when non-Newtonian rheologies are allowed, shear thinning (which reduces apparent viscosity) and nonlinear convective terms (which usually produce the opposite effect) compete in a manner that strongly depends on annular eccentricity. General conclusions are not possible. Fortunately, the time required to compute flows with arbitrary time-varying pipe reciprocation and rotation is no longer than that for stationary pipe, but it was necessary to develop a scheme that is numerically stable for all prescribed motions. Our formulation solves nonlinearly coupled partial differential equations for axial and azimuthal momentum on curvilinear mesh systems that are custom-fitted to the eccentric annulus.

To speed the computations, a coarse mesh such as that in Figure 1.10 is hardcoded. This coarse curvilinear mesh still provides greater physical resolution than is possible using simpler rectangular and circular systems. The ability to edit inner- and outer-circle contours is not supported, and neither is our prior modeling of centrifugal effects arising from borehole axis curvature. The menu designed to access "Transient 2D" capabilities is shown in Figure 1.24, which does allow circles having arbitrary eccentricity and fluid flows with general Herschel-Bulkley yield stress rheologies.

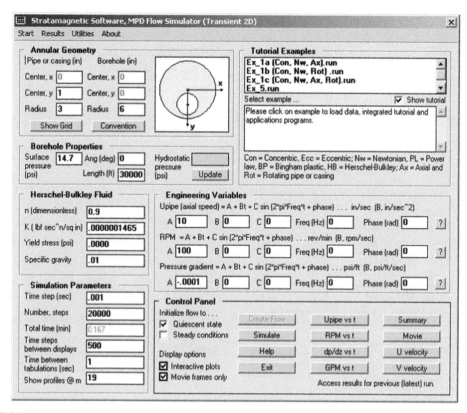

FIGURE 1.24

"Transient 2D" menu interface.

The "Start" menu, as seen in Figure 1.25, for "Transient 2D" resembles that for "Steady 2D." The complementary "Steady 2D" and "Transient 3D multiphase" solvers are accessible from this drop-down menu. The upper left portion of the screen, as before, is reserved for annular geometry definition, with only minor changes to coordinate conventions. Immediately below is a simple utility for calculating hydrostatic pressure at the drillbit using as inputs the surface choke pressure, hole angle, drillstring length, and specific gravity.

At the center left are input text boxes for Herschel-Bulkley fluid rheology that now include an entry for specific gravity. We again emphasize the role of fluid density. For nonrotating flows and general initial conditions, the effects of density disappear asymptotically at large times, assuming, of course, that the drillstring is not reciprocating. In this regard, the unsteady algorithm may be used to solve for steady flows; fast solutions, in fact, are possible as the density grows smaller, since small inertia leads to rapid equilibration (e.g., small specific gravities mean values of 0.01). For rotating flows, density effects never disappear. To calculate steady rotating flows employing the unsteady solver, actual specific gravities must be used. These rules will be apparent from theory.

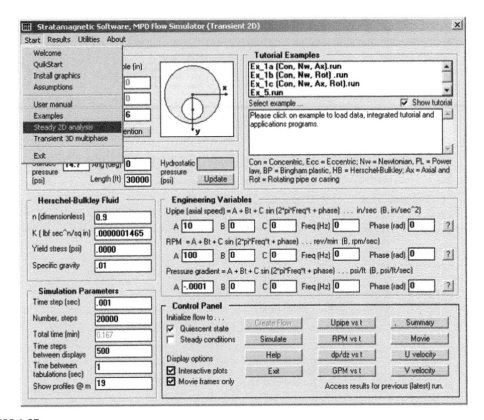

FIGURE 1.25

Transient 2D "Start" menu.

The bottom left of the menu in Figure 1.25 supports time step control; additional parameters (e.g., number of time steps, total time) are self-explanatory. Our transient simulator integrates the governing partial differential equations in time by advancing the discretized form of the equations one time step Δt at a time. Since mesh sizes associated with our curvilinear coordinates are hardcoded, the only parameter available to the user for both accuracy and numerical stability control is Δt. The selected value is constant for the entire duration of the simulation, although in future updates to the algorithm dynamically changing step sizes based on local flow gradients will be used to optimize the integrations. Detailed computed results are available, as indicated in Figures 1.26 and 1.27.

Time step selection represents the most critical decision-making part of the simulation process undertaken by the user. For nonrotating applications, large time step sizes, say 0.01 to 0.05 seconds, might be justifiable and useful, based on validations of the type discussed later in this book. When the inner pipe or casing rotates, the numerical integrations destabilize, and time steps that are 0.001 second or smaller may be necessary. As noted earlier, steady solutions for flows with rotation cannot presently be computed using the purely steady formulation. A limited number of papers

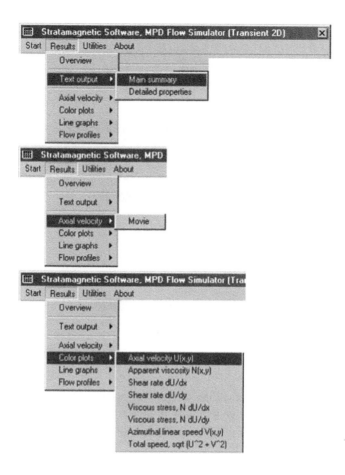

FIGURE 1.26

Transient "Results" menu.

have been published on the subject, but without discussion of stability and computing times, and to this author's knowledge, none of the methods have been offered for commercial use.

Steady flows can be computed as the large-time limit of a transient calculation, provided, of course, that the pipe or casing moves without time-dependent change. Again, small Δt's will generally be necessary. To support this application, the integration scheme is optimized for speed and minimal memory resource requirements. For the present software release, up to 10,000,000 time steps are permitted. If a time step of 0.001 second is assumed, this simulates 10,000 seconds or almost three hours of continuous rig operation.

When a transient partial differential equation is integrated in time, problem-specific boundary and initial conditions are required to constrain and start the solution process. With regard to boundary conditions, we assume that fluid at the outer annular boundary adheres to the boundary and

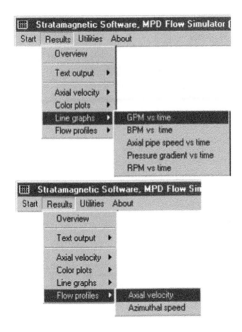

FIGURE 1.27

Transient "Results" menu, more entries.

does not move. Fluid similarly adheres at the drillpipe or casing and moves with whatever transient axial reciprocation or azimuthal rotation motions are prescribed at the inner surface.

We permit two types of initial conditions: completely quiescent and steady flow. By steady conditions, we refer to steady flows without rotation. Thus, we can model the start-up of annular flow from rest, as well as transient reciprocation and rotation starting from nonrotating drill-ahead flowing conditions. Nonrotating flowing initial conditions are actually chosen for special software development reasons. If steady flowing conditions are permitted, they, of course, must first be computed internally. This process must be both fast and stable numerically to ensure a user-friendly experience. If the pipe is nonrotating, the only momentum equation that enters is the axial one, and its steady-state solution can be quickly calculated by assuming small specific gravities and large time steps, since the final solution is independent of fluid density. This is the strategy taken.

If, on the other hand, rotating flow is allowed, the azimuthal momentum additionally enters. Its solution requires actual specific gravities, which may be large; extremely small times steps are necessary for stability, and the user experience requires excessive intervention. It is possible, obviously, for a rotating drillstring to slow down or accelerate, and this possibility is permitted, as discussed later.

The engineering variables menu shown in Figure 1.28 provides a high degree of modeling flexibility. For drillpipe or casing axial speed and rotational rate, and for the applied pressure gradient, assumptions of the general functional form $A + Bt + C \sin (2\pi f + \phi)$ are built into the solver, where A, B, C, f, and ϕ are constants. The units required are indicated next to the input formulas.

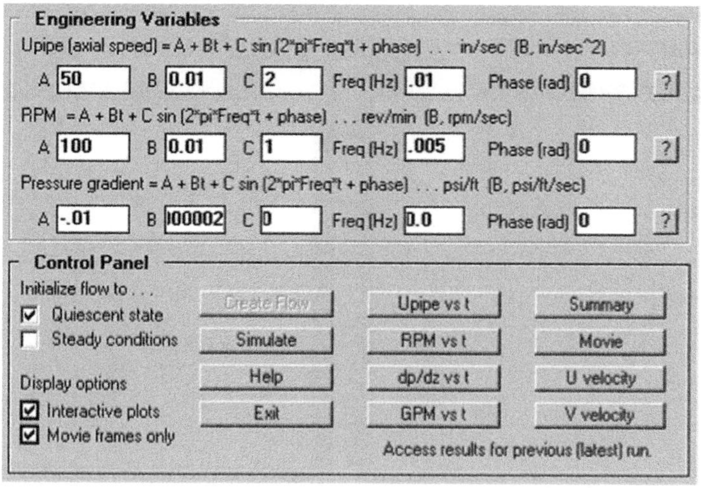

FIGURE 1.28

Engineering variables definition.

Clicking on the question mark to the right of each expression shown in Figure 1.28 produces a line graph of the proposed auxiliary condition (the time scale may be changed from the "Simulation Parameters" menu by altering the time step or total time). Examples are shown in Figure 1.29. Almost all commercial drilling and cementing hydraulics simulators, to this author's knowledge, allow constant pressure gradient at best, but do not support reciprocation or rotation.

The menus shown support general transient specifications; moreover, all three inputs may be transient simultaneously. General reciprocation and rotation capabilities were incorporated by user request, since quantitative methods to assess the effects of axial drillstring vibrations and torsional stick-slip were deemed of interest. Very often, one wants to specify volumetric flow rate and compute pressure gradient. While this is supported in our "Steady 2D" simulator option, it is not possible here because the required transient calculations are very time consuming. However, specification of time-dependent axial pressure gradient, say to model pump ramp-up and ramp-down for swab-surge applications, does to some extent provide the intuitive feeling needed for pressure gradient and flow rate relationships.

Finally, we describe the importance of the "Display options" check boxes at the bottom left of the "Control Panel" interface in Figure 1.25. If "Interactive plots" is not checked, the transient integrations continue at maximum speed and display a simple status note on-screen indicating elapsed time, instantaneous volumetric flow rate, and percentage completed. If it is checked, interactive graphic results are shown at user-defined intervals so that the simulation can be monitored in detail. Results include axial and azimuthal velocity profiles at the top and bottom of the annulus, plus "planar velocity plots" like those in Figure 1.12.

These color plots are saved in a movie file for playback at the end of the simulation—for example, refer to the "Results" menu in Figures 1.26 and 1.27. Three frames from a typical movie are

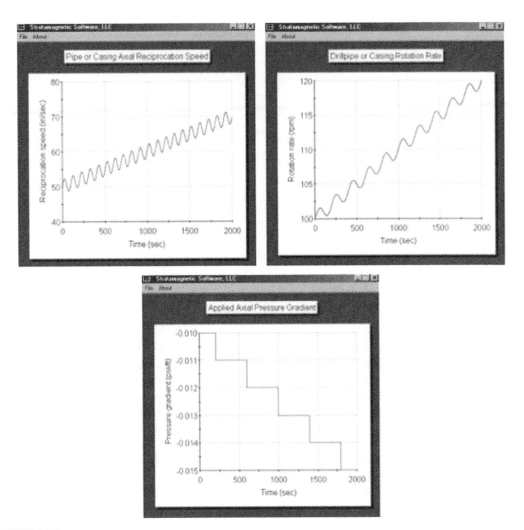

FIGURE 1.29

Example transient reciprocation, rotation, and dp/dz functions.

captured in Figure 1.30. If the box "Movie frames only" is checked, line graphs for velocity profile results are not displayed but movie frames are as they become available. One can, for instance, watch an initially quiescent nonrotating flow evolve in time, with a spatially uniform velocity changing into one with an axial velocity maximum at the wide part of the annulus. If the pipe or casing is made to rotate, this maximum then moves azimuthally in agreement with known computational results from other references. When the simulation ends, a "gpm versus time" plot, as in Figure 1.31, is provided for user insight into volumetric flow rate and unsteady behavior.

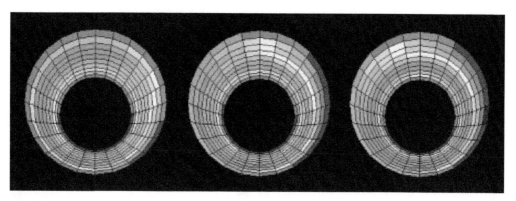

FIGURE 1.30

Example frames from axial velocity movie.

SECTION 1.4

MPD Flow Simulator: Transient, Three-Dimensional, Multiphase Flow

In Sections 1.2 and 1.3, we introduced the "Steady 2D" and "Transient 2D" simulators for non-Newtonian, single-phase annular flow. Aside from obvious differences in "steady" versus "transient," it is important to remember that purely steady formulations for rotating flow are numerically unstable, at least for now, and that rotating pipe or casing flows can only be treated on a transient basis at the present time. Steady rotating solutions are obtained by evaluating transient runs asymptotically to large times. Depending on rotational rate, fluid density, and viscosity, which in turn dictate time step sizes, calculation times may take anywhere from ten seconds to several minutes.

In this section, we discuss the simulator for transient, three-dimensional, multiphase problems. There are two practical problems that we focus on. The first deals with managed pressure drilling and, in particular, the operating scenario sketched in Figure 1.5. Here, a general pumping schedule of non-Newtonian fluid flows is permitted—for example, fluid A pumped for time t_A at a flow rate of GPM_A, followed by fluid B pumped for time t_B under a rate GPM_B, and so on. This fluid travels down the drillpipe, through the drillbit, and finally up the eccentric borehole annulus. We wish to determine the pressure profile along the borehole and particularly the pressure at the bit for all instants in time.

To make calculations tractable, it is assumed that a typical fluid slug length in the pipe or annulus is great compared to the pipe or annular diameter. If so, the locations of all fluid interfaces can be determined kinematically on a volume basis alone without rheological considerations. Then, at any given time t_n, with the volumetric flow rate $Q(t_n)$ known, either of the simulators "Steady 2D" and "Transient 2D" or any of the many specialized flow solvers available in their respective utilities menus can be used to determine the pressure gradient within any fluid slug. Since the lengths of all slugs are also available kinematically, the available set of pressure gradients can be integrated spatially, starting with the pressure value known at the surface choke to provide the pressure distribution along the borehole at that instant in time.

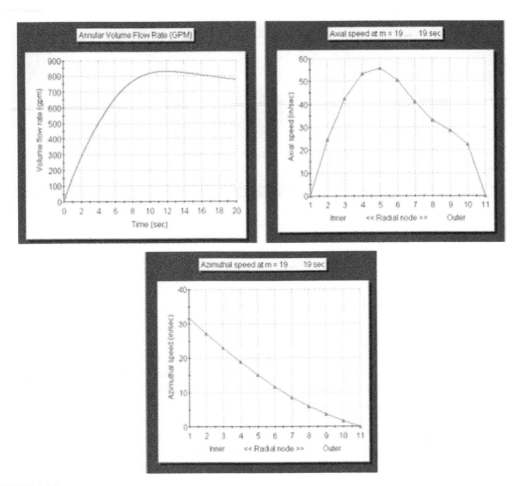

FIGURE 1.31

Typical line plots, "gpm versus time" (*left*), and axial and rotational velocity versus radial location (*center* and *right*).

It is important to understand that this methodology applies to concentric or eccentric annuli, to pipe or casing that may or may not be rotating, and to all fluid rheologies with or without yield stress. Moreover, pressure gradients associated with contiguous fluid slugs may be very different and, in general, may be discontinuous in their values. It is important to realize that pressures at the bit are important in modern ultra-deepwater applications, where drillers have to navigate narrow pressure windows to avoid safety problems. The calculations just discussed support the need for "constant pressure at the bit." They also provide decision options for pressure control using, for instance, changes in pump rate or mud rheology and weight, adjustment to surface pressures at the choke, or pressure variation by altering drillpipe rotational rate.

A second important application is cementing. Here it is also necessary to determine pressure profiles, first to obtain the pressures needed for pumping. The velocity fields and pressures at any location and at any time are available using the procedure previously outlined. Here the details of interfacial mixing may be important, and, if they are desired, calculations may be performed at the user's option. These details require solutions to the coupled velocity and species concentration equations that model both convection and diffusion processes. Calculations for pressure require minutes, primarily arising from problem setup, which have not yet been entirely automated.

More specifically, once the underlying pressure gradients for the different fluid slugs are available from calculations using the "building block" flow solvers, the pressure profile calculation itself requires only seconds. On the other hand, optional calculations for interfacial mixing details (e.g., thickness of different diffusion zones about the annular cross section, time scales required in setting them up, degree of mixing) may require several minutes to an hour, depending on the amount of resolution needed and the time scale of the computation. As in "Transient 2D," the borehole axis here is assumed to be straight.

The main menu shown in Figure 1.32 contains "Start," "Track (Interfaces)," "Zoom3D," and "Utilities" submenus. The "Start" menu shown in Figure 1.33 provides access to generic functions: for graphics installation (needed only once), user manual access, movie playback, access to "Steady

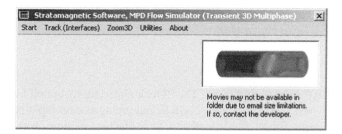

FIGURE 1.32

Main "Transient, 3D, multiphase" menu.

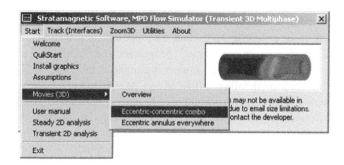

FIGURE 1.33

Start menu, with access to movies and other simulators.

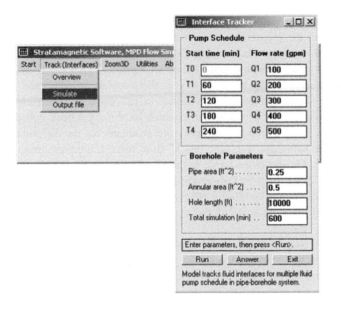

FIGURE 1.34

Interface tracker software (macroscopic properties).

2D" and "Transient 2D" solvers, and so on. The "Track (Interfaces)" option launches the "Interface Tracker" shown in Figure 1.34 with pump schedule and piping system input boxes. Clicking "Run" leads to numerical results for all fluid interfaces as a function of time, which can be viewed by clicking "Answer" or "Output file" under "Track (Interfaces)," as shown in Figure 1.35.

The "Zoom3D" menu hosts calculations for interfacial mixing details. As shown in Figure 1.36, there are two options: the first a "Newtonian mixtures (no rotation)" simulator and the second a "Non-Newtonian mixtures, rotating" option for general fluid rheologies with pipe or casing rotation. The former provides fast, automated calculations, taking advantage of mathematical simplifications offered by Newtonian fluids, while the latter, with more general capabilities, handles all fluid types for concentric or eccentric holes that may contain nonrotating or rotating pipe. While Figures 1.34 and 1.35 deal with "macroscopic" properties, the "Zoom3D" menu options provide "microscopic" solutions. The "Newtonian" option highlighted in Figure 1.36 launches the solver shown in Figure 1.37. Execution details are offered in Chapter 9.

The "Zoom3D" option highlighted in Figure 1.38, "Non-Newtonian mixtures, rotating," again, handles general non-Newtonian rheologies with pipes that may be rotating. All of the applications in this section allow high annular eccentricity. The highlighted option launches the two executable applications shown in Figures 1.39 and 1.40.

The main menu in Figure 1.39 contains time integration input boxes and the "Simulate" button. Prior to clicking "Simulate," the "Pump Schedule" must be defined via Figure 1.40. This screen

```
Multifluid Interface History.txt - Notepad                         _ |□| x|
File  Edit  Format  View  Help
Pump Schedule, Interface Tracking ...

  100 gpm:     0 min < T <   60 min
  200 gpm:    60 min < T <  120 min
  300 gpm:   120 min < T <  180 min
  400 gpm:   180 min < T <  240 min
  500 gpm:             T >  240 min

  Drillpipe area (ft^2):  0.250E+00
  Annular area   (ft^2):  0.500E+00
  Borehole length (ft):   0.100E+05
  Time simulation (min):      600

   ELAPSED  TIME    FLOW       Drillpipe Fluid Interface (feet)
   Minutes  Hours   GPMs    z(1)    z(2)    z(3)    z(4)    z(5)
.
.
.
      0      0.0    100       0       0       0       0       0
      1      0.0    100      53       0       0       0       0
      2      0.0    100     106       0       0       0       0
      3      0.1    100     160       0       0       0       0
      4      0.1    100     213       0       0       0       0
      5      0.1    100     267       0       0       0       0
```

FIGURE 1.35

Detailed numerical interface position results (macroscopic).

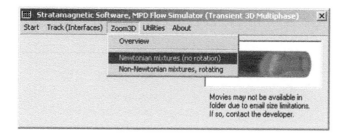

FIGURE 1.36

Newtonian mixtures, no rotation (microscopic properties).

requires inlet and outlet properties and also pressure gradient inputs for the different flow rates assumed.

The pressure gradients referred to in the previous paragraph are shown in Figure 1.41 and apply accordingly as the flows are concentric, rotating, Herschel-Bulkley, and so on. They are quickly launched by checking the option boxes at the bottom of the screen in Figure 1.40.

For completeness, the "Concentric steady flow" modules offered are listed in Figure 1.42, and the corresponding programs are shown in Figure 1.43. Their underlying math models and functions are described elsewhere in this book. For convenience, the "Steady 2D" and "Transient 2D"

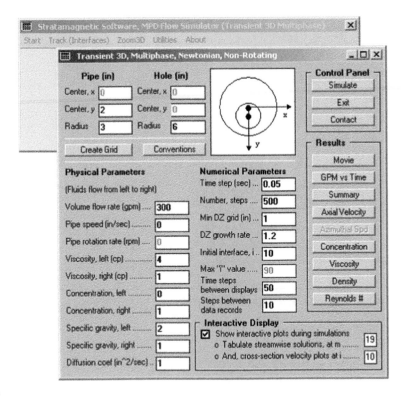

FIGURE 1.37

Newtonian mixtures, no rotation (microscopic).

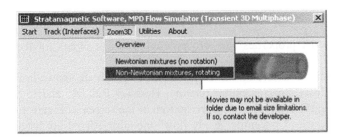

FIGURE 1.38

Non-Newtonian mixtures, rotating (microscopic properties).

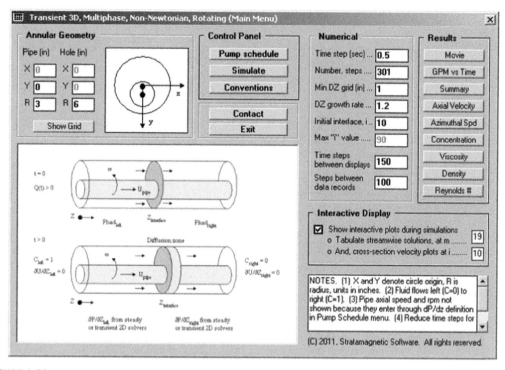

FIGURE 1.39

Non-Newtonian mixtures, rotating (microscopic, main menu).

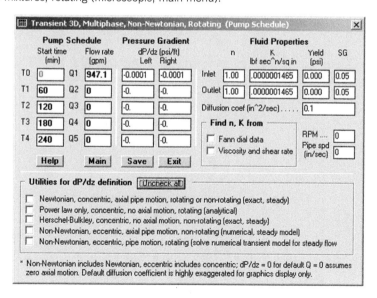

FIGURE 1.40

Non-Newtonian mixtures, rotating (pump schedule).

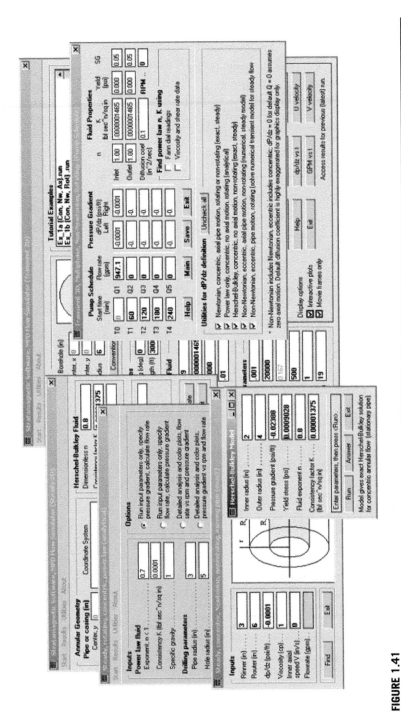

FIGURE 1.41

Pressure gradient utilities.

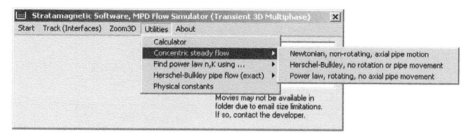

FIGURE 1.42

Concentric steady flow modules.

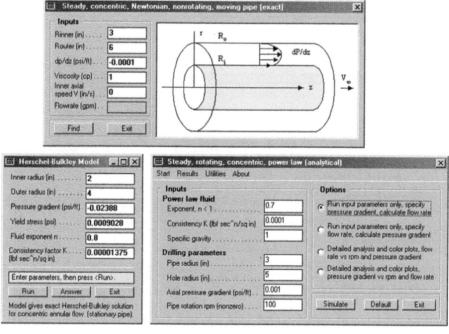

FIGURE 1.43

Three concentric steady flow programs.

annular flow functions are repeated in Figures 1.44 and 1.45 as reminders. The programs in Figure 1.46 support definitions of n and K in Power law applications. Finally, the model in Figure 1.47 provides access to exact Herschel-Bulkley *pipe* flow problems in applications where pressure profiles in the drillpipe are required.

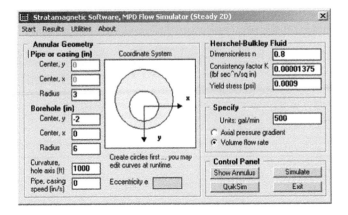

FIGURE 1.44

"Steady 2D" *annular* flow module.

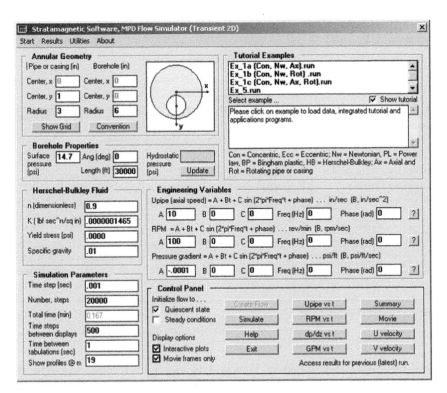

FIGURE 1.45

"Transient 2D" *annular* flow module.

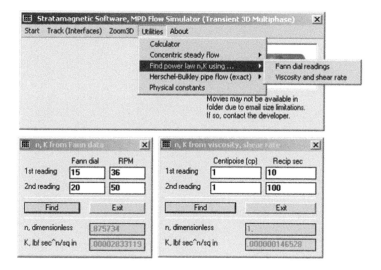

FIGURE 1.46

Finding n and K, utility programs.

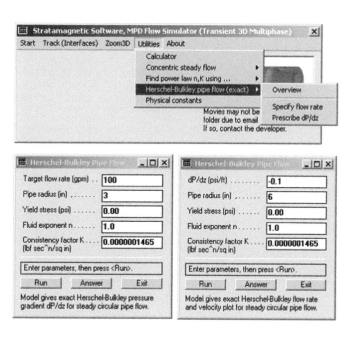

FIGURE 1.47

Exact Herschel-Bulkley *pipe* flow modules, for both forward and inverse calculations.

General Theory and Physical Model Formulation

Well-formulated mathematical models and accurate solutions invariably require a detailed and deep understanding of the partial differential equations underlying fluid flow. Many formulations exist—steady, transient, single-phase, multiphase, Newtonian, non-Newtonian, rotating versus nonrotating, and so on—and then in different coordinate systems, such as rectangular, cylindrical, and curvilinear grid. In this chapter, we develop the general mathematical formulations used in this book, which are to be solved using advanced numerical techniques developed especially for the eccentric annular flows encountered in modern drilling and cementing.

This book introduces a number of formulations important to modern problems in drilling and cementing in deviated wells, and emphasizes inputs and postprocessed quantities important to managed pressure drilling. Our methodologies are "open" to the extent that our models are described and analyzed in complete mathematical detail, subjected to detailed testing, and then validated against one another wherever their input parameters overlap. While many ideas and numerical approaches have been evaluated over the course of our research, only those that have passed our rigorous tests—and that provide fast and stable computational results—are discussed in this book and retained in the final software product for general dissemination.

We stress that a great deal of information can be obtained by studying the form of the equations, even without solution. Many of the fluid flow properties cited in this book were developed simply from visual examination of the equations, and only then were detailed algorithms designed to extract numerical details. Importantly, these properties were used to guide the development of our algorithms and also provided good checkpoints to ensure programming accuracy. In fact, many of the properties casually mentioned in our discussions throughout the book arose from qualitative analysis; to provide the reader with their scientific bases, their trains of thought are documented in Examples 2.1 and 2.2. This chapter develops the main ideas behind our work in annular flow, while Chapter 3 addresses numerical approaches that are key to the solutions of all formulations addressed in this book. Detailed calculated examples are given throughout.

EXAMPLE 2.1

Newtonian Flow Circular Cylindrical Coordinates

In this first example, we study simple Newtonian flows for which the laminar viscosity μ is constant. In practice, viscosity depends on pressure and temperature, but we restrict ourselves to simpler processes for which these dependencies do not arise—by "constant," we imply that viscosity is not affected by the size or shape of the vessel or by the applied pressure gradient or the flow rate, and that its value can be measured unambiguously in a simple viscometer—properties not applicable to flows of non-Newtonian fluids. In particular, we will explore the properties of Newtonian flows written in circular cylindrical coordinates—simple visual inspections of the equations *do* lead to interesting and important conclusions.

The so-called Navier-Stokes equations that apply are given in standard textbooks (e.g., Schlichting, 1968). When "r," "θ," and "z" are radial, azimuthal, and axial coordinates; v_r, v_θ, and v_z are Eulerian velocities in these directions; F_r, F_θ, and F_z are body forces in the same directions; ρ is the constant fluid density; p is pressure; and t is time, the following general partial differential equations can be derived.

Momentum equation in r:

$$\rho\{\partial v_r/\partial t + v_r \partial v_r/\partial r + v_\theta/r \partial v_r/\partial\theta - v_\theta^2/r + v_z \partial v_r/\partial z\} = F_r - \partial p/\partial r$$
$$+ \mu\{\partial^2 v_r/\partial r^2 + 1/r \partial v_r/\partial r - v_r/r^2 + 1/r^2 \partial^2 v_r/\partial\theta^2 - 2/r^2 \partial v_\theta/\partial\theta + \partial^2 v_r/\partial z^2\}$$

$$(2.1)$$

Momentum equation in θ:

$$\rho\{\partial v_\theta/\partial t + v_r \partial v_\theta/\partial r + v_\theta/r \, \partial v_\theta/\partial\theta + v_r v_\theta/r + v_z \partial v_\theta/\partial z\} = F_\theta - 1/r \, \partial p/\partial\theta$$
$$+ \mu\{\partial^2 v_\theta/\partial r^2 + 1/r \, \partial v_\theta/\partial r - v_\theta/r^2 + 1/r^2 \, \partial^2 v_\theta/\partial\theta^2 + 2/r^2 \, \partial v_r/\partial\theta + \partial^2 v_\theta/\partial z^2\}$$

$$(2.2)$$

Momentum equation in z:

$$\rho\{\partial v_z/\partial t + v_r \partial v_z/\partial r + v_\theta/r \, \partial v_z/\partial\theta + v_z \, \partial v_z/\partial z\} = F_z - \partial p/\partial z$$
$$+ \mu\{\partial^2 v_z/\partial r^2 + 1/r \, \partial v_z/\partial r + 1/r^2 \, \partial^2 v_z/\partial\theta^2 + \partial^2 v_z/\partial z^2\}$$

$$(2.3)$$

Mass continuity equation:

$$\partial v_r/\partial r + v_r/r + 1/r \, \partial v_\theta/\partial\theta + \partial v_z/\partial z = 0 \qquad (2.4)$$

The preceding define four equations for the four unknowns v_r, v_θ, v_z, and p. General solutions to these nonlinearly coupled partial differential equations do not exist. We emphasize that, while the formulation just given is written in circular cylindrical coordinates, it does apply to flows past noncircular geometries (in principle the flow through a star-shaped duct, for instance, can be solved, although in practice the solution would be extremely awkward). Understanding this, we ask what general conclusions can be drawn for concentric versus eccentric annular flows. For the remainder of this section, we will ignore the effects of externally imposed body forces (gravity, electric charge, etc.).

Concentric, steady, two-dimensional flows without influx

We first address the most commonly formulated problem—namely, concentric annular flows without azimuthal dependence, so that $\partial/\partial\theta = 0$ (this does not require that $v_\theta = 0$); flows without fluid influx or outflux, for which $v_r = 0$; and then those for which the problem is steady, so that $\partial/\partial t = 0$; finally, we invoke the restriction to purely two-dimensional flows whose properties do not vary

Example 2.1   **49**

from one cross section to the next, so that $\partial/\partial z = 0$. When these conditions are satisfied, the foregoing momentum equations reduce to Equations 2.5, 2.6, and 2.7, while Equation 2.4 for mass conservation is identically satisfied.

Momentum equation in r:

$$\partial p/\partial r = \rho v_\theta^2/r \tag{2.5}$$

Momentum equation in θ:

$$\partial^2 v_\theta/\partial r^2 + 1/r\,\partial v_\theta/\partial r - v_\theta/r^2 = 0 \tag{2.6}$$

Momentum equation in z:

$$\partial^2 v_z/\partial r^2 + 1/r\,\partial v_z/\partial r = (1/\mu)\partial p/\partial z \tag{2.7}$$

We will provide mathematical and software solutions to these later, but for now we emphasize their general properties. The linear azimuthal velocity field v_θ is determined by solving Equation 2.6 subject to constant values at the radial boundaries. At the inner pipe or casing surface, the speed is determined by rotational speed and radius, while at the outer annular wall, the speed is zero. Notice that the solution for v_θ does not involve $\partial p/\partial z$. In other words, the azimuthal motion is simply one induced by "dragging" at the inner pipe surface.

Now consider the solution for axial velocity found by the solution to Equation 2.7 subject to constant speeds at the radial boundaries (e.g., a zero or nonzero translational speed at the inner surface and zero at the outer wall). The solution does not involve the rotational speed and includes μ and the applied pressure gradient $\partial p/\partial z$ only to the extent that they appear in the lumped form $(1/\mu)\,\partial p/\partial z$.

In conclusion, the azimuthal motion does not affect axial flow, and axial motion does not influence azimuthal flow: The two are dynamically independent. Only when v_θ is available is Equation 2.5 used and then only in computing a radial pressure gradient that arises from centrifugal effects. It is remarkable that such general properties can be derived simply by visual inspection without any knowledge of partial differential equations.

Eccentric, steady, two-dimensional flow

Now let us repeat this analysis without the assumption calling for concentric flow; that is, we no longer assume that $\partial/\partial\theta = 0$. In doing so, we may deal with cross sections that contain eccentric circles, but the eccentric annuli may well contain asymmetric washouts at the outer contour and arbitrary cuttings beds at the bottom contour. We will again assume that $\partial/\partial t = \partial/\partial z = 0$, but no longer require $v_r = 0$. Thus, we have

Momentum equation in r:

$$\begin{aligned}
\rho\{v_r\,\partial v_r/\partial r + v_\theta/r\,\partial v_r/\partial\theta - v_\theta^2/r\} = &-\partial p/\partial r \\
&+ \mu\{\partial^2 v_r/\partial r^2 + 1/r\,\partial v_r/\partial r - v_r/r^2 + 1/r^2\partial^2 v_r/\partial\theta^2 - 2/r^2\,\partial v_\theta/\partial\theta\}
\end{aligned} \tag{2.8}$$

Momentum equation in θ:

$$\rho\{v_r\, \partial v_\theta/\partial r + v_\theta/r\, \partial v_\theta/\partial\theta + v_r v_\theta/r\} = -1/r\, \partial p/\partial\theta$$
$$+ \mu\{\partial^2 v_\theta/\partial r^2 + 1/r\, \partial v_\theta/\partial r - v_\theta/r^2 + 1/r^2\, \partial^2 v_\theta/\partial\theta^2 + 2/r^2\, \partial v_r/\partial\theta\} \tag{2.9}$$

Momentum equation in z:

$$\rho\{v_r\partial v_z/\partial r + v_\theta/r\, \partial v_z/\partial\theta\} = -\partial p/\partial z$$
$$+ \mu\{\partial^2 v_z/\partial r^2 + 1/r\, \partial v_z/\partial r + 1/r^2\, \partial^2 v_z/\partial\theta^2\} \tag{2.10}$$

Mass continuity equation:

$$\partial v_r/\partial r + v_r/r + 1/r\, \partial v_\theta/\partial\theta = 0 \tag{2.11}$$

These remain four coupled partial differential equations in four unknowns, whereas Equations 2.6 and 2.7 are uncoupled, linear, ordinary differential equations. Hence, the solutions are difficult to obtain. Now, we have not yet specified an annular geometry, nor have we defined the r−θ coordinate system that applies to the problem. Nonetheless, we can assume in a dimensionless sense that $v_\theta \gg v_r$ so that v_r can be ignored in a first approximation. Our main focus is the resulting momentum equation in z, which now takes the form

$$\partial^2 v_z/\partial r^2 + 1/r\, \partial v_z/\partial r + 1/r^2\, \partial^2 v_z/\partial\theta^2 \approx (1/\mu)\partial p/\partial z + (\rho/\mu)(v_\theta/r)\partial v_z/\partial\theta \tag{2.12}$$

This should be compared with the earlier result in Equation 2.7; that is, $\partial^2 v_z/\partial r^2 + 1/r\, \partial v_z/\partial r = (1/\mu)\, \partial p/\partial z$. The left side now includes an additional term "$1/r^2\, \partial^2 v_z/\partial\theta^2$." In fact, the left-hand operator can be written in the more familiar form "$\partial^2 v_z/\partial x^2 + \partial^2 v_z/\partial y^2$" using rectangular coordinates, from which we recognize the standard Laplace operator. However, it is the right side that is extremely interesting. No longer is the effective pressure gradient simply given by the constant value $(1/\mu)\, \partial p/\partial z$. Instead, this term is modified by the correction $(\rho/\mu)(v_\theta/r)\, \partial v_z/\partial\theta$, which we emphasize is proportional to the fluid density and the inner pipe rotational rate, and is inversely proportional to viscosity.

What are the physical consequences of this modification? In the concentric problem, the total volumetric flow rate could be determined by integrating the product of v_z (from Equation 2.7) and "$2\pi r\, dr$" over the annular domain. The result is proportional to $(1/\mu)\, \partial p/\partial z$, with the constant of proportionality depending only on geometry. The flow rate does not depend on rotational speed. When eccentricity is permitted, however, the effects of pipe rotation *are* coupled nonlinearly. In addition, the correction to $(1/\mu)\, \partial p/\partial z$ now depends on the lumped parameter "mud weight × rpm/viscosity" in a nontrivial manner.

It is important to note that the correction depends on the spatial coordinates r and θ, as well as the yet to be determined solution $v_z(r, \theta)$: It is spatially variable, and volumetric flow rate will no longer depend on $(1/\mu)\, \partial p/\partial z$ alone. Thus, the effective pressure gradient changes from what we have in the concentric case. The flow rate will generally be different, and computations show that, for the same μ, $\partial p/\partial z$, and rpm, the effect of eccentricity is a strong reduction in flow. Note that this conclusion is obtained for a Newtonian fluid having constant viscosity. When non-Newtonian effects are considered, the competing effects of shear thinning will enter. These will be discussed separately in Example 2.2.

Example 2.1 **51**

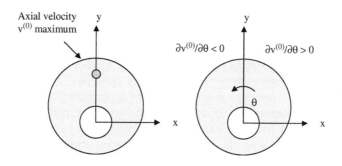

FIGURE 2.1

Location of axial velocity maximum in nonrotating flow.

Furthermore, because the correction also depends on $v_\theta \, \partial v_z/\partial\theta$, we expect that the location of the maximum in axial velocity (in an eccentric annulus with left-right symmetry) found at the wide side along the vertical line of symmetry (e.g., as shown at the left of Figure 2.1) will displace azimuthally, and it does, as our later explanation and all of our subsequent calculations will show.

It suffices to emphasize that eccentricity and rotation effects even for basic Newtonian fluids are extremely subtle. However, simple mathematical constructs can be devised to explore some of these subtleties and to facilitate fast numerical solutions. We explain an important one in the context of Equation 2.12 for v_z, which we rewrite without the subscript "z" for clarity. In mixed coordinates, we have the representation

$$\partial^2 v/\partial x^2 + \partial^2 v/\partial y^2 \approx (1/\mu)\partial p/\partial z + (\rho/\mu)(v_\theta/r)\partial v/\partial\theta \tag{2.13}$$

Now we separate "eccentric, nonrotating" from "eccentric, rotating" effects by isolating the inertia-dependent $(\rho/\mu)(v_\theta/r) \, \partial v/\partial\theta$. In the language of mathematics, we introduce a "regular perturbation expansion" such that $v = v^{(0)} + v^{(1)} + \cdots$ in which the zeroth solution represents leading-order concentric nonrotating effects and the first perturbation to it includes all others. Mathematics books that introduce this subject include the well-known research monographs by Van Dyke (1964), Cole (1968), and Nayfeh (1973). If we next assume that

$$\partial^2 v^{(0)}/\partial x^2 + \partial^2 v^{(0)}/\partial y^2 = (1/\mu)\partial p/\partial z \tag{2.14}$$

then subtraction of Equation 2.14 from Equation 2.13 with the series substitution leads to

$$\partial^2 v^{(1)}/\partial x^2 + \partial^2 v^{(1)}/\partial y^2 \approx (\rho/\mu)(v_\theta/r)\partial v^{(0)}/\partial\theta \tag{2.15}$$

The concentric solution to Equation 2.14, or Equation 2.7, is just the classical Poiseuille pipe flow formula given elsewhere in this book and available in the general literature (e.g., Schlichting (1968)). However, Equation 2.14 applies to eccentric problems too, and its exact numerical solution for arbitrary geometries is a subject of this book and one of the simulators.

But we do not need to solve it to understand its implications. We have shown an eccentric annulus at the left of Figure 2.1 with left-right symmetry. We can imagine that we now have obtained a straight, nonrotating, "out of the page" axial flow solution $v^{(0)}$ applicable to the left diagram.

The location of maximum axial speed is shown at the gray dot. With the θ convention highlighted, it is clear that $\partial v^{(0)}/\partial \theta$ increases at the right of the line of symmetry, while it decreases at the left.

Next, observe that the sign of the azimuthal velocity v_θ in Equation 2.15 cannot change. Thus, $(\rho/\mu)(v_\theta/r)\, \partial v^{(0)}/\partial \theta$, which functions as an effective pressure gradient for the disturbance axial flow $v^{(1)}$, is antisymmetric with respect to the vertical line of symmetry: It subtracts flow on one side and adds at the other. This effective pressure gradient is variable throughout the annular cross section. The driver, which depends on the solution to the azimuthal problem, affects total flow rate in a nontrivial way, although for small values of "mud weight $\times$ rpm/viscosity," it is clear that the solution is proportional to it, with the $v^{(1)}$ field again being antisymmetric. This antisymmetry means that "mud weight $\times$ rpm/viscosity" does not significantly affect total flow rate if it is small. However, when it is large, our symmetry and antisymmetry ideas may break down. It is also clear how, in the presence of unsteady effects, arguments like those offered previously are not possible.

We note that, while we have provided useful discussions on rotation and eccentricity, the numerical solution for steady rotating flows in eccentric domains, even under the assumption of simplified Newtonian flow, has proven to be challenging. A limited number of papers on the subject have been published by several authors, but these have offered few formulation and numerical details and have declined to discuss computing times and numerical stability properties. The author, in fact, has written a steady, rotating flow solver for non-Newtonian eccentric annular flows, which converges for v_z under restrictive conditions. The controlling "mud weight $\times$ rpm/viscosity" parameter, for the larger values characteristic of those parameters used in practical drilling and cementing, always leads to numerical instability.

On the other hand, the perturbation problem for $v_z^{(1)}$ could be solved with unconditional stability; however, the linearization used clearly does not apply physically to high values of "mud weight $\times$ rpm/viscosity." In this book, however, steady-state flows with rotation are successfully solved by integrating the transient equations asymptotically in time until steady conditions are reached using a fast solver.

EXAMPLE 2.2

Shear-Thinning and Non-Newtonian Flow Effects

In the previous example, we studied Newtonian flows for which viscosity always remains constant to focus on the effects of rotation and eccentricity alone. Here we consider non-Newtonian fluids, which generally exhibit shear thinning, but we do not discuss rotation, so that we remove the convective effects of inertia. Whereas before the use of circular cylindrical coordinates facilitated our understanding of pipe rotation, we now introduce rectangular or Cartesian coordinates to assist in our explanations of non-Newtonian viscosity effects. We consider here eccentric annular flows formed by general closed curves (which need not be circular), but for simplicity we restrict ourselves to steady, two-dimensional, single-phase flows. These assumptions are removed later.

The equations for general fluid motions in three dimensions are available from many excellent textbooks (Bird, Stewart, and Lightfoot, 1960; Streeter, 1961; Schlichting, 1968; Slattery, 1981). We cite these without proof. For problems without inner pipe rotation, it turns out that their rectangular form is most suitable in deriving curvilinear coordinate transforms—as we later show, the relevant starting point for rotation effects is cylindrical radial coordinates.

Example 2.2   **53**

Governing equations

Let u, v, and w denote Eulerian fluid velocities, and F_z, F_y, and F_x denote body forces, in the z, y, and x directions, respectively, where (z, y, x) are Cartesian coordinates. Also, let ρ be the constant fluid density and p be the pressure; we denote by S_{zz}, S_{yy}, S_{xx}, S_{zy}, S_{yz}, S_{xz}, S_{zx}, S_{yx}, and S_{xy} the nine elements of the general extra stress tensor $\underline{\underline{S}}$. If t is time and ∂'s represent partial derivatives, the complete equations of motion obtained from Newton's law and mass conservation are

Momentum equation in z:

$$\rho(\partial u/\partial t + u\,\partial u/\partial z + v\,\partial u/\partial y + w\,\partial u/\partial x) = F_z - \partial p/\partial z + \partial S_{zz}/\partial z + \partial S_{zy}/\partial y + \partial S_{zx}/\partial x \quad (2.16)$$

Momentum equation in y:

$$\rho(\partial v/\partial t + u\,\partial v/\partial z + v\,\partial v/\partial y + w\,\partial v/\partial x) = F_y - \partial p/\partial y + \partial S_{yz}/\partial z + \partial S_{yy}/\partial y + \partial S_{yx}/\partial x \quad (2.17)$$

Momentum equation in x:

$$\rho(\partial w/\partial t + u\,\partial w/\partial z + v\,\partial w/\partial y + w\,\partial w/\partial x)$$
$$= F_x - \partial p/\partial x + \partial S_{xz}/\partial z + \partial S_{xy}/\partial y + \partial S_{xx}/\partial x \quad (2.18)$$

Mass continuity equation:

$$\partial u/\partial z + \partial v/\partial y + \partial w/\partial x = 0 \quad (2.19)$$

Simple rheological models

These equations apply to all Newtonian and non-Newtonian fluids. In continuum mechanics, the most common class of empirical models for incompressible, isotropic fluids assumes that $\underline{\underline{S}}$ can be related to the rate of deformation tensor $\underline{\underline{D}}$ by a relationship of the form

$$\underline{\underline{S}} = 2N(\Gamma)\underline{\underline{D}} \quad (2.20)$$

where the elements of $\underline{\underline{D}}$ are

$$D_{zz} = \partial u/\partial z \quad (2.21)$$

$$D_{yy} = \partial v/\partial y \quad (2.22)$$

$$D_{xx} = \partial w/\partial x \quad (2.23)$$

$$D_{zy} = D_{yz} = (\partial u/\partial y + \partial v/\partial z)/2 \quad (2.24)$$

$$D_{zx} = D_{xz} = (\partial u/\partial x + \partial w/\partial z)/2 \quad (2.25)$$

$$D_{yx} = D_{xy} = (\partial v/\partial x + \partial w/\partial y)/2 \quad (2.26)$$

In Equation 2.20, $N(\Gamma)$ is the "apparent viscosity" satisfying

$$N(\Gamma) > 0 \quad (2.27)$$

$\Gamma(z, y, x)$ being a scalar functional of u, v, and w defined by the tensor operation

$$\Gamma = \{2\text{trace}(\underline{\underline{D}} \bullet \underline{\underline{D}})\}^{1/2} \tag{2.28}$$

Unlike the constant laminar viscosity μ in classical Newtonian flow, we will demonstrate that the apparent viscosity depends on the details of the particular problem being considered—for example, the rheological model used, the exact annular geometry occupied by the fluid, the applied pressure gradient, or the net volumetric flow rate. Also, it varies with the position (z, y, x) in the annular domain. Thus, single measurements obtained from viscometers are usually not meaningful in practice. In fact, inferences can be very misleading.

Examples

To fix ideas, consider the simple but important Ostwald-de Waele model for two-parameter "Power law" fluids, for which

$$N(\Gamma) = K\,\Gamma^{n-1} \tag{2.29a}$$

where the "consistency factor" K and the "fluid exponent" n are constants. Such Power law fluids are "pseudoplastic" when $0 < n < 1$, Newtonian when $n = 1$, and "dilatant" when $n > 1$. Most drilling fluids are pseudoplastic.

In the limit ($n = 1$, $K = \mu$), Equation 2.29a reduces to the Newtonian model with $N(\Gamma) = \mu$, where μ is the constant laminar viscosity; in this classical limit, stress is directly proportional to the rate of strain. Only for Newtonian flows is volumetric flow rate a linear function of applied pressure gradient and inversely proportional to μ.

Power law and Newtonian fluids respond instantaneously to applied pressure and stress. But if the fluid behaves as a rigid solid until the net applied stresses have exceeded some known critical yield value, say S_{yield}, then Equation 2.29a can be generalized by writing

$$N(\Gamma) = K\,\Gamma^{n-1} + S_{yield}/\Gamma \ \text{ if } \ \{1/2\ \text{trace}(\underline{\underline{S}} \bullet \underline{\underline{S}})\}^{1/2} > S_{yield}$$
$$\underline{\underline{D}} = 0 \ \text{ if } \ \{1/2\ \text{trace}(\underline{\underline{S}} \bullet \underline{\underline{S}})\}^{1/2} < S_{yield} \tag{2.29b}$$

In this form, Equation 2.29b rigorously describes the Herschel-Bulkley fluid. When the limit ($n = 1$, $K = \mu$) is taken, the first equation becomes

$$N(\Gamma) = \mu + S_{yield}/\Gamma \ \text{ if } \ \{1/2\ \text{trace}(\underline{\underline{S}} \bullet \underline{\underline{S}})\}^{1/2} > S_{yield} \tag{2.29c}$$

This is the Bingham plastic model, where μ is now the "plastic viscosity." For Herschel-Bulkley and Bingham plastic flows in circular pipes, exact analytical solutions can be developed for velocity distribution, plug zone radius, and total flow rate (these limits include Newtonian and Power law fluids). Analogous solutions are available for flows between parallel plates. Exact solutions for concentric annuli are not presently available, but are derived in closed analytical form elsewhere in this book and used to validate numerical flow models.

Example 2.2   55

ILLUSTRATION 2.1

For tutorial purposes, we examine a limit of two-dimensional Power law flows, where the axial velocity $u(y, x)$ does not depend on the axial coordinate z. In the absence of rotation, the velocities v and w in the cross-plane satisfy $v = w = 0$, so the functional Γ or shear rate in Equation 2.29a takes the form

$$\Gamma = [(\partial u/\partial y)^2 + (\partial u/\partial x)^2]^{1/2} \tag{2.30}$$

and Equation 2.29a becomes

$$N(\Gamma) = K[(\partial u/\partial y)^2 + (\partial u/\partial x)^2]^{(n-1)/2} \tag{2.31}$$

The apparent viscosity reduces to the conventional $N(\Gamma) = K (\partial u/\partial y)^{(n-1)}$ formula for one-dimensional, parallel plate, and "slot flow" flows considered in the drilling and cementing literature.

When both independent variables y and x for the cross section are present, as in the case for eccentric annular flow, significant mathematical difficulty arises. For one, the ordinary differential equation for annular velocity in simple concentric geometries becomes a partial differential equation. And whereas the former requires boundary conditions at two points, the partial differential equation requires no-slip boundary conditions imposed along two arbitrarily closed curves.

The nonlinearity of the governing equation and the irregular annular geometry only compound these difficulties. Despite such complexities, the resulting problem is simple in a sense. The momentum equations for v and w vanish identically and that for mass conservation implies that $u = u(y,x)$ only. The single remaining equation is

$$\partial S_{zy}/\partial y + \partial S_{zx}/\partial x = \partial P/\partial z = \text{constant} \tag{2.32}$$

where the constant pressure gradient $\partial P/\partial z$ is prescribed. This is to be compared to the simpler Equation 2.7. Since $\underline{\underline{S}} = 2N\underline{\underline{D}}$, Equation 2.32 reduces to

$$\partial(N \, \partial u/\partial y)/\partial y + \partial(N \, \partial u/\partial x)/\partial x = \partial P/\partial z \tag{2.33}$$

Substitution of Equation 2.31 shows that Equation 2.33 can be written as a nonlinear Poisson equation—that is, as Equation 2.34—in the form

$$\begin{aligned}\partial^2 u/\partial y^2 + \partial^2 u/\partial x^2 = [\partial P/\partial z + (1-n)N(\Gamma)(u_y{}^2 u_{yy} \\ + 2u_y u_x u_{yx} + u_x{}^2 u_{xx})/(u_y{}^2 + u_x{}^2)]/N(\Gamma)\end{aligned} \tag{2.34}$$

which is to be compared with Equation 2.14. This equation, together with extensions for rotation and complicated rheological effects, is solved exactly in our software models. Our only purpose in writing it down explicitly here is to provide a "live" example showing why nonlinear effects are complicated.

The Newtonian limit with $n = 1$ reduces Equation 2.34 to the classical Poisson equation $\partial^2 u/\partial y^2 + \partial^2 u/\partial x^2 = (1/\mu) \, \partial P/\partial z$, with several important properties. For example, doubling the pressure gradient while doubling the viscosity leaves $u(y,x)$ unchanged: Only the lumped driver $(1/\mu) \, \partial P/\partial z$ appears. And, for instance, doubling $\partial P/\partial z$ with μ constant doubles u everywhere, a property obvious from simple rescaling. Also, μ is just the quantity measured in a viscometer, and its value remains unchanged for all pressure gradients and flow cross sections.

However, when n is not unity, the complicated terms at the right of Equation 2.34 remain. Casual observation leads us to conclude, for example, that doubling the pressure gradient will do *something*, but exactly what is uncertain. Because the divisor of $\partial P/\partial z$ is not just a constant "μ" but a complicated function that, because it depends on the as yet unknown solution $u(y,x)$, the so-called "apparent viscosity" is unknown. In fact, it will vary from case to case, and it will depend on the applied pressure gradient plus the size and shape of the vessel, and it will be variable throughout the flow cross section. Hence, we have the origin of the terms "shear thickening" and "shear thinning." Shear-thickening and shear-thinning fluids are non-Newtonian because their viscosities increase or decrease, respectively, as the applied shearing stress increases. "Silly Putty" is shear thickening, while ketchup is shear thinning.

ILLUSTRATION 2.2

As a second tutorial example, consider the case of steady helical flow with $\partial/\partial z = 0$ along the hole axis (refer to the discussion in Bird, Stewart, and Lightfoot (2002) for an example). In this case, the shear rate satisfies the formula $\Gamma = d\gamma/dt = \{(r\, d(v_\theta/r)/dr)^2 + (dv_z/dr)^2\}^{1/2} = \{(r\, d\Omega/dr)^2 + (dv_z/dr)^2\}^{1/2}$, where the usual rotational rate is defined by $\Omega = v_\theta/r$. An extension of this expression was used in Bird et al. to study non-Newtonian pipe flows with combined axial and azimuthal flow where both velocity fields are coupled. It is also used in Equation 5.47 in Chapter 5, where the effects of rotation in concentric annuli with Power law fluids are considered.

Note on Mass Density

It can be seen from Equation 2.32 that fluid density ρ completely disappears in this steady flow without rotation. However, it is important that, from Example 2.1, density remains significant when the flow rotates because the nonlinear convective terms do not vanish. It also goes without saying that density effects are all important in transient analysis because inertia is important. We will demonstrate later that steady flows can be computed from unsteady algorithms using small densities for rapid convergence—but this strategy is applicable only when there is no underlying pipe rotation.

Only n and K (and not "μ") are "absolutes" for Power law flow modeling; they can be obtained from viscometer measurements. The foregoing difficulties apply not just to Power law fluids but to all non-Newtonian fluids, with or without yield stress. When yield stresses are present, other complications arise—for example, the inability to identify a priori the size and shape of the plug zone means that such problems cannot be solved for practical annular geometries. We do, fortunately, offer a rigorous solution to this problem later.

In summary, we offer several general principles from the discussions of Examples 2.1 and 2.2. In particular,

- In Newtonian flow, viscosity is a constant of the motion (barring changes due to pressure and temperature) that is unambiguously determined from viscometer measurement.
- In nonrotating Newtonian flow, the lumped quantity $(1/\mu)\, \partial p/\partial z$ controls the dynamics, and changes to it will proportionally change u(y,z) everywhere. Thus, faster testing with inexpensive fluids, together with simple arithmetic extrapolation, can be used in engineering design.
- For concentric annuli in steady Newtonian rotating flow, azimuthal velocities do not depend on the pressure gradient, and axial flows are unaffected by rotation: The two are dynamically uncoupled.
- Annular eccentricity introduces changes to the applied pressure gradient that are variable throughout the flow domain (the velocity likewise scales differently at different cross-sectional locations) when rotation is allowed. Their magnitudes are proportional to the product "density × rpm/viscosity." This effect generally decreases the flow rate (as rotational speed increases) for a fixed pressure gradient—this nontrivial modification applies even to simple Newtonian fluids without shear thinning.
- Non-Newtonian fluids (even without rotation and three-dimensionality) exhibit shear-thickening and shear-thinning properties. In a concentric annulus with a rotating inner pipe, drilling fluid viscosity will decrease because of azimuthal motion, so that net flow rate increases relative to the nonrotating case, assuming that pressure gradient is fixed. Complications arise when this is countered by the effects of eccentricity—computational methods are required to determine the exact balances between the two.
- Non-Newtonian flows in eccentric borehole annuli with rotation will exhibit shear-dependent changes to viscosity, plus changes to applied pressure gradient that depend on rotational speed, fluid density, and viscosity (the "apparent viscosity" now varies throughout the flow domain). Simple rescaling arguments cannot be used to deduce flow properties for u(y,z) because the governing equations are extremely complicated in form.
- For non-Newtonian flows, laboratory testing and extrapolation are not possible because of the foregoing complications—hence, the only recourse for prediction and job planning is full-scale testing with actual nonlinear fluids or, alternatively, detailed computational fluid-dynamics analysis.

Example 2.2   **57**

Field and laboratory examples

Figures 2.2 and 2.3, together with the related discussions, are obtained from correspondence with John Lofton, of Chevron, to whom the author is grateful. Figure 2.2 provides a "pressure-while-drilling" (or PWD) log from a field run. PWD logs provide real-time pressures as they are conveyed to the surface with measurement-while-drilling tools and are essential to drilling safety. Such logs can monitor downhole conditions accurately and supply updates to calibrate software models used for planning.

Lofton writes,

Look at 1600 hrs on 25 April 02. After the connection at 1693' (red arrow), the pump is on (green curve), and the rotary is abruptly increased up to 100 RPM (red curve). The standpipe pressure

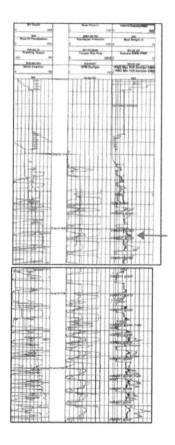

FIGURE 2.2

A "pressure-while-drilling" (PWD) log.

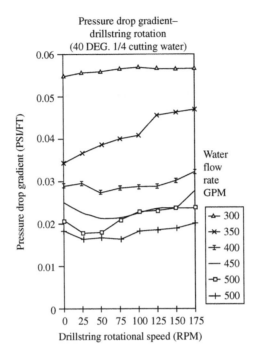

FIGURE 2.3

A laboratory example for a 40-degree well.

(blue curve) spikes—increased pump pressure. The ECD (red and black curves on the far right) both increase. The rotation has increased the pump pressure and the annular friction for the same pump rate. This response seems consistent throughout the PWD log. This is a directional well from a platform with an angle of less than 45 degrees and is using a low-density water-based mud.

I have also looked at broader industry applications—some of which I do not have first hand, on-location experience. There was a study done at the University of Tulsa on the effects of rotation in inclined wellbores. I think it is excellent, honest work: no products to sell, no bias on the outcome. The effects of rotation were investigated at 40, 65, and 90 degrees of inclination. The annular pressure was monitored with rotation at each of these inclinations. The results at 40 degrees were similar to those of the PWD log above and reflect my experience in the field, especially at the lower end of the flow-rates—300 gpm and 350 gpm.

Results for 65 and 90 degrees were more erratic, with some resulting in reduced pressure gradients—possibly because of hole geometry changes due to unflushed cuttings.

The last comment on pronounced rotation effects at lower volumetric flow rates is especially significant. Lower flow rates point toward higher values of the dimensionless azimuthal to axial velocity ratio, which is a good indicator of rotation coupling to the overall flow. Many drillers have

Example 2.3   **59**

also indicated cuttings transport problems in larger-diameter holes—large diameters are precisely the ones with smaller annular velocities. These two observations support the use of rotating pipe models in planning drilling jobs.

EXAMPLE 2.3

Curvilinear Grid Formulation for Highly Eccentric Annular Flows with General Non-Newtonian Fluids *without* Rotation

Here we consider eccentric annular flows formed by general closed curves (which need not be circular) in order to introduce the methodology. For simplicity, we restrict ourselves to steady, two-dimensional, single-phase flows of non-Newtonian without rotary pipe movement, although constant-speed axial translation is permitted. The effects of pipe or casing rotation require a different formalism that is developed in Example 2.4, one that builds on the present introductory work. Two-phase, three-dimensional flow extensions are covered later in this book. There is some redundancy between Example 2.2 and the following exposition; however, this is retained for completeness and clarity. The equations for general fluid motions in three spatial dimensions are available from many excellent textbooks (e.g., Bird, Stewart, and Lightfoot, 1960; Streeter, 1961; Schlichting, 1968; Slattery, 1981), and we cite these without proof.

Governing equations

Let u, v, and w denote Eulerian fluid velocities, and F_z, F_y, and F_x denote body forces in the z, y, and x directions, respectively, where (z, y, x) are Cartesian coordinates. Also, let ρ be the constant fluid density and p be the pressure; we denote by S_{zz}, S_{yy}, S_{xx}, S_{zy}, S_{yz}, S_{xz}, S_{zx}, S_{yx}, and S_{xy} the nine elements of the general extra stress tensor $\underline{S}$. If t is time and ∂'s represent partial derivatives, the complete transient equations of motion obtained from Newton's law and mass conservation are

Momentum equation in z:

$$\rho(\partial u/\partial t + u\,\partial u/\partial z + v\,\partial u/\partial y + w\,\partial u/\partial x)$$
$$= F_z - \partial p/\partial z + \partial S_{zz}/\partial z + \partial S_{zy}/\partial y + \partial S_{zx}/\partial x \qquad (2.35)$$

Momentum equation in y:

$$\rho(\partial v/\partial t + u\,\partial v/\partial z + v\,\partial v/\partial y + w\,\partial v/\partial x)$$
$$= F_y - \partial p/\partial y + \partial S_{yz}/\partial z + \partial S_{yy}/\partial y + \partial S_{yx}/\partial x \qquad (2.36)$$

Momentum equation in x:

$$\rho(\partial w/\partial t + u\,\partial w/\partial z + v\,\partial w/\partial y + w\,\partial w/\partial x)$$
$$= F_x - \partial p/\partial x + \partial S_{xz}/\partial z + \partial S_{xy}/\partial y + \partial S_{xx}/\partial x \qquad (2.37)$$

Mass continuity equation:

$$\partial u/\partial z + \partial v/\partial y + \partial w/\partial x = 0 \qquad (2.38)$$

Rheological flow models

The equations just given apply to all Newtonian and non-Newtonian fluids. In continuum mechanics, the most common class of empirical models for incompressible, isotropic fluids assumes that $\underline{\underline{S}}$ can be related to the rate of deformation tensor $\underline{\underline{D}}$ by an expression of the form

$$\underline{\underline{S}} = 2\,N(\Gamma)\,\underline{\underline{D}} \qquad (2.39)$$

where the elements of $\underline{\underline{D}}$ are

$$D_{zz} = \partial u/\partial z \qquad (2.40)$$

$$D_{yy} = \partial v/\partial y \qquad (2.41)$$

$$D_{xx} = \partial w/\partial x \qquad (2.42)$$

$$D_{zy} = D_{yz} = (\partial u/\partial y + \partial v/\partial z)/2 \qquad (2.43)$$

$$D_{zx} = D_{xz} = (\partial u/\partial x + \partial w/\partial z)/2 \qquad (2.44)$$

$$D_{yx} = D_{xy} = (\partial v/\partial x + \partial w/\partial y)/2 \qquad (2.45)$$

In Equation 2.39, $N(\Gamma)$ is the "apparent viscosity" satisfying

$$N(\Gamma) > 0 \qquad (2.46)$$

$\Gamma(z, y, x)$ being a scalar functional of u, v, and w defined by the tensor operation

$$\Gamma = \{2\ \text{trace}(\underline{\underline{D}} \bullet \underline{\underline{D}})\}^{1/2} \qquad (2.47)$$

Unlike the constant laminar viscosity in classical Newtonian flow, the apparent viscosity depends on the details of the particular problem being considered—for example, the rheological model used, the exact annular geometry occupied by the fluid, the applied pressure gradient, or the net volumetric flow rate. Also, it varies with the position (z, y, x) in the annular domain. Thus, single measurements obtained from viscometers may not be meaningful in practice.

Power law fluids

As a model to fix ideas, we will focus briefly on one practical but important simplification. Our discussion later applies to general fluids with yield stresses. For now, the Ostwald-de Waele model for two-parameter "Power law" fluids assumes

$$N(\Gamma) = K\,\Gamma^{n-1} \qquad (2.48a)$$

where the "consistency factor" K and the dimensionless "fluid exponent" n are constants.

Such Power law fluids are "pseudoplastic" when $0 < n < 1$, Newtonian when $n = 1$, and "dilatant" when $n > 1$. Most drilling fluids are pseudoplastic. In the limit ($n = 1$, $K = \mu$), Equation 2.48a reduces to the Newtonian model with $N(\Gamma) = \mu$, where μ is the constant laminar viscosity; in this limit, stress is directly proportional to the rate of strain. Only for Newtonian flows is total volumetric flow rate a linear function of applied pressure gradient.

Example 2.3   **61**

Herschel-Bulkley yield stress fluids

Power law and Newtonian fluids respond instantaneously to applied pressure and stress. But if the fluid behaves as a rigid solid until the net applied stresses have exceeded some known critical yield value, say S_{yield}, then Equation 2.48a can be generalized by writing

$$N(\Gamma) = K \, \Gamma^{n-1} + S_{yield}/\Gamma \ \text{ if } \ \{1/2 \ \text{trace}(\underline{\underline{S}} \bullet \underline{\underline{S}})\}^{1/2} > S_{yield}$$

$$\underline{\underline{D}} = 0 \ \text{ if } \ \{1/2 \ \text{trace}(\underline{\underline{S}} \bullet \underline{\underline{S}})\}^{1/2} < S_{yield} \qquad (2.48b)$$

In this form, Equation 2.48b rigorously describes the general Herschel-Bulkley fluid. When the limit ($n = 1$, $K = \mu$) is taken, the first equation becomes

$$N(\Gamma) = \mu + S_{yield}/\Gamma \ \text{ if } \ \{1/2 \ \text{trace}(\underline{\underline{S}} \bullet \underline{\underline{S}})\}^{1/2} > S_{yield} \qquad (2.48c)$$

This is the Bingham plastic model, where μ is now the "plastic viscosity." For Herschel-Bulkley and Bingham plastic flows in circular pipes, exact analytical solutions can be developed for velocity distribution, plug zone radius, and total flow rate (these limits include Newtonian and Power law fluids), and are given elsewhere in this book. Analogous solutions are available for flows between parallel plates. Exact solutions for concentric annuli do not appear to be available, but are derived in closed analytical form later in this book; these are used to validate various numerical eccentric flow models.

Conventionally, until now, eccentric annular flows containing fluids with nonzero yield stresses have been more difficult to analyze, both mathematically and numerically, than those marked by zero yield. This is so because there coexist "dead" (or "plug") and "shear" flow regimes with distinct internal boundaries that must be determined as part of the solution. This "plug versus no-plug" transition introduces a type of nonlinearity in the formulation, which exists even for "n = 1" Bingham plastics; Herschel-Bulkley nonlinearities associated with fractional n's make matters worse. Fluids with yield stresses complicate the grid generation problem because distinct but unknown internal boundaries exist.

In particular, even if a plug zone's size and shape are known, the transition contour itself defines a coordinate curve: Radial-like lines approaching from either side and crossing it have slope discontinuities at intersections, and the underlying conservation laws have to be rederived with such properties in mind. In dealing with yield stress fluids, the existence of a second isolated domain has long impeded flow-modeling efforts, and solutions for complicated annular domains have been impossible. Fortunately, this problem has been addressed and solved from a different perspective using new but mathematically rigorous methods. The approach, described at the end of this example, applies to all fluids with and without yield stresses.

Borehole configuration

Our configuration is shown in Figure 2.4. A drillpipe (or casing) and borehole combination is inclined at an angle α relative to the ground, with $\alpha = 0°$ for horizontal and $\alpha = 90°$ for vertical wells. Here "z" denotes any point within the annular fluid; section AA is a cut taken normal to the

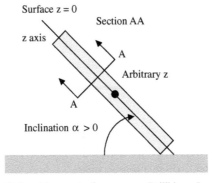

FIGURE 2.4

Borehole configuration.

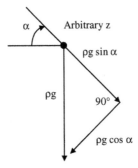

FIGURE 2.5

Gravity vector components.

local z axis. Figure 2.5 resolves the vertical body force due to gravity at "z" into components parallel and perpendicular to the axis, while Figure 2.6 provides a detailed picture of the cross section at section AA.

Now we specialize the previous equations to downhole flows. In Figures 2.4, 2.5, and 2.6, we have aligned z, which increases downward, with the axis of the borehole. The axis may be inclined, varying from $\alpha = 0°$ for horizontal holes to 90° for vertical holes. The plane of the variables (y,x) is perpendicular to the z axis, and (z, y, x) are mutually orthogonal Cartesian coordinates. The body forces due to the gravitational acceleration g can be resolved into components

$$F_z = \rho \, g \, \sin \alpha \qquad (2.49)$$

$$F_x = -\rho \, g \, \cos \alpha \qquad (2.50)$$

$$F_y = 0 \qquad (2.51)$$

Example 2.3   **63**

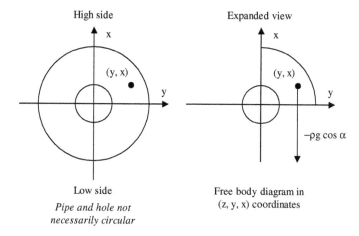

High side

Expanded view

(y, x)

(y, x)

$-\rho g \cos \alpha$

Low side

Pipe and hole not necessarily circular

Free body diagram in (z, y, x) coordinates

FIGURE 2.6

Gravity vector components.

If we now assume that the drillpipe does not rotate, the resulting flow can only move in a direction parallel to the borehole axis. This requires that the velocities v and w vanish. Axial translation is still permissible. Therefore,

$$v = w = 0 \qquad (2.52)$$

Since the analysis applies to constant density flows, we obtain

$$\partial \rho / \partial t = 0 \qquad (2.53)$$

Equations 2.38, 2.52, and 2.53 together imply that the axial velocity u(y, x, t) does not depend on z. And, if we further confine ourselves to steady laminar flow (that is, to flows driven by axial pressure gradients that do not vary in time), we find that

$$u = u(y, x) \qquad (2.54)$$

depends at most on two independent variables—namely, the cross-sectional coordinates y and x. For a concentric drillpipe and borehole, it is more convenient to collapse y and x into a radial coordinate $r = (x^2 + y^2)^{1/2}$, for which we later provide a complete analysis. For general eccentric flows, the lack of similar algebraic transformations drives the use of grid generation methods. Substitution of Equations 2.52 and 2.54 into Equations 2.35, 2.36, and 2.37 leads to

$$0 = \rho\, g \sin \alpha - \partial p / \partial z + \partial S_{zy} / \partial y + \partial S_{zx} / \partial x \qquad (2.55)$$

$$0 = -\partial p / \partial y \qquad (2.56)$$

$$0 = -\rho\, g \cos \alpha - \partial p / \partial x \qquad (2.57)$$

If we introduce, without loss of generality, the pressure separation of variables,

$$P = P(z, x) = p - z\, \rho\, g \sin \alpha + x\, \rho\, g \cos \qquad (2.58)$$

we can replace Equations 2.55, 2.56, and 2.57 by the single equation

$$\partial S_{zy}/\partial y + \partial S_{zx}/\partial x = \partial P/\partial z = \text{constant} \tag{2.59}$$

where the constant pressure gradient $\partial P/\partial z$ is prescribed. Recall the definitions of the deformation tensor elements given in Equations 2.40 through 2.45 and the fact that $\underline{S} = 2N\underline{D}$ to rewrite Equation 2.59 as

$$\partial(N \, \partial u/\partial y)/\partial y + \partial(N \, \partial u/\partial x)/\partial x = \partial P/\partial z \tag{2.60}$$

Early approach, viscosity derivatives omitted

The general "after 2010" method available at the end of this example applies to fluid flows with and without yield stresses. In the next few paragraphs, though, we summarize for completeness the approach used from 1990 to 2010, since a number of algorithms developed and distributed during that time frame required the simplifications discussed here. We will explain their need and the reasons for their deficiencies and provide the needed fixes.

Now, for two-dimensional flows whose velocities do not depend on the axial coordinate z and that further satisfy the nonrotating flow assumption, we have $v = w = 0$. The functional Γ in Equation 2.48a takes the simple form

$$\Gamma = [(\partial u/\partial y)^2 + (\partial u/\partial x)^2]^{1/2} \tag{2.61}$$

so that Equation 2.48b becomes

$$N(\Gamma) = K \, \Gamma^{n-1} + S_{yield}/\Gamma \ \ \text{if} \ \ \{1/2 \, \text{trace}(\underline{S} \bullet \underline{S})\}^{1/2} > S_{yield}$$

$$\Gamma = [(\partial u/\partial y)^2 + (\partial u/\partial x)^2]^{1/2} \tag{2.62}$$

$$\underline{D} = 0 \ \ \text{if} \ \ \{1/2 \, \text{trace}(\underline{S} \bullet \underline{S})\}^{1/2} < S_{yield}$$

The apparent viscosity reduces to the conventional $N(\Gamma) = K \, (\partial u/\partial y)^{(n-1)}$ formula for one-dimensional, parallel plate, and "slot flow" flows considered in the literature in the Power law limit. When both independent variables y and x for the cross section are present, as for eccentric annular flow, significant mathematical difficulty arises. For one, the ordinary differential equation for annular velocity in simple concentric geometries becomes a partial differential equation. And whereas the former requires boundary conditions at two points, the partial differential equation requires no-slip boundary conditions imposed along two arbitrarily closed curves. The nonlinearity of the governing equation and the irregular annular geometry only compound these difficulties.

We illustrate the decades-old problem by returning to our Power law example, for which the apparent viscosity function $N(\Gamma)$ is given exactly by the nonlinear equation

$$N(\Gamma) = K[(\partial u/\partial y)^2 + (\partial u/\partial x)^2]^{(n-1)/2} \tag{2.63}$$

Note that Equation 2.60 (that is, $\partial \, (N \, \partial u/\partial y)/\partial y + \partial \, (N \, \partial u/\partial x)/\partial x = \partial P/\partial z$) and Equation 2.63 make up the entire system to be solved along with general no-slip velocity boundary conditions at drill-pipe and borehole surfaces. This formulation also allows constant-velocity axial pipe movement.

Example 2.3   **65**

It is important, for the purposes of numerical analysis, to recognize how the net result can be written as a nonlinear Poisson equation:

$$\partial^2 u/\partial y^2 + \partial^2 u/\partial x^2 = [\partial P/\partial z + (1-n)N(\Gamma)$$
$$(u_y^2 u_{yy} + 2u_y u_x u_{yx} + u_x^2 u_{xx})/(u_y^2 + u_x^2)]/N(\Gamma) \tag{2.64}$$

In this form, conventional solution techniques for elliptic equations were employed at first. These include iterative techniques as well as direct inversion methods. The nonlinear terms in the square brackets, for example, were evaluated using the latest values in a successive approximations scheme.

Also, various algebraic simplifications were used at different times. For some values of n, particularly those near unity, these nonlinear terms represented negligible higher-order effects if the "1-n" terms were small in a dimensionless sense compared with pressure gradient effects. For small n, the second derivative terms on the right side may be unimportant, since such flows contain flat velocity profiles. In most of the early work, the principal effects of nonlinearity were modeled using the simpler and more stable Poisson model in Equation 2.65, one that is not unlike the classical equation for Newtonian flow. The apparent viscosity that acts in concert with the driving pressure gradient was still variable, nonlinear, and dependent on both geometry and rate.

$$\partial^2 u/\partial y^2 + \partial^2 u/\partial x^2 \approx N(\Gamma)^{-1} \partial P/\partial z \tag{2.65}$$

This approximation was used because the additional terms in Equation 2.64 were numerically unstable. This is not a constant viscosity model because N is still nonlinear and variable throughout the flow cross section; only its spatial derivatives were ignored. The model, which appeared to be unconditionally stable, importantly retained the strong influence of local geometry on annular velocity (e.g., low bottom speeds in eccentric holes regardless of rheology or flow rate), and extensive comparisons with detailed numerical models and laboratory data suggested that the results were reasonable. Recent extensions, however, now allow us to keep Equation 2.60 in its entirety, enabling fast and stable calculations even for problems with yield stress. Mathematical and numerical details are offered later in this chapter.

Additional postprocessing formulas

Once the solution for the velocity field is available, additional formulas are evaluated to provide useful physical information. Borehole temperature is sometimes important in drilling and cementing. Our momentum model then requires a coupled solution to a temperature partial differential equation with convective and conductive terms. In the general case, a source term accounts for heat generation by internal friction and heat may flow to and from formation boundaries. External and internal sources of heat may affect local fluid viscosity, since n and K depend on temperature.

For problems that are not isothermal, the existence of a steady flow is not guaranteed. The estimation of heat source strength from velocity gradients, while most likely unimportant, is nonetheless discussed here because the cumulative effects of distributed sources over large time scales may have a pronounced effect on the flow if this frictional heat is not dissipated into the boundary. For completeness, we therefore give the expression for the "dissipation function"—that is, the distributed heat source term—depends on local velocity gradients.

We noted that when the temperature $T(z, y, x, t)$ is important, a partial differential equation with conductive and convective terms couples to our momentum equations. This energy equation contains a positive definite quantity Φ called the "dissipation function" that represents the distributed source term responsible for local heat generation. In general, it takes the form

$$
\begin{aligned}
\Phi(z, y, x) = {} & S_{zz}\partial u/\partial z + S_{yy}\partial v/\partial y + S_{xx}\partial w/\partial x \\
& + S_{zy}(\partial u/\partial y + \partial v/\partial z) + S_{zx}(\partial u/\partial x + \partial w/\partial z) \\
& + S_{yx}(\partial v/\partial x + \partial w/\partial y)
\end{aligned}
\tag{2.66}
$$

Applying assumptions consistent with the foregoing analysis, we obtain

$$
\Phi = N(\Gamma)\{(\partial u/\partial y)^2 + (\partial u/\partial x)^2\} > 0
\tag{2.67}
$$

where, as before, we use Equation 2.63 for the apparent viscosity in its entirety. Equation 2.67 shows that velocity gradients, not magnitudes, contribute to temperature increases.

In other computations, we provide values of local viscous stresses and their corresponding shear rates. These stresses are the rectangular components

$$
S_{zy} = N(\Gamma)\partial u/\partial y
\tag{2.68}
$$

$$
S_{zx} = N(\Gamma)\partial u/\partial x
\tag{2.69}
$$

The shear rates in these equations are $\partial u/\partial y$ and $\partial u/\partial x$, respectively. These quantities are useful for several reasons. They are physically important in estimating the efficiency with which fluids in deviated wells remove cuttings beds having specified mechanical properties. From the numerical analysis point of view, they allow checking of computed solutions for physical consistency (e.g., high values at solid surfaces, zeros within plug flows) and required symmetries.

We next discuss mathematical issues regarding computational grid generation and numerical solution. These ideas are highlighted because we solve the complete boundary value problem, satisfying no-slip velocity conditions exactly, without simplifying the annular geometry.

Boundary-conforming, curvilinear grid generation

In many engineering problems, a judicious choice of coordinate systems simplifies calculations and brings out the salient physical features more transparently than otherwise. For example, the use of cylindrical coordinates for single-well problems in petroleum engineering leads to elegant "radial flow" results that are useful in well testing. Cartesian grids, on the other hand, are preferred in simulating oil and gas flows from rectangular fields.

The annular geometry modeling considered here is aimed at eccentric flows with cuttings beds, arbitrary borehole wall deformations, and unconventional drill collar or casing-centralizer cross sections. Obviously, simple coordinate transforms are not readily available to handle arbitrary domains of flow. Without resorting to crude techniques, for instance, applying boundary conditions along mean circles and squares, or invoking "slot flow" assumptions, there has been no real flow modeling alternative until the first publication of *Borehole Flow Modeling*.

Example 2.3   **67**

Fortunately, results from differential geometry allow us to construct "boundary-conforming, curvilinear coordinates" that are natural for physical modeling and computation. These general techniques extend classical ideas on conformal mapping. They have accelerated progress in simulating aerospace flows past airfoils and cascades, and are only beginning to be applied in the petroleum industry. Thompson, Warsi, and Mastin (1985) provide an excellent introduction to the subject.

To those familiar with conventional analysis, it may seem that the choice of (y, x) coordinates in Equation 2.65 is "unnatural." After all, in the limit of a concentric annulus, the equation does not reduce to a radial "r-only" formulation. But our use of such coordinates was motivated by the new gridding methods that, like classical conformal mapping, are founded on Cartesian coordinates. The approach, developed in detail in Example 3.1 (in Chapter 3) in essence requires us to first solve a set of nonlinearly coupled, second-order partial differential equations. In particular,

$$(y_\eta^2 + x_\eta^2)y_{\xi\xi} - 2(y_\xi y_\eta + x_\xi x_\eta)y_{\xi\eta} + (y_\xi^2 + x_\xi^2)y_{\eta\eta} = 0 \tag{2.70}$$

$$(y_\eta^2 + x_\eta^2)x_{\xi\xi} - 2(y_\xi y_\eta + x_\xi x_\eta)x_{\xi\eta} + (y_\xi^2 + x_\xi^2)x_{\eta\eta} = 0 \tag{2.71}$$

are considered with special mapping conditions related to the annular geometry. These are no simpler than the original flow equations, and arguably worse, since there are now two more equations, but they importantly introduce a first step that does not require solution on complicated domains.

Equations 2.70 and 2.71 are, importantly, solved on simple rectangular (ξ, η) grids. Once the solution is obtained, the results for $x(\xi, \eta)$ and $y(\xi, \eta)$ are used to generate the metric transformations needed to reformulate the physical equations for u in (ξ, η) coordinates. The flow problem is then solved in these rectangular computational coordinates using standard numerical methods. These new coordinates implicitly contain all the details of the input geometry, providing fine resolution in tight spaces as needed. To see why, we now briefly describe the boundary conditions used in the mapping. Figures 2.7(a) and 2.7(b) indicate how a general annular region would map into a rectangular computational space under the proposed scheme.

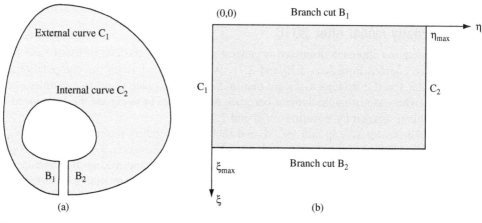

FIGURE 2.7

(a) Irregular physical (y, x) plane. (b) Rectangular computational plane.

Again, the idea rests with special computational coordinates (ξ, η). A discrete set of "user-selected" physical coordinates (y, x) along curve C_1 in Figure 2.7(a) is specified along the straight line $\eta = 0$ in Figure 2.7(b). Similarly, (y, x) values obtained from curve C_2 in Figure 2.7(a) are specified along $\eta = \eta_{max}$ in Figure 2.7(b). Values for (y, x) chosen along "branch cuts" $B_{1,2}$ in Figure 2.7(a) are required to be single-valued along edges $\xi = 0$ and $\xi = \xi_{max}$ in Figure 2.7(b).

With (y, x) prescribed along the rectangle of Figure 2.7(b), Equations 2.70 and 2.71 for $y(\xi, \eta)$ and $x(\xi, \eta)$ can be numerically solved. Once the solution is obtained, the one-to-one correspondences between all physical points (y, x) and computational points (ξ, η) are known. The latter is the domain chosen for numerical computation for annular velocity. Finite difference representations of the no-slip conditions "u = 0" that apply along C_1 and C_2 of Figure 2.7(a) are very easily implemented in the rectangle of Figure 2.7(b). At the same time, the required modifications to the governing equation for u(y, x) are modest. For example, the simplified Equation 2.65 becomes

$$(y_\eta^2 + x_\eta^2)u_{\xi\xi} - 2(y_\xi y_\eta + x_\xi x_\eta)u_{\xi\eta}$$
$$+ (y_\xi^2 + x_\xi^2)u_{\eta\eta} = (y_\xi x_\eta - y_\eta x_\xi)^2 \partial P / \partial z / N(\Gamma) \tag{2.72}$$

whereas the result for Equation 2.64 requires additional terms. For Equation 2.72 and its exact counterpart, the velocity terms in the apparent viscosity $N(\Gamma)$ of Equation 2.63 transform according to

$$u_y = (x_\eta u_\xi - x_\xi u_\eta)/(y_\xi x_\eta - y_\eta x_\xi) \tag{2.73}$$

$$u_x = (y_\xi u_\eta - y_\eta u_\xi)/(y_\xi x_\eta - y_\eta x_\xi) \tag{2.74}$$

These relationships are also used to evaluate the dissipation function. Again, we emphasize that in our solution for velocity, Equation 2.72 is importantly solved in rectangular computational coordinates (ξ, η). We leave the details to be developed in Chapter 3, and for now continue with our development of broad and basic concepts.

Exact viscosity model after 2010

The grid generation approach discussed is general and rigorous, and fast methods have been developed to solve coupled Equations 2.70 and 2.71. We have transformed the computational problem for the annular speed u from an awkward one in the physical plane (y, x) to a simpler one in (ξ, η) coordinates, where the irregular domain becomes rectangular. In doing so, we introduced the intermediate problem dictated by Equations 2.70 and 2.71.

When solutions for $y(\xi, \eta)$ and $x(\xi, \eta)$ and their corresponding metrics are available, Equation 2.72, which is slightly more complicated than the original Equation 2.65, can be solved conveniently using existing "rectangle-based" methods without compromising the annular geometry. Although complicated, containing more algebraic terms in the host equation for velocity, the resulting system does allow faster computing because fewer equations are actually used in the iteration process. For iterative methods, fewer equations mean much faster convergence.

The assumptions behind Equation 2.65, used in earlier work and introduced purely for numerical stability, are physically unacceptable for problems where apparent viscosity varies significantly

Example 2.3   **69**

within the flow domain. The solution ultimately developed, allowing us to retain all apparent viscosity derivatives that lead to almost unconditional stability, was simple. In particular, recall that the complete equation in rectangular coordinates, Equation 2.60,

$$\partial(N \, \partial u/\partial y)/\partial y + \partial(N \, \partial u/\partial x)/\partial x = \partial P/\partial z$$

can be expanded in the form

$$\partial^2 u/\partial y^2 + \partial^2 u/\partial x^2 + N_y/N \, \partial u/\partial y + N_x/N \, \partial u/\partial x = 1/N \, \partial P/\partial z \qquad (2.75)$$

so that $\partial^2 u/\partial y^2 + \partial^2 u/\partial x^2 = 1/N \, \partial P/\partial z - N_y/N \, \partial u/\partial y - N_x/N \, \partial u/\partial x$. For practical reasons, in the past it was simplest to modify the old solution algorithm by replacing "$1/N \, \partial P/\partial z$" with "$1/N \, \partial P/\partial z - N_y/N \, \partial u/\partial y - N_x/N \, \partial u/\partial x$." The new terms were evaluated using the latest available solutions for u in an iterative semi-explicit scheme, or by solving a fully implicit scheme in which the new terms were determined at the same time as those in $\partial^2 u/\partial y^2 + \partial^2 u/\partial x^2$. Both of these approaches often led to unstable or nonconvergent results.

It can be shown, after detailed numerical testing, that the "obvious" central differencing used in our prior numerical approach—that is,

$$(1/N \, \partial N/\partial x)_i \approx 1/N_i(N_{i+1} - N_{i-1})/(x_{i+1} - x_{i-1}) \qquad (2.76)$$

is numerically unstable, while the less obvious approach utilizing

$$(1/N \, \partial N/\partial x)_i = \{\partial(\ln N)/\partial x\}_i \approx \{(\ln N)_{i+1} - (\ln N)_{i-1}\}/(x_{i+1} - x_{i-1}) \qquad (2.77)$$

is very stable. This stable discretization is used and the new terms are all retained in curvilinear coordinate form. Since Equation 2.60 is nonlinear, von Neumann analyses are impossible, but hundreds of practical simulations have demonstrated the value of the logarithmic representation. Equation 2.60, together with the finite differencing in Equation 2.77, represents the new methodology used in this book for fluids with and without yield stress.

Equations 2.70 and 2.71 were solved by rewriting them as a single vector equation for $x + iy$, employing simplifications from complex variables, and discretizing the end equation using second-order accurate formulas. The finite difference equations were then reordered so that the coefficient matrix is sparse, banded, and computationally efficient. Finally, the "successive line over relaxation" (SLOR) method was used to obtain the solution in an implicit and iterative manner. The SLOR scheme is unconditionally stable on a linearized von Neumann basis and is quickly performed. Mesh generation requires 1 to 2 seconds of computing time on typical computers.

Once the transformations for $y(\xi, \eta)$ and $x(\xi, \eta)$ are available for a given annular geometry, Equation 2.60 can be solved any number of times for different applied pressure gradients, volumetric flow rate constraints, or fluid rheology models, without recomputing the mapping. Because Equation 2.60 is similar to Equations 2.70 and 2.71, the same procedure is used for its solution. These iterations converge quickly and stably because the meshes used were smooth. When solutions for the velocity field $u(\xi, \eta)$ are available, these also requiring just seconds, simple inverse mapping relates each computed "u" with its unique image in the physical (y, x) plane. With $u(y, x)$ and its spatial derivatives known, postprocessed quantities like $N(\Gamma)$, S_{zy}, S_{zx}, their shear rates, and apparent viscosities are easily calculated and displayed in physical (y, x) coordinates.

Drilling and production engineers recognize that flow properties in eccentric annuli correlate to some extent with annular position (e.g., low bottom speeds regardless of rheology). Our text-based graphical display software projects u(y, x) and all postprocessed quantities on the annular geometry. This facilitates quick, efficient visual correlation of computed physical properties or inferred characteristics (e.g., "cuttings transport efficiency" and "stuck pipe probability") with annular geometry. These highly visual outputs and sophisticated color graphics, together with the speed and stability of the scheme, promote an understanding of annular flow in an interactive, real-time manner.

Extensions for yield stress fluids

We return now to fluids with nonzero yield stresses. In general, there exist internal boundaries separating "dead" (or "plug") and flowing "shear" regimes. These unknown boundaries must be obtained as part of the solution. Previously we indicated that these boundaries must be found as part of the complete solution process. But even if they were known explicitly, numerical solutions would be no easier: Slope discontinuities or "kinks" affect coordinate lines crossing the fluid interface, and special methods must be developed to model sudden changes in slope (the standard finite difference formulas used in numerical analysis require a function and its derivatives to be continuous).

The basic problem actually arises from the method devised by mathematicians to "simplify" combined plug and shear flows using Equation 2.48b. The resulting "if-then" model is responsible for creating two unnatural domains that must be related by additional auxiliary conditions. We remedy the problem as follows. In reality, flows do not suddenly change from flowing to nonflowing or vice versa: A steep but continuous transition prevails. Consider, for example, the conventional Herschel-Bulkley viscoplastic model, which includes Bingham plastics as a special limit when the model exponent "n" is unity. As in Equation 2.48b, this requires that

$$\tau = \tau_0 + K(d\gamma/dt)^n, \quad \text{if } \tau > \tau_0 \tag{2.78a}$$

$$d\gamma/dt = 0, \quad \text{otherwise} \tag{2.78b}$$

where τ is the shear stress, τ_0 is the yield stress, K is the consistency factor, n is the exponent, and $d\gamma/dt$ is the shear rate. This model is far from perfect. Both Herschel-Bulkley and Bingham plastic models, for instance, predict infinite viscosities in the limit of vanishing shear rate, a fact that often leads to numerical instabilities. In addition, the behavior is not compatible with the conservation laws that govern many complex flows.

An alternative to Equations 2.78a and 2.78b is the continuous function suggested by Souza Mendez and Dutra (2004):

$$\tau = \{1 - \exp(-\eta_0 \, d\gamma/dt/\tau_0)\}\{\tau_0 + K(d\gamma/dt)^n\} \tag{2.79}$$

which applies *everywhere* in the problem domain. The corresponding apparent viscosity is

$$\eta = \tau/(d\gamma/dt) \tag{2.80}$$

$$= \{1 - \exp(-\eta_0 d\gamma/dt/\tau_0)\}\{\tau_0/(d\gamma/dt) + K(d\gamma/dt)^{n-1}\} \tag{2.81}$$

Example 2.3   **71**

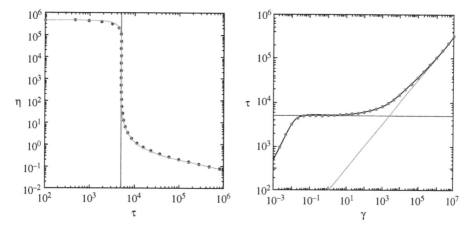

FIGURE 2.8

Extended Herschel-Bulkley law.

Sketches for Equations 2.80 and 2.81 are given in Figure 2.8. We describe this as the "extended Herschel-Bulkley" model (our terminology). For infinite shear rates, we recover $\tau = \tau_0 + K \, (d\gamma/dt)^n$. For low shear rates, a simple Taylor expansion leads to

$$\eta \approx \{\eta_0(d\gamma/dt)/\tau_0\}\{\tau_0/(d\gamma/dt) + K(d\gamma/dt)^{n-1}\} \tag{2.82a}$$

$$\approx \eta_0 \tag{2.82b}$$

where it is clear now that the parameter η_0 represents a very high plug zone viscosity.

The use of Equations 2.79, 2.80, and 2.81 in numerical algorithms simplifies both formulation and coding, since internal boundaries and plug domains do not need to be determined as part of the solution. A single constitutive law (as opposed to the use of both Equations 2.48a and 2.48b) applies everywhere; moreover, the continuous function assumed also possesses continuous derivatives and automatically provides rapid transitions across boundaries separating plug and sheared flows. Standard finite difference formulas then apply.

Also, the use of Equation 2.81 integrates naturally with the method behind Equation 2.77. Essentially, the apparent viscosity N is stored in a Fortran function statement that is called and updated as needed. We emphasize that $N(\Gamma) = K \, [(\partial u/\partial y)^2 + (\partial u/\partial x)^2]^{(n-1)/2}$, as given in Equation 2.63, applies to unidirectional flows without changes in the z direction (this too must be reexpressed in curvilinear coordinates).

From the programming and software perspective, there is no distinction between zero-yield and yield stress fluids. In a practical computer program, the plug zone viscosity might be assumed, for example, anywhere from 100 to 1,000 cp. In fact, we choose high values of η_0 that additionally stabilize the numerical integration schemes used. This strategy is applied throughout our work, both to our iterative relaxation schemes for steady-state problems and to our transient integration schemes for more complicated managed pressure drilling formulations. Finally, our "extended" model is *not* the "generalized Herschel-Bulkley" (GHB) relation in Becker et al. (2003). That model merely provides one additional curve-fitting parameter via $(\tau/\tau_{ref})^m = (\tau_0/\tau_{ref})^m + (\mu_\infty \, d\gamma/dt/\tau_{ref})^n$. While useful, the GHB model does not address plug zone size and shape issues. Thus, it is not used or considered in this book.

EXAMPLE 2.4

Curvilinear Grid Formulation for Eccentric Annular Flows with General Non-Newtonian Fluids *with* Rotation

In the previous example, we introduced three strategies relevant to modeling non-Newtonian flows in highly eccentric annuli: (a) stable numerical modeling of variable apparent viscosities and their spatial derivatives; (b) extended representations for Herschel-Bulkley yield stress fluids to compute internal boundaries naturally; and (c) curvilinear grid generation to accommodate irregular geometric details such as hole eccentricity, cuttings beds, and washouts. Implementation details are reserved for Chapter 3.

In this example, we further develop the formalism to address problems with inner pipe or casing rotation, which arise frequently in managed pressure drilling and effective cementing and completions. The problem is all the more important because drillstring rotation can be used to actively control downhole pressure in modern managed pressure applications, thus providing another option for well management besides mud rheology, pump rate, and surface choke adjustment.

We have seen from the foregoing example how the use of rectangular or Cartesian coordinates provides a natural starting point for detailed geometric modeling of highly eccentric holes. This technique was first discussed in the author's *Borehole Flow Modeling* (1992) and *Computational Rheology* (2001), but in the context of nonrotating flows. When pipe rotation is permitted, cylindrical coordinates provide obvious first advantages in describing inner circle steady or transient movements, but ultimately the resulting description must be rewritten to rectangular coordinates for transfer to curvilinear grid representation using the method of Example 2.3. Thus, rotation problems involve levels of algebraic manipulations in three successive coordinate systems: cylindrical, rectangular, and curvilinear. Actually, two more "hidden" coordinate transforms apply that are related to computer screen displays; however, these represent issues outside of fluid dynamics and are not discussed.

Again, we start with circular cylindrical coordinates so that steady or transient rotational rates can be easily prescribed at the circular pipe or casing surface. The general non-Newtonian rheological equations for unsteady single-phase flow with and without yield stress are given in references previously cited and are listed below without proof. The momentum equations in the "r," "θ," and "z" directions are, respectively,

$$
\begin{aligned}
\rho(\partial v_r/\partial t + v_r \, \partial v_r/\partial r + v_\theta/r \, \partial v_r/\partial \theta - v_\theta^2/r + v_z \, \partial v_r/\partial z) \\
= F_r - \partial P/\partial r + 1/r \, \partial(r \, S_{rr})/\partial r + 1/r \, \partial S_{r\theta}/\partial \theta - 1/r \, S_{\theta\theta} + \partial S_{rz}/\partial z
\end{aligned}
\tag{2.83}
$$

$$
\begin{aligned}
\rho(\partial v_\theta/\partial t + v_r \partial v_\theta/\partial r + v_\theta/r \, \partial v_\theta/\partial \theta + v_r v_\theta/r + v_z \, \partial v_\theta/\partial z) \\
= F_\theta - 1/r \, \partial P/\partial \theta + 1/r^2 \, \partial(r^2 S_{\theta r})/\partial r + 1/r \, \partial S_{\theta\theta}/\partial \theta + \partial S_{\theta z}/\partial z
\end{aligned}
\tag{2.84}
$$

$$
\begin{aligned}
\rho(\partial v_z/\partial t + v_r \, \partial v_z/\partial r + v_\theta/r \, \partial v_z/\partial \theta + v_z \, \partial v_z/\partial z) \\
= F_z - \partial P/\partial z + 1/r \, \partial(r \, S_{zr})/\partial r + 1/r \, \partial S_{z\theta}/\partial \theta + \partial S_{zz}/\partial z
\end{aligned}
\tag{2.85}
$$

Example 2.4   **73**

where F denotes body forces, while the equation for mass conservation takes the form

$$\partial v_r/\partial r + v_r/r + 1/r\,\partial v_\theta/\partial\theta + \partial v_z/\partial z = 0 \tag{2.86}$$

In the preceding, v_r, v_θ, and v_z are radial, azimuthal, and axial velocity components, respectively. Again, we have

$$\underline{\underline{S}} = 2\,N(\Gamma)\underline{\underline{D}} \tag{2.87}$$

denoting the deviatoric stress tensor, $N(\Gamma)$; the apparent viscosity function, Γ; the shear rate; and, now, the deformation tensor, $\underline{\underline{D}}$, whose elements are defined by

$$D_{rr} = \partial v_r/\partial r \tag{2.88a}$$

$$D_{\theta\theta} = 1/r\,\partial v_\theta/\partial\theta + v_r/r \tag{2.88b}$$

$$D_{zz} = \partial v_z/\partial z \tag{2.88c}$$

$$D_{r\theta} = D_{\theta r} = \tfrac{1}{2}[r\,\partial(v_\theta/r)/\partial r + 1/r\,\partial v_r/\partial\theta] \tag{2.88d}$$

$$D_{\theta z} = D_{z\theta} = \tfrac{1}{2}(\partial v_\theta/\partial z + 1/r\,\partial v_z/\partial\theta) \tag{2.88e}$$

$$D_{rz} = D_{zr} = \tfrac{1}{2}(\partial v_r/\partial z + \partial v_z/\partial r) \tag{2.88f}$$

These equations are later solved analytically in Example 5.6 (in Chapter 5) for steady concentric rotation with axial flow for Power law fluids. Other limits are addressed in this book as well. For the purposes of modeling eccentric flow with inner pipe rotation, our ultimate objective being the use of transformed curvilinear coordinates, we compare Equations 2.60 and 2.85 to obtain the composite but exact axial flow model:

$$\begin{aligned}\rho(\partial u/\partial t + v_r\,\partial u/\partial r + v_\theta/r\,\partial u/\partial\theta + u\,\partial u/\partial z)\\ = F_z - \partial P/\partial z + \partial(N\,\partial u/\partial y)/\partial y + \partial(N\,\partial u/\partial x)/\partial x\end{aligned} \tag{2.89}$$

We introduce a coordinate system centered with the circular pipe or casing throughout this book. Then, for steady flows, we have $\partial/\partial t = 0$, while the two-dimensional flow assumption states that $\partial/\partial z = 0$. If we further assume that radial velocities are small compared with those in the axial and azimuthal directions, a premise that is valid so long as massive fluid influxes or outfluxes are ruled out, we have in the absence of body forces

$$\rho v_\theta/r\,\partial u/\partial\theta \approx -\partial P/\partial z + \partial(N\,\partial u/\partial y)/\partial y + \partial(N\,\partial u/\partial x)/\partial x \tag{2.90}$$

where we have used the notation $u = v_z$ to be consistent with Example 2.3.

Equation 2.90 is just the extension of Equation 2.13 to handle non-Newtonian rheologies. The simpler model was developed in the context of Newtonian rotating, eccentric annular flows. Here we rewrite the left side to x and y coordinates using "$x = r\cos\theta$ and $y = r\sin\theta$," and rewrite the reexpressed azimuthal derivative in rectangular coordinates according to the chain rule results $\partial/\partial\theta = -y\,\partial/\partial x + x\,\partial/\partial y$ and $\partial/\partial r = (x/r)\,\partial/\partial x + (y/r)\,\partial/\partial y$. The resulting x–y equation is then transformed to curvilinear coordinates as explained in Example 2.3, and the same x–y methods for steady and transient flow apply.

Equation 2.90 still contains a velocity v_θ that plays the role of a variable coefficient. As we indicated in Example 2.1, its magnitude is proportional to fluid density and to azimuthal speed and, here also, inversely proportional to apparent viscosity. In the most general case, the nonlinearly coupled system formed by Equations 2.84 and 2.90 must be solved. However, for rotating annular flows, simplifications are possible. We first reconsider Equation 2.84 in the form

$$
\begin{aligned}
\rho(\partial v_\theta/\partial t &+ v_r\,\partial v_\theta/\partial r + v_\theta/r\,\partial v_\theta/\partial\theta + v_r v_\theta/r + v_z\,\partial v_\theta/\partial z) \\
&= -1/r\,\partial P/\partial\theta + 1/r^2\,\partial(r^2\,S_{\theta r})/\partial r + 1/r\,\partial S_{\theta\theta}/\partial\theta + \partial S_{\theta z}/\partial z \\
&\approx -1/r\,\partial P/\partial\theta + N(\Gamma)(\partial^2 v_\theta/\partial r^2 + 1/r\,\partial v_\theta/\partial r - v_\theta/r^2 + 1/r^2\,\partial^2 v_\theta/\partial\theta^2 \\
&\quad + 2/r^2\,\partial v_r/\partial\theta + \partial^2 v_\theta/\partial z^2)
\end{aligned}
\tag{2.91}
$$

Applying the assumptions used for Equation 2.90, plus the fact that the azimuthal flow is primarily "dragged" by the inner surface in the sense shown in Example 2.1, so that $\partial/\partial\theta$ on the left side and the induced pressure gradient on the right are both small compared to axial effects, we have an approximate linear partial differential equation without explicit pressure dependence obtained as

$$
\partial^2 v_\theta/\partial r^2 + 1/r\,\partial v_\theta/\partial r + 1/r^2\,\partial^2 v_\theta/\partial\theta^2 - v_\theta/r^2 \approx 0
\tag{2.92}
$$

which must be solved subject to $v_\theta = 0$ at the outer wall and $v_\theta = R\omega$ at the circular pipe surface, where R is the pipe radius and ω is the rotational rate (for transient flows, the $\rho\partial v_\theta/\partial t$ term is retained).

Equation 2.92, we emphasize, must be solved in curvilinear coordinates consistently with the x–y form of Equation 2.90. The required form is straightforwardly obtained. For this purpose, we note that $x = r\cos\theta$ and $y = r\sin\theta$ imply that $r = (x^2 + y^2)^{1/2}$ and $\partial^2 v_\theta/\partial r^2 + 1/r\,\partial v_\theta/\partial r + 1/r^2 \partial^2 v_\theta/\partial\theta^2 = \partial^2 v_\theta/\partial x^2 + 1/r^2\,\partial^2 v_\theta/\partial y$, so we have a simpler $\partial^2 v_\theta/\partial x^2 + 1/r^2\,\partial^2 v_\theta/\partial y - v_\theta/(x^2 + y^2) = 0$.

In summary, the steady, rotating flow formulation valid for non-Newtonian fluids with or without yield stress solves the coupled system

$$
\partial(N\,\partial u/\partial y)/\partial y + \partial(N\,\partial u/\partial x)/\partial x \approx \partial P/\partial z + \rho v_\theta/r\,\partial u/\partial\theta
\tag{2.93}
$$

$$
\partial^2 v_\theta/\partial x^2 + 1/r^2\,\partial^2 v_\theta/\partial y - v_\theta/(x^2 + y^2) = 0
\tag{2.94}
$$

subject to $u = u_0$ (a constant positive, zero, or negative axial pipe movement speed) and $v_\theta = R\omega$ at the inner pipe, plus $u = v_\theta = 0$ at the outer annular boundary. Note that the apparent viscosity function depends on both u and v_θ and embodies shear-thinning and shear-thickening effects, so the two are coupled nonlinearly. Again, Equations 2.93 and 2.94 are written in x–y coordinates but are transformed to boundary-conforming, curvilinear coordinates for solution. A controlling parameter in numerical stability is $\rho\omega/\mu$, where μ is proportional to the average apparent viscosity. The required details related to geometric transformations are explained in Chapter 3.

Numerical Analysis and Algorithm Development Strategies

In Chapter 1, we provided an overview of all of the simulation functions addressed by the new annular flow technologies; in Chapter 2, we presented the overall theoretical framework for boundary-conforming, curvilinear grid approaches to the modeling of non-Newtonian fluids (with and without yield stress) with drillpipe or casing that may be stationary or rotating. This framework applies to both the single-phase flows treated in Chapters 1 through 7 and their multiphase extensions in Chapters 8 and 9.

In the present chapter, mathematical details required in implementing curvilinear grids are described without the usual recourse to differential geometry, an abstract subject area covered in specialized graduate-level courses. Importantly, in Example 3.1, these advanced ideas are developed using basic calculus, so they are understandable to practicing engineers and extendable by software practitioners without much difficulty. We show how arbitrary annular flow "donut" regimes are converted to simple rectangular domains for fast computation. We also demonstrate how the nonlinear mesh generation equations, whose numerical solutions are often slow and unstable, can be rapidly computed using an unconditionally stable algorithm.

Example 3.2 outlines how the same methods can be used to model noncircular (nonannular) ducts, an application that is important in studying hole enlargement in drilling engineering and pipeline clogging in underwater applications. Practical applications to solids deposition modeling are addressed in Example 3.3, and numerous calculated results are offered for both Newtonian and non-Newtonian flow. Finally, Example 3.4 introduces the subject of finite difference analysis using a solution for classical Hagen-Poiseuille pipe flow in circular pipes, and rapidly progresses to state-of-the-art curvilinear grid–based methods for non-Newtonian fluids where velocities and apparent viscosities are found iteratively.

EXAMPLE 3.1

Grid Generation for Eccentric Annular Flow

We introduce abstract (but useful and important) ideas in coordinate transformations, typically presented from a differential geometry perspective but here reworked using only elementary calculus to clearly convey the ideas. Suppose we wish to express a function f(x, y) in terms of convenient independent variables ξ and η. If the transformations $x = x(\xi, \eta)$ and $y = y(\xi, \eta)$ are available, direct substitution allows us to rewrite $f(x, y)$ in the form

$$f(x, y) = F(\xi, \eta) \tag{3.1}$$

In Equation 3.1, the functional relation $F(\xi, \eta)$ between ξ and η is generally different from the relation $f(x, y)$ connecting x and y. Derivatives of $f(x, y)$ with respect to x and y are easily related to derivatives of $F(\xi, \eta)$ taken with respect to ξ and η. By applying the chain rule of calculus, we have

$$F_\xi = f_x x_\xi + f_y y_\xi \tag{3.2}$$

$$F_\eta = f_x x_\eta + f_y y_\eta \tag{3.3}$$

where subscripts, along with ∂'s, will be used to indicate partial derivatives. Equations 3.2 and 3.3 can be algebraically solved for f_x and f_y to yield

$$f_x = (y_\eta F_\xi - y_\xi F_\eta)/J \tag{3.4}$$

$$f_y = (x_\xi F_\eta - x_\eta F_\xi)/J \tag{3.5}$$

where

$$J(\xi, \eta) = x_\xi y_\eta - x_\eta y_\xi \tag{3.6}$$

is known as the Jacobian of the transformation. For reasons that will be apparent later, we will refer to this Jacobian as "big jay."

Most boundary value problems occurring in mathematical physics involve second-order partial differential equations. To express such equations in (ξ, η) coordinates, transformations similar to those in Equations 3.4 and 3.5 are therefore needed for f_{xx}, f_{xy}, and f_{yy}. Throughout this presentation, f and F are considered to be sufficiently smooth, so it is possible to interchange the order of differentiation between any two independent variables. By "smooth," we mean that sudden discontinuities are not expected in physical solutions. This is assured, for instance, if kinks do not appear in the cross-sectional geometry. Application of the chain rule to Equations 3.2 and 3.3 leads to

$$\begin{aligned} F_{\xi\xi} &= f_x x_{\xi\xi} + x_\xi(f_{xx}x_\xi + f_{xy}y_\xi) + f_y y_{\xi\xi} + y_\xi(f_{yx}x_\xi + f_{yy}y_\xi) \\ &= x_{\xi\xi}f_x + y_{\xi\xi}f_y + x_\xi^2 f_{xx} + y_\xi^2 f_{yy} + 2x_\xi y_\xi f_{xy} \end{aligned} \tag{3.7}$$

Similarly,

$$F_{\eta\eta} = x_{\eta\eta}f_x + y_{\eta\eta}f_y + x_\eta^2 f_{xx} + y_\eta^2 f_{yy} + 2x_\eta y_\eta f_{xy} \tag{3.8}$$

and

$$F_{\eta\xi} = x_{\eta\xi}f_x + y_{\eta\xi}f_y + x_\eta x_\xi f_{xx} + y_\eta y_\xi f_{yy} + (x_\eta y_\xi + x_\xi y_\eta)f_{xy} \tag{3.9}$$

Now, let us rewrite Equations 3.7, 3.8, and 3.9, treating the functions f_{xx}, f_{xy}, and f_{yy} as algebraic unknowns on the left-hand side of a three-by-three system. That is, we write the foregoing equations as follows:

Example 3.1   **77**

$$x_\xi^2\, f_{xx} + \qquad\qquad 2x_\xi y_\xi\, f_{xy} + \; y_\xi^2\, f_{yy} = F_{\xi\xi} - x_{\xi\xi}\, f_x - y_{\xi\xi}\, f_y \qquad\qquad (3.10)$$

$$x_\eta^2\, f_{xx} + \qquad\qquad 2x_\eta y_\eta\, f_{xy} + \; y_\eta^2\, f_{yy} = F_{\eta\eta} - x_{\eta\eta}\, f_x - y_{\eta\eta}\, f_y \qquad\qquad (3.11)$$

$$x_\eta x_\xi f_{xx} + (x_\eta y_\xi + x_\xi y_\eta)\, f_{xy} + y_\eta y_\xi\, f_{yy} = F_{\eta\xi} - x_{\eta\xi}\, f_x - y_{\eta\xi}\, f_y \qquad\qquad (3.12)$$

In this form, the solutions for f_{xx}, f_{xy}, and f_{yy} can be easily obtained using determinants. However, we need not write down individual solutions, since we will not need to use them in our applications. But we will make use of the rectangular Laplace operator $f_{xx} + f_{yy}$, which, in curvilinear coordinates, takes the form

$$f_{xx} + f_{yy} = (\alpha\, F_{\xi\xi} - 2\beta F_{\xi\eta} + \gamma F_{\eta\eta})/J^2 + [(\alpha\, x_{\xi\xi} - 2\beta x_{\xi\eta} + \gamma x_{\eta\eta})(y_\xi F_\eta - y_\eta F_\xi)$$
$$+\, (\alpha\, y_{\xi\xi} - 2\beta y_{\xi\eta} + \gamma y_{\eta\eta})(x_\eta F_\xi - x_\xi F_\eta)]/J^3 \qquad\qquad (3.13)$$

where the Greek letter coefficients represent the nonlinear functions

$$\alpha = x_\eta^2 + y_\eta^2 \qquad\qquad (3.14)$$

$$\beta = x_\xi x_\eta + y_\xi y_\eta \qquad\qquad (3.15)$$

$$\gamma = x_\xi^2 + y_\xi^2 \qquad\qquad (3.16)$$

Thompson's Mapping

So far, we have not imposed any constraints on the functions $x = x(\xi, \eta)$ and $y = y(\xi, \eta)$, or their inverses $\xi = \xi(x, y)$ and $\eta = \eta(x, y)$. One well-known transformation is Thompson's mapping, originally developed to solve the Navier-Stokes equations for viscous flows past planar airfoils in aerospace applications; see, for example, Thompson (1984); Thompson, Warsi, and Mastin (1985); and Tamamidis and Assanis (1991). This method was modified for use in Chin (1992, 2001) to study steady non-Newtonian flows in eccentric annuli and noncircular pipes. In this approach, $\xi(x, y)$ and $\eta(x,y)$ are *defined* as solutions to the elliptic equations

$$\xi_{xx} + \xi_{yy} = P^*(\xi, \eta) \qquad\qquad (3.17)$$

$$\eta_{xx} + \eta_{yy} = Q^*(\xi, \eta) \qquad\qquad (3.18)$$

where P^* and Q^* are functions chosen (by ingenious guesswork) to control local grid density.

We will explain the exact motivation behind the Thompson approach later. For now, however, we ask, "What equations govern $x = x(\xi, \eta)$ and $y = y(\xi, \eta)$ given Equations 3.17 and 3.18?" At this point, it is helpful to understand that Equation 3.13 holds for *any* function f. That is, for any prescribed set of transformations, Equation 3.13 can be viewed as a source of useful identities. Let us take $f(x, y) = \xi(x, y)$, in which case $F(\xi, \eta) = \xi$; then $F_\eta = 0$, and all second derivatives of F with respect to ξ and η vanish. Substitution in Equation 3.13 and replacement of the resulting Laplacian of ξ with respect to x and y using Equation 3.17 lead to

$$-y_\eta(\alpha\, x_{\xi\xi} - 2\beta\, x_{\xi\eta} + \gamma\, x_{\eta\eta}) + x_\eta(\alpha\, y_{\xi\xi} - 2\beta\, y_{\xi\eta} + \gamma\, y_{\eta\eta}) = P^* J^3 \qquad\qquad (3.19)$$

Similarly, consider $f(x, y) = \eta(x, y)$, so that $F(\xi, \eta) = \eta$. It follows that $F_\xi = 0$ and that all second derivatives of F with respect to ξ and η vanish. Substitution in Equation 3.13 and replacement of the Laplacian of η with respect to x and y using Equation 3.18 lead to

$$y_\xi(\alpha\, x_{\xi\xi} - 2\beta\, x_{\xi\eta} + \gamma\, x_{\eta\eta}) - x_\xi(\alpha\, y_{\xi\xi} - 2\beta\, y_{\xi\eta} + \gamma\, y_{\eta\eta}) = Q^* J^3 \qquad (3.20)$$

If we now regard $(\alpha\, x_{\xi\xi} - 2\beta\, x_{\xi\eta} + \gamma\, x_{\eta\eta})$ and $(\alpha\, y_{\xi\xi} - 2\beta\, y_{\xi\eta} + \gamma\, y_{\eta\eta})$ as algebraic unknowns in a simple two-by-two system, Equations 3.19 and 3.20 can be solved, thus yielding Thompson's well-known elliptic equations

$$\alpha x_{\xi\xi} - 2\beta x_{\xi\eta} + \gamma\, x_{\eta\eta} + J^2(P^* x_\xi + Q^* x_\eta) = 0 \qquad (3.21)$$

$$\alpha y_{\xi\xi} - 2\beta y_{\xi\eta} + \gamma\, y_{\eta\eta} + J^2(P^* y_\xi + Q^* y_\eta) = 0 \qquad (3.22)$$

Interestingly, we have derived these relationships using basic calculus, without recourse to more esoteric notions from differential geometry. Equations 3.21 and 3.22 are nonlinearly coupled because the coefficients α, β, and γ in Equations 3.14 through 3.16 depend on both $x(\xi, \eta)$ and $y(\xi, \eta)$.

Some reciprocity relations

For practical reasons, we will need to convert results between physical $x-y$ and computational $\xi-\eta$ planes. Thus, reciprocity relationships are needed. Let us return to general considerations and for now refrain from invoking Thompson's assumptions. In particular, we examine the general transformations

$$x = x(\xi, \eta) \qquad (3.23)$$

$$y = y(\xi, \eta) \qquad (3.24)$$

From elementary calculus, the total differentials dx and dy are given by

$$x_\eta d\eta + x_\xi d\xi = dx \qquad (3.25)$$

$$y_\eta d\eta + y_\xi d\xi = dy \qquad (3.26)$$

Equations 3.25 and 3.26 can be solved in terms of $d\xi$ and $d\eta$, thus leading to the relationships

$$d\eta = -y_\xi dx/J + x_\xi dy/J \qquad (3.27)$$

$$d\xi = +y_\eta dx/J - x_\eta dy/J \qquad (3.28)$$

where the "big jay" Jacobian is given by Equation 3.6. Now, we can similarly consider the inverse transformation. If we write

$$\eta = \eta(x, y) \qquad (3.29)$$

$$\xi = \xi(x, y) \qquad (3.30)$$

Example 3.1   **79**

it follows that

$$d\eta = \eta_x dx + \eta_y dy \qquad (3.31)$$

$$d\xi = \xi_x dx + \xi_y dy \qquad (3.32)$$

Comparison of Equation 3.27 with Equation 3.31, and of Equation 3.28 with Equation 3.32, leads to

$$\eta_x = -y_\xi / J \qquad (3.33)$$

$$\eta_y = x_\xi / J \qquad (3.34)$$

$$\xi_x = y_\eta / J \qquad (3.35)$$

$$\xi_y = -x_\eta / J \qquad (3.36)$$

On the other hand, we might have proceeded from the definitions for the total differentials $d\xi$ and $d\eta$, and reconsidered Equations 3.31 and 3.32 as

$$\eta_x dx + \eta_y dy = d\eta \qquad (3.37)$$

$$\xi_x dx + \xi_y dy = d\xi \qquad (3.38)$$

Equations 3.37 and 3.38 can be solved algebraically for dx and dy to give

$$dx = -\xi_y d\eta / j + \eta_y d\xi / j \qquad (3.39)$$

$$dy = +\xi_x d\eta / j - \eta_x d\xi / j \qquad (3.40)$$

where the "little jay" Jacobian satisfies

$$j(x, y) = \xi_x \eta_y - \xi_y \eta_x \qquad (3.41)$$

Comparison of Equation 3.25 with Equation 3.39, and of Equation 3.26 with Equation 3.40, leads to

$$x_\eta = -\xi_y / j \qquad (3.42)$$

$$x_\xi = \eta_y / j \qquad (3.43)$$

$$y_\eta = \xi_x / j \qquad (3.44)$$

$$y_\xi = -\eta_x / j \qquad (3.45)$$

Finally, comparison of Equation 3.33 with Equation 3.45, Equation 3.34 with Equation 3.43, Equation 3.35 with Equation 3.44, and lastly, Equation 3.36 with Equation 3.42 leads to

$$J(\xi, \eta) j(x, y) = 1 \qquad (3.46)$$

or

$$(x_\xi y_\eta - x_\eta y_\xi)(\xi_x \eta_y - \xi_y \eta_x) = 1 \qquad (3.47)$$

It is important to understand that the equations obtained in this section are generally valid, regardless of Thompson's or any other transformations. They allow us to move conveniently between quantities expressed in the physical (x, y) and computational (ξ, η) planes.

Conformal mapping limits

Conformal mapping is a powerful technique used to transform simple harmonic solutions into those applicable to more complicated shapes. Here, we explore its general properties and attempt to understand conformal mapping from a mathematical viewpoint. Usually, methods from complex variables analysis are used to introduce the following concepts, but as in our above treatment for coordinate transformations, the basic results can be developed using only elementary calculus. We *now* formally reintroduce the Cauchy-Riemann conditions:

$$\xi_x = \eta_y \tag{3.48}$$

$$\eta_x = -\xi_y \tag{3.49}$$

Let us differentiate Equation 3.48 with respect to x and Equation 3.49 with respect to y; elimination of the cross-derivative term between the two results leads to Equation 3.50. A similar procedure yields Equation 3.51.

$$\xi_{xx} + \xi_{yy} = 0 \tag{3.50}$$

$$\eta_{xx} + \eta_{yy} = 0 \tag{3.51}$$

Equations 3.50 and 3.51 are both elliptic; they are, in fact, *exactly* Thompson's Equations 3.17 and 3.18, but with $P^* = Q^* = 0$. Since $\xi(x, y)$ and $\eta(x, y)$ satisfy Laplace's equation, they are said to be harmonic. And because harmonic functions are generally obtained as real and imaginary parts of complex analytical functions, Equations 3.50 and 3.51 are usually derived more elegantly using complex variables methods. The latter are also used to derive "free" solutions to equations like "$( )_{xx} + ( )_{yy} = 0$."

That is, if solutions for $\phi_{xx} + \phi_{yy} = 0$ are known, solutions to a related $\psi_{xx} + \psi_{yy} = 0$ can be deduced. In reservoir engineering, ϕ might represent Darcy pressure, in which case ψ would describe streamlines. Of course, in real-world problems that satisfy more complicated partial differential equations, conformal mapping methods cannot be used—in fact, none of the flow models used in annular flow modeling satisfy Laplace's equation. We therefore address the use of curvilinear coordinate transformations for general boundary value problems.

To understand the implications of Equations 3.48 and 3.49 in transformed coordinates, we turn to our reciprocity relations. If the ξ_x and η_y in Equation 3.48 are replaced by their equivalents using our Equations 3.43 and 3.44, and if η_x and ξ_y in Equation 3.49 are replaced by their equivalents using Equations 3.42 and 3.45, we obtain

$$y_\eta = x_\xi \tag{3.52}$$

$$y_\xi = -x_\eta \tag{3.53}$$

Example 3.1   **81**

which imply, using the same procedure we have described, that

$$x_{\xi\xi} + x_{\eta\eta} = 0 \qquad (3.54)$$

$$y_{\xi\xi} + y_{\eta\eta} = 0 \qquad (3.55)$$

Thus, $x(\xi, \eta)$ and $y(\xi, \eta)$ are likewise harmonic, but in the variables ξ and η. Equations 3.54 and 3.55 are simpler than Equations 3.21 and 3.22, with $P^* = Q^* = 0$. Reciprocity shows that there exists a duality between physical and mapped planes, and vice versa, for conformal transformations; that is, Equations 3.50 and 3.51 are mirror images of Equations 3.54 and 3.55. One might have anticipated this type of reversibility, but it is not directly evident from Equations 3.21 and 3.22. Equations 3.54 and 3.55 are consistent with Thompson's original Equations 3.21 and 3.22.

Use of the Cauchy-Riemann conditions in the transformed plane—that is, Equations 3.52 and 3.53—in Equations 3.14 through 3.16 leads to $\alpha = \gamma$ and $\beta = 0$. In this presentation, our grid generation discussions include derivations for results of broad theoretical interest. However, due to resource limitations in ongoing research, our applications will be restricted to $P^* = Q^* = 0$. We stress that Equations 3.54 and 3.55 are linear, unlike Equations 3.21 and 3.22. However, they do *not* generally uncouple for true conformal mappings, as they might superficially suggest, since x and y cannot be arbitrarily specified along boundaries: To be conformal, x and y must satisfy Equations 3.52 and 3.53 everywhere.

Solutions to mesh-generation equations

We show how our geometrical transforms are useful in solving boundary value problems. To explain the issues clearly, we consider an elementary reservoir flow application. Commercial simulators calculate pressures, velocities, and other properties on rectangular grids. Again, their x−y coordinate lines do not conform to the irregular curves defining actual boundaries; also, high grid densities imposed near wells imply similarly high densities far away, where such resolution is unnecessary. This results in large, inefficient computing domains containing dead flow and large matrices. Sometimes, coarse meshes are used everywhere, together with high-density "corner point" modeling to provide grid refinement close to wells. However, many refrain from their usage because cross-derivative terms in the transformed flow equations, which increase computing time, are incorrectly ignored in the matrix inversion for numerical expediency.

Boundary conditions

Although the industry focuses on Cartesian meshes, more effective boundary-conforming, curvilinear grids *can* be generated, adapting to both far-field and near-field boundaries. We now reiterate the basic ideas because they are essential to understanding the method, but here they focus on the boundary conditions needed to supplement Thompson's equations. Suppose that a transform $\xi = \xi(x, y)$, $\eta = \eta(x, y)$ exists that maps the irregular, doubly connected domain defined by the general well and far-field reservoir boundaries of Figure 3.1 into the singly connected rectangle of Figure 3.2.

Physically insignificant branch cuts B_1 and B_2 have been introduced, which will be discussed. Such a mapping effectively allows calculations to be performed on more desirable high-resolution grids like the one in Figure 3.3. It is clear that more meaningful flow models can be formulated using "ξ, η" coordinates; improved flow description is possible, with fewer grids and less matrix inversion.

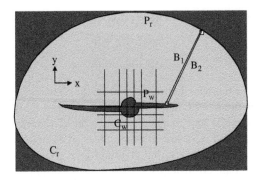

FIGURE 3.1

Irregular domain with inefficient rectangular meshes (application for reservoir flow into a fractured well). *Source:* From Chin (2002).

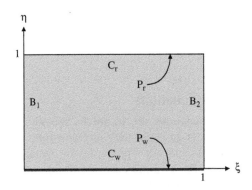

FIGURE 3.2

Irregular domain mapped to rectangular $\xi - \eta$ computational space.

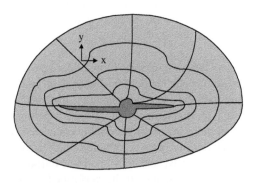

FIGURE 3.3

Physical domain in boundary-conforming coordinates.

Example 3.1   **83**

Now, it is known from complex variables that conformal transformations satisfy linear Laplace equations in x and y, but classical methods unfortunately do not explain *how* the mappings are obtained.

Thompson, again, developed a novel approach. Rather than dealing with $\xi = \xi(x, y)$ and $\eta = \eta(x, y)$ directly, the method equivalently considers inverse functions $x = x(\xi, \eta)$ and $y = y(\xi, \eta)$ satisfying two *nonlinearly* coupled equations, Equations 3.21 and 3.22:

$$(x_\eta^2 + y_\eta^2)x_{\xi\xi} - 2(x_\xi x_\eta + y_\xi y_\eta)x_{\xi\eta} + (x_\xi^2 + y_\xi^2)x_{\eta\eta} = 0 \qquad (3.56)$$

$$(x_\eta^2 + y_\eta^2)y_{\xi\xi} - 2(x_\xi x_\eta + y_\xi y_\eta)y_{\xi\eta} + (x_\xi^2 + y_\xi^2)y_{\eta\eta} = 0 \qquad (3.57)$$

where ξ and η are independent variables. How are these used to create mappings? Suppose that contour C_W in Figure 3.1 is to map into $\eta = 0$ of Figure 3.2. The user first discretizes C_W in Figure 3.1 by penciling along it a sequence of dots chosen to represent the curve. If these are selected in an orderly, say clockwise, fashion, they define the direction in which ξ increases. Along $\eta = 0$, values of x and y are known (e.g., from measurement on graph paper) as functions of ξ. Similarly, x and y values along C_r are known as functions of ξ on $\eta = 1$ of Figure 3.2. These provide the boundary conditions for Equations 3.56 and 3.57, which are augmented by single-valuedness constraints at arbitrarily chosen branch cuts B_1 and B_2.

In Thompson's and similar approaches, Equations 3.56 and 3.57 are discretized by finite differences and solved by point or line relaxation, starting with guesses for the dependent variables x and y. The problem is linearized by approximating all nonlinear coefficients using values from earlier iterations. Typically, several updates to Equation 3.56 are taken, followed by updates to Equation 3.57, with this cycling process, often unstable, repeated until convergence. Variations of the approach are known, with 100×100 mesh systems in the $\xi-\eta$ plane requiring minutes of computing time on typical personal computers. Once $x = x(\xi, \eta)$ and $y = y(\xi, \eta)$ are solved and tabulated as functions of ξ and η, physical coordinates are generated. First, η is fixed; for each node ξ along this η, the computed values of (x, y) pairs are successively plotted in the x–y plane to produce the required closed contour. This procedure is repeated for all values of η until the entire family of closed curves is obtained, with limit values $\eta = 0$ and $\eta = 1$ again describing C_w and C_r. Orthogonals are constructed by repeating the procedure, with the η and ξ roles reversed.

This process provides the mapping only. The partial differential equation governing the physical problem must be transformed into (ξ, η) coordinates and solved. For instance, in reservoir simulation, Darcy's pressure equation must be expressed in terms of ξ, η and solved; in aerodynamics, the Navier-Stokes equations are solved. Thompson's simplification lies not in the transformed host equation, which may contain mixed derivatives and variable coefficients, but in the computational domain itself, because it takes on a rectangular form amenable to simple numerical solution. While the transformed equation is more complicated, fewer equations are actually needed to represent the complete flow, thus leading to rapid convergence and much faster solutions.

Fast iterative solutions

This section describes the solution to the mapping equations—solutions to the transformed momentum equations, discussed elsewhere in this book, depend on the engineering model under consideration. Existing solution methods solving $x(\xi, \eta)$ and $y(\xi, \eta)$ stagger the solutions for Equations 3.56 and 3.57. For example, crude solutions are used to initialize the coefficients of Equation 3.56, and improvements to $x(\xi, \eta)$ are obtained. These are used to evaluate the coefficients of Equation 3.57, in order to obtain an improved $y(\xi, \eta)$; then attention turns to Equation 3.56 again, and so on, until convergence is achieved. Various means are used to implement these iterations—for example, point SOR, line SLOR, line SOR with explicit damping, alternating-direction-implicit, and multigrid, with varying degrees of success. Often these schemes diverge computationally. In any event, this staggering introduces different artificial time levels while iterating. However, classic numerical analysis suggests that faster convergence and improved stability are possible by reducing their number.

A new approach to rapidly solving the mesh equations was developed by this author using a simple idea. This has since been validated and extended. Consider $z_{\xi\xi} + z_{\eta\eta} = 0$, for which $z_{i,j} \approx (z_{i-1,j} + z_{i+1,j} + z_{i,j-1} + z_{i,j+1})/4$ holds on constant grid systems. This well-known averaging law motivates the *recursion formula* $z_{i,j}^{n} = (z_{i-1,j}^{n-1} + z_{i+1,j}^{n-1} + z_{i,j-1}^{n-1} + z_{i,j+1}^{n-1})/4$ often used to illustrate and develop multilevel iterative solutions; an approximate, and even trivial, solution can be used to initialize the calculations, and correct nonzero solutions are always produced from non-zero boundary conditions.

But the well-known Gauss-Seidel method is fastest: As soon as a new value of $z_{i,j}$ is calculated, its previous value is discarded and overwritten by the new value. This speed is accompanied by low memory requirements, since there is no need to store both n and $n-1$ level solutions: Only a single array, $z_{i,j}$ itself, is required in programming. Our approach to Equations 3.56 and 3.57 was motivated by the following idea. Rather than solving for $x(\xi, \eta)$ and $y(\xi, \eta)$ in a staggered, leapfrog manner, is it possible to *simultaneously* update x and y in a similar "once-only" manner? Are convergence rates significantly increased? What formalism permits us to solve in Gauss-Seidel fashion? What are the programming implications?

Complex variables are used in harmonic analysis problems; for example, the real and imaginary parts of an analytical function $f(z)$, where $z = x + iy$, provide solutions satisfying Laplace's equation. Here we use complex analysis differently. We *define* a dependent variable z by $z(\xi, \eta) = x(\xi, \eta) + iy(\xi, \eta)$, and then add Equation 3.56 plus i times Equation 3.57, in order to obtain the result $(x_\eta^2 + y_\eta^2)z_{\xi\xi} - 2(x_\xi x_\eta + y_\xi y_\eta)z_{\xi\eta} + (x_\xi^2 + y_\xi^2)z_{\eta\eta} = 0$. Now, the complex conjugate of z is $z^*(\xi, \eta) = x(\xi, \eta) - iy(\xi, \eta)$, from which we find that $x = (z + z^*)/2$ and $y = -i(z - z^*)/2$. Substitution produces the simple and equivalent one-equation result

$$(z_\eta z_\eta^*)z_{\xi\xi} - (z_\xi z_\eta^* + z_\xi^* z_\eta)z_{\xi\eta} + (z_\xi z_\xi^*)z_{\eta\eta} = 0 \qquad (3.58)$$

This form yields significant advantages. First, when z is declared a complex variable in a Fortran program, Equation 3.58 represents, for all practical purposes, a *single* equation in $z(\xi, \eta)$. There is no need to leapfrog between x and y solutions now, since a single formula analogous to the classical model $z_{i,j} = (z_{i-1,j} + z_{i+1,j} + z_{i,j-1} + z_{i,j+1})/4$ is easily written for the $z_{i,j}$ related to Equation 3.58 using second-order central differences. Because both x and y are simultaneously resident in computer memory, the extra time level present in staggered schemes is eliminated, as in the Gauss-Seidel method. In thousands of test simulations conducted using point and line relaxation, convergence times are shorter by factors of two to three, with convergence rates far exceeding those obtained for cyclic solutions between $x(\xi, \eta)$ and $y(\xi, \eta)$. Convergence appears to be unconditional, monotonic, and stable. Because Equation 3.58 is nonlinear, von Neumann tests for exponential stability and traditional estimates for convergence rates do not apply, but the evidence for stability and convergence, while empirical, remains very strong and convincing.

Example 3.1   85

On Laplacian transformations

We have introduced expedient ways to solve Equations 3.56 and 3.57 computationally for $x(\xi, \eta)$ and $y(\xi, \eta)$. These mappings are not generally conformal. The fast solution method for Equation 3.58 is important because the properties afforded by conformal transformations are not useful for physical problems governed by models other than Laplace's equation. As noted earlier, even the simplest engineering rheology problems satisfy (usually nonlinear) inhomogeneous equations that are far more complicated in structure.

For example, under conformal transformation, it is possible to show that the $x-y$ Laplacian satisfies $f_{xx} + f_{yy} = (F_{\xi\xi} + F_{\eta\eta})/J(\xi, \eta)$. In physical problems where "$f_{xx} + f_{yy} = 0$," it follows similarly that "$F_{\xi\xi} + F_{\eta\eta} = 0$," since J is never zero—this invariance maps simple solutions into more difficult ones for complicated shapes, providing, in effect, "free" mathematical solutions. In the problems addressed in this book, the governing partial differential equations are not as elementary; for instance, the equation "$\rho \partial u/\partial t = -\partial p/\partial z + \mu(\partial^2 u/\partial x^2 + \partial^2 u/\partial y^2)$" is needed to describe the simplest transient Newtonian flows, with non-Newtonian motions satisfying even more complicated models. This means that any advantages inherent in conformal mapping are lost by virtue of the more involved physics. Thus, Equations 3.56 and 3.57 are entirely adequate for solving such problems, provided the Laplacian $f_{xx} + f_{yy}$ is replaced by its appropriately transformed value. In fact, if Equations 3.56 and 3.57 are used to simplify Equation 3.13, we obtain the remarkable result that

$$f_{xx} + f_{yy} = (\alpha F_{\xi\xi} - 2\beta F_{\xi\eta} + \gamma F_{\eta\eta})/J^2 \qquad (3.59)$$

As an example, the steady, isotropic reservoir flow problem for liquids solving $p_{xx} + p_{yy} = 0$ for the Darcy pressure amounts to $\alpha P_{\xi\xi} - 2\beta P_{\xi\eta} + \gamma P_{\eta\eta} = 0$ in transformed coordinates. But since $p = p_{well}$ at the inner borehole contour and $p = p_\infty$ at the far-field boundary are constants, noting that the well may contain fractures and other geometric anomalies, and that the far field may be highly irregular, derivatives with respect to ξ must vanish, leaving as the governing equation a simplified $P_{\eta\eta} = 0$ whose general analytical solution takes the form $P(\eta) = (p_\infty - p_{well})\eta + p_{well}$.

This means that the solution to a superset of pressure problems can be expressed in terms of a single geometric mapping! The solution to the grid generation problem thus provides the general solution to the steady reservoir flow problem in a very elegant manner. This fact is used in the reservoir engineering book of Chin (2002) to develop numerical solutions for flows in reservoirs with very complicated geometries (e.g., a reservoir having the shape of Texas; see Figure 1.2 in Chapter 1).

When the governing partial differential equation is more complicated than $p_{xx} + p_{yy} = 0$, then, of course, more specialized techniques are required. This book develops the required methods for non-Newtonian eccentric annular flows with general fluid properties under steady and transient conditions. In three-dimensional problems with constant cross sections, the same geometric mapping applies in all cross sections and J is independent of z. Then the general Laplacian transforms according to

$$f_{xx} + f_{yy} + f_{zz} = (\alpha F_{\xi\xi} - 2\beta F_{\xi\eta} + \gamma F_{\eta\eta})/J^2 + F_{zz} \qquad (3.60)$$

This result will prove extremely useful later in this book when we address transient multiphase flow in three dimensions. But mappings can be developed that are much more general. For instance, a very long borehole with geometric anomalies and variations in the axial direction (that are measurable by caliper logs) can be modeled by creating mappings at periodic axial distances, resulting in a Jacobian with z dependence. The corresponding flow equations likewise contain the modified J, and, needless to say, the numerical solution is far more complicated.

The more general models developed for the applications treated in this book will be shown to reduce to simpler ones for steady and transient, two-dimensional, single-phase flows, and will also be shown to be numerically consistent, as computed asymptotic results taken for large times and axial distances reduce to known results obtained in earlier chapters. The extraordinary degree of cross-checking undertaken for the model development reported in this book ensures that our formulations are consistent and correct in areas where assumptions overlap. This provides a degree of user confidence needed for job planning in real-world field applications.

EXAMPLE 3.2

Mappings for Flows in Singly Connected Ducts

In our annular flow discussions, where the domain of interest lies between the pipe and the borehole wall, we deal, in a mathematical sense, with "donuts," as further suggested in Figure 3.3, which shows a fractured well in a petroleum reservoir (that is, "donuts," as in "coffee and donuts"). Domains with "holes" such as these are known as "doubly connected" regions; two holes, for example, lead to those that are "triply connected."

A wealth of material on connectivity is available in mathematical topology. In our more mundane work, we also deal with "singly connected" domains such as those in Figure 3.4. These are important in modeling flows in complicated ducts, which in petroleum engineering include pipeline cross sections with clogging debris, boreholes with substantial clogging where the bottom completely fills with cuttings and the hole is no longer annular, and so on. To generalize the idea of single connectivity, the possibilities in Figure 3.4 are available for thought.

In Figure 3.1, we introduced "branch cuts" (across which special conditions were invoked for single-valuedness) to transform a doubly connected region into one that is singly connected. The transforms for problems such as those in Figure 3.4 are more easily obtained because branch cuts are unnecessary. Once $x-y$ point values at A, B, C, D, and all intermediate points are assigned (e.g., by interpolating from graph paper sketches), they can be directly imposed as boundary conditions at the edges of the rectangle in Figure 3.5. Then the previous solution process for "z" applies. The resulting generalized duct flow computer model was easily developed from the annular flow work for a pipeline application. In this case, we wished to dynamically couple a debris growth model that ultimately clogs the flow cross section in order to study wax and hydrate buildup in cold subsea environments. We discuss the solids deposition modeling strategy before providing example computed results.

EXAMPLE 3.3

Solids Deposition Modeling and Applications

What is "solids deposition modeling" and what is its role in pipe or annular flow dynamics? Although numerous studies have been directed, for instance, at wax deposition and hydrate formation, none have addressed the dynamic interaction between the solids deposition process and the

Example 3.3   **87**

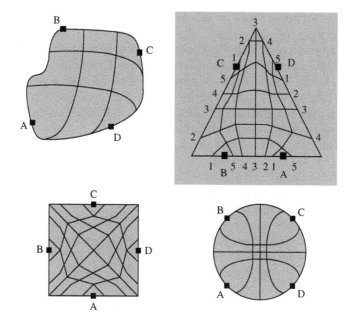

FIGURE 3.4

Singly connected "pipe flow" domains.

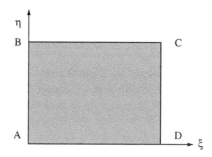

FIGURE 3.5

Mapped computational domain.

velocity field imparted by the flowing non-Newtonian fluid. The latter serves dual functions: It assists with solid particle placement, but, at the same time, the viscous stress field associated with it tends to remove particles that have adhered to solid surfaces.

Until now, determining the velocity field alone has proven difficult, if not impossible: Nonlinear flow equations must be solved for geometric domains that are far from ideal in shape. However, the methods developed in this book for annular and general duct flow permit fast and robust solutions and also efficient postprocessing and visual display for quantities like apparent viscosity, shear rate, and viscous stress. In this sense, "half" of the problem has been resolved, and in this chapter we address the remaining half.

To understand the overall philosophy, it is useful to return to the problem of mudcake formation and erosion, and of cuttings transport, considered in detail in Chapter 5. As we will find, the plugging or cleaning of a borehole annulus can be a dynamic, time-dependent process. For example, the inability of the low-side flow to remove cuttings results in debris bed formation, when cuttings combine with mudcake to form mechanical structures. Forced filtration of drilling mud into the formation compacts these beds, and individual particle identities are lost: The resulting beds, characterized by well-defined yield stresses, alter the shape of the borehole annulus and the properties of the flow.

But the bed can be eroded or removed, provided the viscous stress imparted by the flowing mud in the modified annulus exceeds the yield value. If this is not possible, plugging will result and stuck pipe is possible. On the other hand, alternative remedial actions are possible. The driller can change the composition of the mud to promote more effective cleaning, increase the volumetric flow rate, or both. Successfully doing so erodes cuttings accumulations and, ideally, promotes dynamic "self-cleaning" of the hole.

In a sense, developing a new "constitutive relation"—for example, postulating Newtonian or Power law properties and deriving complementary flow equations—is simpler than designing solids deposition models. The mathematical process needed to "place" stress-strain relations in momentum differential equation form is more straightforward than the cognitive process required to understand every step of a new physical phenomenon (e.g., wax deposition or hydrate formation). In this section, we introduce a philosophy behind modeling solids deposition, and as a first step develop a simple model for mudcake and cuttings bed buildup over porous rock. We emphasize that there are no simple answers: Each problem is unique, and the developmental process is very iterative.

Mudcake buildup on porous rock

Borehole annuli are lined with slowly thickening mudcake that, over large time scales, will reduce cross-sectional size. However, dynamic equilibrium is usually achieved because erosive forces in the flow stream limit such thickening. As a first step in understanding this process, growth in the absence of erosion must be characterized, but even this requires a detailed picture of the physics. The reader should carefully consider the steps needed in designing deposition models, taking this example as a model.

Since the permeability of the formation greatly exceeds that of mudcake, and the thickness of mudcake is small compared with the borehole radius, we can model cake growth in the idealized lineal flow test setup in Figure 3.6. We consider a one-dimensional experiment where mud, in essence a suspension of clay particles in water, is allowed to flow through filter paper. Initially, the flow rate is rapid. But as time progresses, solid particles (typically 6 percent to 40 percent by volume for light to heavy muds), such as barite, are deposited onto the surface of the paper, forming a mudcake that, in turn, retards the passage of mud filtrate by virtue of the resistance to flow that the cake provides.

We therefore expect the filtrate volumetric flow rate and cake growth rate to decrease with time, while the filtrate volume and cake thickness continue to increase, but ever more slowly. These qualitative ideas can be formulated precisely because the problem *is* based on well-defined

Example 3.3   **89**

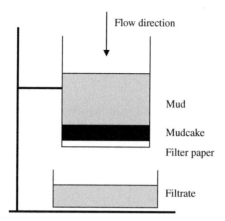

Flow direction

Mud

Mudcake

Filter paper

Filtrate

FIGURE 3.6

Simple laboratory mudcake buildup experiment.

physical processes. For one, the composition of the homogeneous mud during this filtration does not change: Its *solid fraction* is always constant. Second, the flow *within* the mudcake *is* a Darcy flow and is therefore governed by the equations used by reservoir engineers. The only problem, though, is the presence of a *moving boundary*—namely, the position interface separating the mudcake from the mud that ultimately passes through it and continually adds to its thickness. The physical problem, therefore, is a *transient* process that requires somewhat different mathematics than that taught in fundamental partial differential equations courses.

Mudcakes in reality may be compressible; that is, their mechanical properties may vary with applied pressure differential. We will be able to draw upon reservoir engineering methods developed for subsidence and formation compaction later. For now, a simple constitutive model for incompressible mudcake buildup—that is, the filtration of a fluid suspension of solid particles by a porous but rigid mudcake—can be constructed from first principles. First, let $x_c(t) > 0$ represent cake thickness as a function of the time, where $x_c = 0$ indicates zero initial thickness. Also, let V_s and V_l denote the volumes of solids and liquids in the mud suspension, and let f_s denote the *solid fraction* defined by $f_s = V_s/(V_s + V_l)$. Since this does not change throughout the filtration, its time derivative must vanish.

If we set $df_s/dt = (V_s + V_l)^{-1}dV_s/dt - V_s(V_s + V_l)^{-2}(dV_s/dt + dV_l/dt) = 0$, we can show that $dV_s = (V_s/V_l)\,dV_l$. But since, separately, $V_s/V_l = f_s/(1 - f_s)$, it follows that $dV_s = \{f_s/(1 - f_s)\}dV_l$. This is, essentially, a *conservation of species* law for the solid particles making up the mud suspension, and does not as yet embody any assumptions related to mudcake buildup. Frequently, we might note, the drilling fluid is thickened or thinned in the process of making hole; if so, the equations derived here should be reworked with $f_s = f_s(t)$ and its corresponding time-dependent pressure drop.

To introduce the mudcake dynamics, we observe that the total volume of solids dV_s deposited on an elemental area dA of filter paper during an infinitesimal time dt is $dV_s = (1 - \phi_c)\,dA\,dx_c$, where ϕ_c is the mudcake porosity. During this time, the volume of filtrate flowing through our filter paper screen is $dV_l = |v_n|\,dA\,dt$, where $|v_n|$ is the Darcy velocity of the filtrate through the cake and past the paper. We now set our two expressions for dV_s equal in order to form $\{f_s/(1 - f_s)\}$

$dV_1 = (1 - \phi_c) \, dA \, dx_c$, and replace dV_1 with $|v_n| \, dA \, dt$, so that we obtain $\{f_s/(1 - f_s)\} \, |v_n| \, dA \, dt = (1 - \phi_c) \, dA \, dx_c$.

Now it is seen that the dA's cancel, and we are led to a generic equation governing mudcake growth. In particular, the cake thickness $x_c(t)$ satisfies the ordinary differential equation

$$dx_c(t)/dt = \{f_s/\{(1 - f_s)(1 - \phi_c)\}\} |v_n| \qquad (3.61a)$$

At this point, we assume a one-dimensional, constant-density, single *liquid* flow. For such flows, the constant Darcy velocity is $(k/\mu)(\Delta p/L)$, where $\Delta p > 0$ is the usual "delta p" pressure drop through the core of length L, assuming that a Newtonian approximation applies. The corresponding velocity for the present problem is $|v_n| = (k/\mu)(\Delta p/x_c)$, where k is the mudcake permeability and μ is a mean filtrate viscosity. Substitution in Equation 3.61a leads to

$$dx_c(t)/dt = \{kf_s\Delta p/\{\mu(1 - f_s)(1 - \phi_c)\}\}/x_c \qquad (3.61b)$$

If the mudcake thickness is infinitesimally thin at $t = 0$, with $x_c(0) = 0$, Equation 3.61b can be integrated, with the result that

$$x_c(t) = \sqrt{[\{2kf_s\Delta p/\{\mu(1 - f_s)(1 - \phi_c)\}\}t]} > 0 \qquad (3.61c)$$

This demonstrates that cake thickness in a lineal flow grows with time the same as $\sqrt{t}$. However, it grows ever more slowly because increasing thickness means increasing resistance to filtrate through-flow, the source of the solid particulates required for mudcake buildup; consequently, filtrate buildup also slows.

To obtain the filtrate production volume, we combine $dV_1 = |v_n|dA \, dt$ and $|v_n| = (k/\mu)(\Delta p/x_c)$ to form $dV_1 = (k\Delta p dA/\mu) \, x_c^{-1} dt$. Using Equation 3.61c, we find $dV_1 = (k\Delta p dA/\mu)[\{2kf_s\Delta p/\{\mu(1 - f_s)(1 - \phi_c)\}\}]^{-1/2}(t)^{-1/2} \, dt$. Direct integration, assuming zero filtrate initially, yields

$$V_1(t) = 2(k\Delta p dA/\mu)[\{2kf_s\Delta p/\{\mu(1 - f_s)(1 - \phi_c)\}\}]^{-1/2}(t)^{1/2}$$
$$= \sqrt{\{2k\Delta p(1 - f_s)(1 - \phi_c)/(\mu f_s)\}} \sqrt{t} \, dA \qquad (3.61d)$$

This correctly reproduces the common observation that filtrate volume increases in time as $\sqrt{t}$. The mudcake deposition model in Equation 3.61c, at this point, is credible and is significant in that it explicitly highlights the roles of the individual parameters k, f_s, Δp, μ, and ϕ_c.

Now, along the walls of general boreholes that are not necessarily circular, containing drillpipes that need not be concentric, the "$x_c(t)$" in Equation 3.61c applies at each location; of course, "$x_c(t)$" must be measured in a direction perpendicular to the local surface area. This thickness increases with time by the same amount everywhere; consequently, the hole area decreases and the annular geometry changes, with more pronounced curvature. At the same time, drilling fluid is flowing parallel to the borehole axis. This flow, generally non-Newtonian, must be calculated using the methods developed in this book. The mechanical yield stress τ_y of the formed cake, which must be separately determined in the laboratory, is an important physical constant of the system. If the stress τ imparted by the fluid is less than τ_y, a very simple deposition model might allow Equation 3.61c to proceed "as is."

However, if $\tau > \tau_y$ applies locally, one might postulate, instead of Equation 3.61c, an "erosion model,"

$$dx_c(t)/dt = f(\ldots) \qquad (3.62)$$

Example 3.3   **91**

where the function "f < 0" might depend on net flow rate, gel level, weighting material characteristics, and the magnitude of the difference "$\tau - \tau_y$." In unconsolidated sands penetrated by deviated wells, "f" may vary azimuthally, since gravity effects at the top of the hole differ from those at the bottom. And in highly eccentric annuli, mudcake at the low side may be thicker than high-side cake because lower viscous stress levels are less effective in cake removal.

Again, the mudcake buildup and removal process is time dependent, and very dynamic, at least computationally. In the present example, we conceptually initialize calculations with a given eccentric annulus, possibly contaminated by cake, and calculate the non-Newtonian flow characteristics associated with this initial state. Equations 3.61c and 3.62 are applied at the next time step to determine modifications to the initial shape. Then flow calculations are repeated, with the entire process continuing until some clear indicator of hole equilibrium is achieved. The hole may tend to plug, in which case remedial planning is suggested, or it may tend to remain open.

In any event, the development of deposition and erosion models such as those in Equations 3.61c and 3.62 requires a detailed understanding of the physics and consequently calls for supporting laboratory experiments. As this example for mudcake deposition shows, it *is* possible to formulate phenomenological models analytically when the "pieces of the puzzle" are well understood, as we have for the "$\sqrt{t}$" model governing mudcake growth.

By the same token, it should be clear that in other areas of solids deposition modeling—for example, accumulation of produced fines, wax buildup, and hydrate plug formation in pipelines—"simple answers" are not yet available. More than likely, the particular models used will depend on the reservoir in question and will probably change throughout the life of the reservoir. For this reason, the present chapter focuses on generic questions and attempts to build a sound research approach and modeling philosophy for workers entering the field. At the present time, much of the published research on wax deposition and hydrate formation focuses on fundamental processes like crystal growth and thermodynamics. An experimental database providing even qualitative information is not yet available for detailed model development. Nonetheless, we can speculate on how typical models may appear and comment on the mathematical forms in which they can be expressed.

Depositional mechanics

In this section, we introduce the reader to basic ideas in different areas of solids deposition and transport by fluid flow, if only to highlight common physical processes and mathematical methods. By far, the most comprehensive literature is found in sedimentary transport and slurry movement, specialties that have been well developed in civil engineering over decades of research. Anderson (1961) and Kapfer (1973) provide an excellent introduction to established techniques. These references, in fact, motivated the cuttings transport research in Chapter 5. Concepts and results from these and related works are covered next.

Sedimentary transport

Sediment transport is important to river, shoreline, and harbor projects. The distinction between "cohesive" and "noncohesive" sediments is usually made. For example, clays are cohesive, while sand and gravel in stream beds consist of discrete particles. In cohesive sediments, the resistance to erosion depends primarily on the strength of the cohesive bond between the particles. Variables affecting particle lift-off include parameters like bed shear stress; fluid viscosity; particle size, shape, and mass density; and number density distribution. Different forces are involved in holding grains down and entraining them into the flow. These include gravity, frictional resistance along grain contacts, "cohesiveness" or "stickiness" of clays due to electrochemical attraction, and forces parallel to the bed such as shear stress. The "sediment-transporting capacity" of a moving fluid is the maximum rate at which moving fluid can transport a particular sediment aggregation.

Lift forces are perpendicular to the flow direction and depend on the shapes of individual particles. For example, a stationary spherical grain in a uniform stream experiences no lift, since upper and lower flow fields are symmetric; however, a spinning or "tumbling" spherical grain does experience lift. Flat grains oriented at nonzero angles with respect to the uniform flow also experience lift, whose existence is apparent from asymmetry. Of course, oncoming flows need not be uniform. It turns out that small, heavy particles that have settled in a lighter viscous fluid can resuspend if the mixture is exposed to a shear field. This interaction between gravity and shear-induced fluxes strongly depends on particle size and shape. Note that the above force differs from the lift for airplane wings: Small grains "see" low Reynolds number flows, while much larger bodies operate at high Reynolds numbers. Thus, formulas obtained in different fluid specialties must be carefully evaluated before they are used in deposition modeling. In either case, mathematical analysis is very difficult.

Once lifted into the flow stream, overall movement is dictated by the vertical "settling velocity" of the particle and by the velocity in the main flow. Settling velocity is determined by balancing buoyancy and laminar drag forces, with the latter strongly dependent on fluid rheology. For Newtonian flows, the classic Stokes solution applies; for non-Newtonian flows, analytical solutions are not available. Different motions are possible. Finer silts and clays will more or less float within a moving fluid. On the other hand, sand and gravel are likely to travel close to the bed; those that "roll and drag" along the bottom move by traction, while those that "hop, skip, and jump" move by the process of saltation.

In general, modeling non-Newtonian flow past single stationary particles is difficult, even for the most accomplished mathematicians. Flows past unconstrained bodies are even more challenging. Finally, modeling flows past aggregates of particles is likely to be impossible without additional simplifying statistical assumptions. For these reasons, useful and practical deposition and transport models are likely to be empirical, so scalable laboratory experiments are highly encouraged. Simpler "ideal" flow setups that provide some physical insight into key parameters are likely to be more useful than "practical engineering" examples that include too many interacting variables.

Slurry transport

A large body of literature exists for slurry transport (e.g., coal slurries, slurries in mining applications, slurries in process plants, and so on). A comprehensive review is neither possible nor necessary, since water is the carrier fluid in the majority of references. However, many fundamental ideas and approaches apply. Early references provide discussions on sewage sludge removal, emphasizing prevalent non-Newtonian behavior, while acknowledging that computations are not practical. They also discuss settling phenomena in slurries—for example, the influence of particle size, particle density, and fluid viscosity.

Example 3.3   **93**

"Minimum velocity" formulas are available that, under the assumptions cited, are useful in ensuring clean ducts when the carrier fluid is water. The notion of "critical tractive force," the value of shear stress at which bed movement initiates, is introduced; this concept is important in our discussions of cuttings transport. Both "velocity" and "stress" criteria are used later in this chapter to construct illustrative numerical models of eroding flows. Also, the distinction between transport in closed conduits and open channels is made.

The literature additionally addresses the effects of channel obstructions and the formation of sediment waves; again, restrictions to water as the carrier fluid are required. Numerous empirical formulas for volumetric flow rate that would give clean conduits are available in the literature; however, their applicability to oilfield debris, waxes, and hydrates is uncertain. While we carefully distinguish between velocity and stress as distinctly different erosion mechanisms, we note that in some flows the distinction is less clear. At times, for example, the decrease in bed shear stress is primarily a function of decreasing flow velocity.

Waxes and paraffins: Basic ideas

As hot crude flows from reservoirs into cold pipelines, with low temperatures typical under deep subsea conditions, wax crystals may form along solid surfaces when wall temperatures drop below the "cloud point" or "wax appearance temperature." Crystals may grow in size until the wall is fully covered, with the possibility of encapsulating oil in the wax layers. Wax deposition can grow preferentially on one side of the pipe due to gravity segregation. As wax thickness builds, the pressure drop along the pipe must be increased to maintain constant flow rate, and power requirements increase. Constant pressure processes yield decreasing flow rates.

Paraffin deposition can be controlled through various means. Insulation and direct pipe heating will reduce exposure to the cold environment. Mechanical pigging is possible. Chemical inhibitors can also be used. For example, surfactants or dispersants alter the ability of wax particles to adhere to each other or to pipe wall surfaces; in the language of sedimentary transport, they become less cohesive and behave more like discrete entities. Biochemical methods, for instance, use bacteria to control wax growth.

In this book, we will address the effect of nonlinear fluid rheology and noncircular duct flow in facilitating wax erosion. The "critical tractive force" ideas developed in slurry transport, extended in Chapter 5 to cuttings removal, again apply to bed-like deposits. Recent authors, for example, introduce "critical wax tension" analogously, defined as the critical shear force required to remove a unit thickness of wax deposit; the exact magnitude depends on oil composition, wax content, temperature, buildup history, and aging.

More complications

Paraffin deposition involves thermodynamics, but other operational consequences arise that draw from all physical disciplines.

- Electro-kinetic effects may be important with heavy organic constituents. Potential differences along the conduit may develop due to the motion of charged particles; these induce alterations in colloidal particle charges downstream that promote deposition. That is, electrical charges in the crude may encourage migration of separated waxes to the pipe wall.

- In low flow rate pipelines, certain waxes sink because of gravity and form sludge layers at the low side. Also, density segregation can lead to recirculating flows of the type modeled in Chapter 5.
- For lighter waxes, buoyancy can cause precipitated wax to collect at the top of the pipe. (In the simulations performed in this chapter, no distinction is made between "top" and "bottom," since our "snapshots" can be turned "upside down.")
- Deposited wax will increase wall roughness and therefore increase friction, thus reducing pipe-line flow capacity.
- Suspended particulates such as asphaltenes, formation fines, corrosion products, silt, and sand may encourage wax precipitation, acting as nuclei for wax separation and accumulation. Wax particles so separated may not necessarily deposit along walls; they may remain in suspension, altering the rheology of the carrier fluid, affecting its ability to "throw" particles against pipe walls or to remove wax deposits by erosion.
- Although significant deposition is unlikely under isothermal conditions—that is, when pipeline crude and ocean temperatures are in equilibrium—wall deposits may nonetheless form. Pipe roughness, for instance, can initiate stacking, leading to local accumulations that may further grow.

Wax precipitation in detail

Waxy crude may contain a variety of light and intermediate hydrocarbons (e.g., paraffins, aromatics, naphthenic, wax, heavy organic compounds) and low amount of resins, asphaltenes, and organo-metallics. Wax in crudes consists of paraffin (C18−C36) and naphthenic (C30−C60) hydrocarbons. These wax components exist in various states (i.e., gas, liquid, or solid) depending on temperature and pressure. When wax freezes, crystals are formed. Those formed from paraffin wax are known as "macrocrystalline" while those originating from naphthenes are "microcrystalline."

When the temperature of waxy crude is decreased, the heavier fractions in wax content appear first. The "cloud point" or "wax appearance temperature" is the temperature below which the oil is saturated with wax. Deposition occurs when the temperature of the crude falls below the cloud point. Paraffin will precipitate under slight changes in equilibrium conditions, causing loss of solubility of the wax in the crude. Wax nucleation and growth may occur along the pipe surface and within the bulk fluid. Precipitation within the fluid causes its viscosity to increase and alters the non-Newtonian characteristics of the carrier fluid. Increases in frictional drag may initiate pumping problems and higher overall pipe pressures. Note that the carrier fluid is rarely a single-phase flow. More often than not, wax deposition occurs in three-phase oil, water, and gas flow, over a range of gas-oil ratios, water cuts, and flow patterns, which can vary significantly with pipe inclination angle.

Wax deposition control

The most direct means of control, though not necessarily the least inexpensive, targets wall temperature by insulation or heating, possibly through internally heated pipes. But the environment is far from certain. Some deposits do not disappear on heating and are not fully removed by pigging. Crudes may contain heavy organics like asphaltenes and resin, which may not crystallize upon cooling and may

Example 3.3   **95**

not have definite freezing points; these interact with wax differently and may prevent or enhance wax crystal formation. Solvents provide a different alternative. However, those containing benzene, ethyl benzene, toluene, and so on are encountering increased opposition from regulatory and environmental concerns. The problems are acute for offshore applications; inexpensive and environmentally friendly control approaches with minimal operational impact are desired.

Wax growth on solid surfaces, under static conditions, is believed to occur by molecular diffusion. Behind most deposition descriptions are liquid phase models and equations of state, with the exact composition of the wax phase determined by the model and the physical properties of the petroleum fractions. We do not attempt to understand the detailed processes behind wax precipitation and deposition in this section. Instead, we focus on fluid-dynamical modeling issues, demonstrating how non-Newtonian flows can be calculated for difficult "real-world" duct geometries that are less than ideal. The "mere" determination of the flow field itself is significant, since it provides information to evaluate different modes of deposition and to address important remediation issues.

For example, in sediment transport, flow nonuniformities play dual roles: They may "throw" particles onto surfaces, where they adhere, or they can remove buildups by viscous shear. Both effects must be studied using experiments considering the background velocity and stress fields that analysis provides. Modeling approaches hope to establish the hydrodynamic backbone that makes accurate forecasting of these phenomena possible. Is it possible to design a fluid that keeps particles suspended or, perhaps, to understand the conditions under which the flow self-cleans? What are the rheological effects of chemical solvents? Wax can cause crude oil to gel and deposit on tubular surfaces. What shear stresses are required to remove it? Waxy crude oil may gel after a period of shutdown. What levels of pressure are required to initiate start-up of flow?

Modeling dynamic wax deposition

In principle, modeling the dynamic, time-dependent interaction between waxy deposits attempting to grow and duct flows attempting to erode them is similar to, although slightly more complicated than, the mudcake model developed earlier. The *deposition*, or *growth* model, shown conceptually in Figure 3.7(a), consists of two parts: a thermal component in which buildup is driven by temperature gradients and a mechanical component in which velocity "throws" additional particles that have precipitated in the bulk fluid into the wax-lined pipe surface.

This velocity may be coupled to the temperature environment. Various solids convection models are available in the fluids literature, and in general different deposition models are needed in different production scenarios. The competing *erosive* model is schematically shown in Figure 3.7(b), in which we emphasize the role of non-Newtonian fluid stress at the walls; it is similar to our model for cuttings transport removal from stiff beds. Wax yield stress may be determined in the laboratory or inferred from mechanical pigging data (see, for example, Souza Mendes et al., 1999, or related pipeline literature).

Hydrate control

Natural gas production from deep waters can be operationally hampered by pipeline plugging due to gas hydrates. Predicting the effects of pipe hydraulics on hydrate behavior is necessary to achieve optimal hydrate control. As exploration moves offshore, the need to minimize production

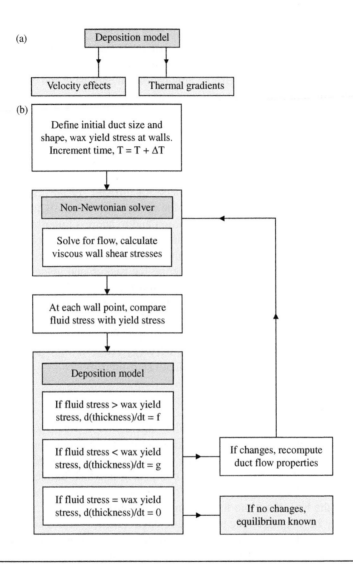

FIGURE 3.7

(a) Conceptual deposition model. (b) Fluid flow and solids deposition model interaction.

facility construction and maintenance costs becomes important. Producers are seeking options that permit the transport of unprocessed fluids miles from wellheads or subsea production templates to central processing facilities located in shallower water. Deepwater, multiphase flowlines can offer cost-saving benefits to operators; consequently, basic and applied research related to hydrate control is an active area of interest.

Hydrate crystallization takes place when natural gas and water come into contact at low temperature and high pressure. Hydrates are "ice-like" solids that form when sufficient amounts of water are available, a "hydrate former" is present, and the proper combinations of temperatures and

Example 3.3   **97**

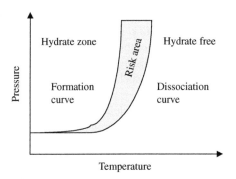

FIGURE 3.8

Hydrate dependence on "P" and "T."

pressures are conducive. Gas hydrates are crystalline compounds that form whenever water contacts the constituents found in natural gas, gas condensates, and oils, at the hydrate formation equilibrium temperatures and pressures, as Figure 3.8 shows. Hydrate crystals can be thought of as integrated networks of hydrogen-bonded, "soccer ball"–shaped ice cages with gas constituents trapped within.

Low seabed temperatures and high pressures can significantly impact the commercial risk of deepwater projects. Hydrates can cause plugging, an unacceptable condition given the inaccessibility of deep subsea pipelines. Hydrate plugging is not new and early on profoundly affected onshore production and flow. But these problems became less severe as hydrate phase equilibrium data became available; these data provided the basis for modern engineering and chemical inhibition procedures using methanol and glycol. Such treatments can be costly in deep water, though, given the quantities of inhibitor required, not to mention expensive storage facilities; but they remain attractive, as recent research has led the way to more effective, low-toxicity compounds as useful alternatives to methanol or glycol. Field and laboratory studies have had some success, but problems remain that must be solved before the industry gains advantages in utilizing these inhibitors.

Operational considerations are also important to hydrate mitigation. Proper amounts of chemicals must arrive at target flowline locations at the required time to control the rate of crystal formation, growth, agglomeration, and deposition. This combined chemical and hydrodynamic control strategy in general multiphase pipeline environments must be effective over extended shutin periods to accommodate a range of potential offshore operating scenarios.

Understanding the effects of chemicals on rheology and flow represents one aspect of the mitigation problem. In pipeline plugging, we are concerned, as noted previously, with the effects of obstructions on pressure drops and flow rates. On the other hand, natural hydrates represent a potentially important source of natural gas, although they can potentially clog pipelines. One possible delivery solution is to convert associated gases into frozen hydrates, which are then mixed with refrigerated crude oil to form slurries, which are in turn pumped through pipelines and into shuttle tankers for transport to shore. By blending ground hydrates with suitable carrier fluids, transportable slurry can be formed that efficiently delivers "gas" to market.

Several questions are immediately apparent. How finely should hydrates be ground? What is the ideal "solids in fluid" concentration? Fineness, of course, influences rheology; the solids that remain affect plugging, and the combination controls delivery economics. And what happens as hydrates convect into higher-pressure pipeline regimes? In any event, we are concerned with the ability to pump the slurry and also with the ability of the slurry to erode hydrate plugs that have formed in the flow path. These considerations require a model that is able to simulate flows in duct geometries that are far from circular. With such a model, we can simulate worst-case conditions and optimize operations.

In this section, we will not focus on the physics and chemistry of hydrate formation, the kinetics of formation and agglomeration, or the physiochemical characterization of the solid constituents. Instead, we will study flows past "hydrate plugs." Wax buildup is "predictable" to the extent that depositions can be found at top and low sides and, all too often, azimuthally about the entire circumference. Hydrates, in contrast, may appear "randomly." For example, they can form as layers separating gas on the top side and water on the low side. In terms of size, hydrate particles may vary from finely dispersed solids to big lumps that stick to the walls of pipelines. Hydrate particle size is nonuniform and follows wide distribution densities. In general, however, large plugs can be found almost anywhere, a situation that challenges non-Newtonian flow modeling in arbitrary ducts. Simulation is important in defining start-up procedures because large plugs are associated with extremely large pressure drops that may be difficult to achieve in practice.

Pipe inclination may play a significant role for denser fluids. Ibraheem et al. (1998) observe that, for their horizontal and 45° positions, predictions may be optimistic, since lift forces, virtual mass effects, and so on are not incorporated and that a two-dimensional model will be necessary. This caution is well justified. In Chapter 5, we show that density stratification can lead to recirculation vortices that plug the pipeline.

Recapitulation

Very subtle questions are possible. Can hydrate pipeline blockages lead to increased flowline pressures that facilitate additional hydrate growth? Can viscous shear stresses develop within a carrier fluid, or perhaps a hydrate slurry, that support "self-cleaning," which in turn eliminates isolated plugs that form? Again, the formalism developed in Figure 3.7(b) for wax removal applies, but now with Figure 3.7(a) replaced by one applicable to hydrate formation. We will show that numerical simulations can be conveniently performed for large, asymmetrically shaped plugs; that is, our grid generation and velocity solvers are truly "robust" in the numerical sense. Thus, it is clear that the simulation methodology also applies to other types of conduits, valves, and fittings that can potentially support hydrate formation.

Modeling concepts and integration

Our mathematical description of time-dependent mudcake buildup, without erosive effects, is relevant to wax buildup under nonisothermal conditions. Recall that once the cake starts building, its incremental growth retards further buildup, since additional resistance impedes fluid filtration. Thus, the rate of cake growth should vary inversely with cake thickness; in fact, we have shown earlier in Equation 3.61b that

$$dx_c(t)/dt = \{kf_s \Delta p / \{\mu(1 - f_s)(1 - \phi_c)\}\}/x_c$$

Example 3.3   **99**

and we showed, in Equation 3.61c that direct integration of "$x_c \, dx_c = \cdots$" leads to "$\frac{1}{2} x_c^2 = \cdots \, t$"—that is, the "$\sqrt{t}$ law":

$$x_c(t) = \sqrt{[\{2kf_s\Delta p / \{\mu(1 - f_s)(1 - \phi_c)\}\}t]} > 0$$

In this section, we introduce some elementary but preliminary ideas, with the hope of stimulating further research. The following illustrative examples were designed to be simple to show how mathematics and physics go hand in hand.

Wax buildup due to temperature differences

Paraphrasing an earlier statement on mudcake growth, "once wax starts building, its incremental growth retards further buildup, since additional insulation impedes heat transfer." Let R_{pipe} denote the inner radius of the pipe, which is constant, and let $R(t) < R_{pipe}$ denote the time-varying radius of the wax-to-fluid interface. In cake buildup, growth rate is proportional to the pressure gradient; here it is proportional to the heat transfer rate, or temperature gradient $(T - T_{pipe})/(R - R_{pipe})$ by virtue of Fourier's law of conduction, with T being the fluid temperature. We therefore write, analogously to Equation 3.61b,

$$dR/dt = \alpha(T - T_{pipe})/(R - R_{pipe}) \tag{3.63}$$

where $\alpha > 0$ is an empirically determined constant. Cross-multiplying leads to $(R - R_{pipe}) \, dR = \alpha(T - T_{pipe}) \, dt$, where $T - T_{pipe} > 0$. Direct integration yields

$$\tfrac{1}{2}(R - R_{pipe})^2 = \alpha(T - T_{pipe})t > 0 \tag{3.64}$$

where we have used the initial condition $R(0) = R_{pipe}$ when $t = 0$.

Hence, according to this simple model, the thickness of the wax will vary as $\sqrt{t}$ under static conditions. Of course, in reality, α may depend weakly on T, crystalline structure, and other factors, and deviations from $\sqrt{t}$ behavior are not unexpected. Furthermore, it is not completely clear that Equation 3.63 in its present form is correct; for example, dR/dt might be replaced by dR^n/dt, but, in any event, guidance from experimental data is necessary. This buildup model treats wax deposition due to thermal gradients, but obviously other modes exist. For general problems in arbitrarily shaped ducts, wax particles, debris, and fines convected with the fluid may impinge against pipe walls at rates proportional to local velocity gradients; or they may deposit at low or high sides by way of gravity segregation, either because they are heavy or because they are buoyant.

Simulating erosion

Again, any model is necessarily motivated by empirical observation, so our arguments are only plausible. As Equations 3.65a and 3.65b show, for non-Newtonian flow in circular pipes, it is generally true that

$$\tau(r) = r \, \Delta p/2L > 0 \tag{3.65a}$$

$$\tau_w = R \, \Delta p/2L > 0 \tag{3.65b}$$

These equations are interesting because they show how shear stress τ must decrease as R decreases: Thus, any wax buildup must be accompanied by lower levels of stress and therefore decreases in the ability to self-clean or erode the wax. The most simplistic erosion model might take the form

$$dR/dt = \beta(\tau - \tau_y) > 0 \tag{3.66}$$

where $\beta > 0$ is an empirical constant, $\tau - \tau_y > 0$, and τ_y is the yield stress of the wax coating. Thus, R increases with time (i.e., the cross-section "opens up") The uncertainties again remain—for example, R can be replaced by R^2. Note that Equations 3.65a and 3.65b do not apply to annular flows.

Deposition and flow field interaction

Our solution to the nonlinear rheology equations on curvilinear meshes is "straightforward" because the problem is at least well defined and tractable numerically. But the same cannot be said for wax or hydrate deposition modeling, since each individual application must be treated on a customized basis. As we have suggested in the above discussions, numerous variables enter, even in the simplest problems. For example, these include particle size, shape, and distribution; cohesiveness; buoyancy; heat transfer; multiphase fluid flow; dissolved wax type; debris content; fluid rheology; pipeline characteristics; surface roughness; insulation; centrifugal force due to bends; volumetric flow rate; and so on.

Nonetheless, when a particular engineering problem is well understood, the dominant interactions can be identified, and integrated fluid flow and wax or hydrate deposition models can be constructed. The following simulations demonstrate different types of integrated models that have been designed to simulate flows in clogging and self-cleaning pipelines. These examples illustrate the broad range of applications that are possible, where the computational "engines" developed for curvilinear grid-based methods have proven invaluable in simulating operational reality.

Detailed calculated examples

In this section, six simulation examples are discussed in detail. These demonstrate how the general duct model can be used to host different types of solids deposition mechanisms. However, the exact "constitutive relations" used are proprietary to the funding companies and cannot be listed here.

SIMULATION 1

Wax Deposition with Newtonian Flow in Circular Duct

In this first simulation set, we consider a unit centipoise Newtonian fluid, flowing in an initially circular duct; in particular, we assume a 6-in. radius so that the cross-sectional area is 113.1 in.^{2}. A family of "smile-shaped" surfaces is selected for the solids buildup boundary family of curves, since wax surfaces are expected to be more curved than flat. This buildup increases with time, and for convenience the final duct cross section is assumed to be an exact semicircle whose area is 113.1/2 or 56.55 in.^{2} A deposition model is invoked, and intermediate "cross-sectional area versus volumetric flow rate" results, assuming an axial pressure gradient of 0.001 psi/ft, at selected time intervals, are given in Figure 3.9.

Example 3.3   101

Area (in·²)	Rate (gpm)
.1129E+03	.7503E+05 (full circle)
.1082E+03	.6931E+05
.1035E+03	.6266E+05
.9882E+02	.5670E+05
.9411E+02	.5090E+05
.8941E+02	.4531E+05
.8470E+02	.3994E+05
.8000E+02	.3483E+05
.7529E+02	.3000E+05
.7059E+02	.2549E+05
.6588E+02	.2132E+05
.6117E+02	.1752E+05
.5647E+02	.1411E+05 (semi-circle)

FIGURE 3.9

Flow rate versus duct area, with dp/dz fixed.

How do we know that computed results are accurate? We selected Newtonian flow for this validation because the Hagen-Poiseuille volumetric flow rate formula (see Chapter 1) for *circular* pipes can be used to check our numbers. This classic solution, assuming dp/dz = 0.001 psi/ft, R = 6 in., and $\mu = 1$ cp, shows that the flow rate is exactly 0.755E + 05, as compared to our 0.750E+05 gpm. The ratio 755/750 is 1.007, thus yielding 0.7 percent accuracy.

Another indicator of accuracy is found in our computation of area. Obviously, the formula "πR^2" applies to our starting shape, which again yields 113.1 in.^{2}. However, we have indicated 112.9 in Figure 3.9, for a 0.2 percent error. Why an error at all? This appears because our general topological analysis never utilizes "πR^2." The formulation is expressed in terms of metrics of the transformations $x(\xi, \eta)$ and $y(\xi, \eta)$.

Therefore, if computed circle areas agree with "πR^2" and volumetric flow rates are consistent with classical Hagen-Poiseuille flow results, our mathematical boundary value problems, numerical analysis, and programming are likely to be correct. The last entry in Figure 3.9 gives our area for the semicircle, which is to be compared with an exact 113.1/2 or 56.55 in.^{2}. From the ratio 56.55/56.47 = 1.001, our "error" of 0.1 percent suggests that the accompanying 0.1411E + 05-gpm rate is also likely to be correct.

Interestingly, from the top and bottom lines of Figure 3.9, it is seen that a 50 percent reduction in flow area, from "fully circular" to "semi-circular," is responsible for a five-fold decrease in volume throughput. This demonstrates the severe consequence of even partial blockage. Because the flow is Newtonian and linear in this example, the conclusion is "scalable" and applicable to all Newtonian flows. That is, it applies to pipes of all radii R, to all pressure gradients dp/dz, and to all viscosities μ.

Why is "scalability" a property of Newtonian flows? To see that this is true, we return to the governing equation "$(\partial^2/\partial x^2 + \partial^2/\partial y^2) u(x, y) = 1/\mu\ dp/dz$" in the duct coordinates (x, y). Suppose that a solution u(x, y) for a given value of "$1/\mu\ dp/dz$" is available. If we replace this by "$C/\mu\ dp/dz$," where C is a constant, it is clear that Cu must solve the modified problem. Similarly, if Q and τ represent total volumetric flow rate and shear stress in the original problem, the corresponding rescaled values are CQ and Cτ. This would not be true if, for example, μ were a nonlinear function of $\partial u/\partial x$ and $\partial u/\partial y$, as in the case of non-Newtonian fluids; and if it were, it is now obvious that μ, or "N(Γ)," in the non-Newtonian flow notation must now vary with x and y because Γ depends on $\partial u/\partial x$ and $\partial u/\partial y$. Interestingly, we have deduced these important properties even without "solving" the differential equation!

Unfortunately, in the case of non-Newtonian fluids, generalizations such as these cannot be made, and each problem must be considered individually. The extrapolations available to linear mathematical analysis are just not available. It is instructive to examine, in detail, the velocity, apparent viscosity, shear rate, viscous shear stress distributions, and so on for the similar sequence of simulations for non-Newtonian flows. Because generalizations cannot be offered, we do not need to quote the exact parameters assumed. Figure 3.10(a−h) provides "time lapse" results for a Power law fluid simulation; note, for example, how apparent viscosities are not constant but, in fact, vary throughout the cross-sectional area of the duct.

Our methodology and software allow us to plot all quantities of physical interest at each time step. Again, these quantities are needed to interpret solids deposition data obtained in research flow loop experiments,

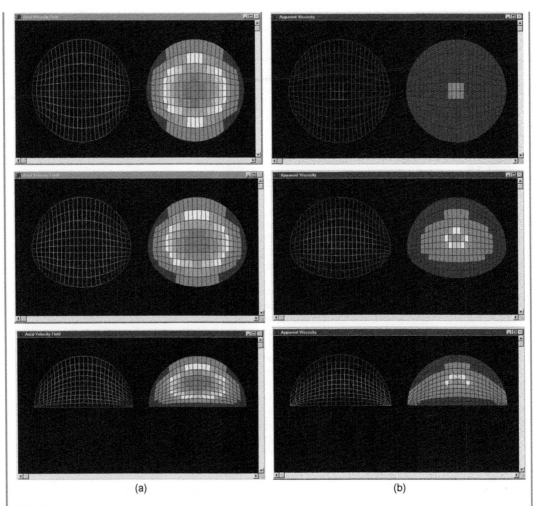

| (a) | (b) |

FIGURE 3.10

(a) Time lapse sequence: axial velocity "U." (b) Time lapse sequence: apparent viscosity, "N(Γ)." (c) Time lapse sequence: viscous stress, "N(Γ) $\partial U/\partial x$." (d) Time lapse sequence: viscous stress, "N(Γ) $\partial U/\partial y$." (e) Time lapse sequence: shear rate, "$\partial U/\partial x$." (f) Time lapse sequence: shear rate, "$\partial U/\partial y$." (g) Time lapse sequence: Stokes' product, "N(Γ)U." (h) Time lapse sequence: dissipation function, "Φ." *Note:* Parts (e) through (h) on pages 104 and 105.

because deposition mechanisms are not very well understood. Due to space limitations, only the first and last "snapshots," plus an intermediate one, are shown; in the final time step, our initially circular duct has become purely semicircular. The varied "snapshots" shown are also instructive because, to the author's knowledge, similar detailed results have never before appeared in the literature.

Example 3.3   **103**

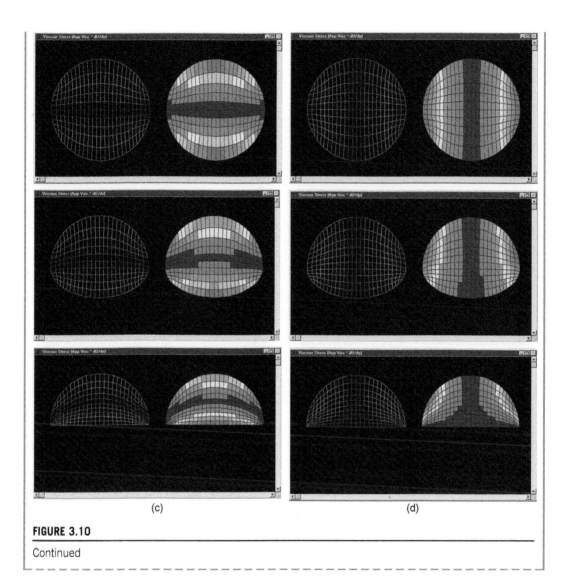

(c)   (d)

FIGURE 3.10

Continued

SIMULATION 2

Hydrate Plug with Newtonian Flow in Circular Duct (Velocity Field)

In this simulation, consider the flow about an isolated but growing "hydrate plug." This model does not offer any geometric symmetry because, in reality, such blockages can form randomly within the duct. Thus, our curvilinear grid algorithms are useful in modeling real flows and determining pressure drops associated with plugs having different shapes. For now, we assume Newtonian flow so that our results are scalable in the sense of the previous example. This is *not* a limitation of the solver, which handles very nonlinear, non-Newtonian fluids. A Newtonian

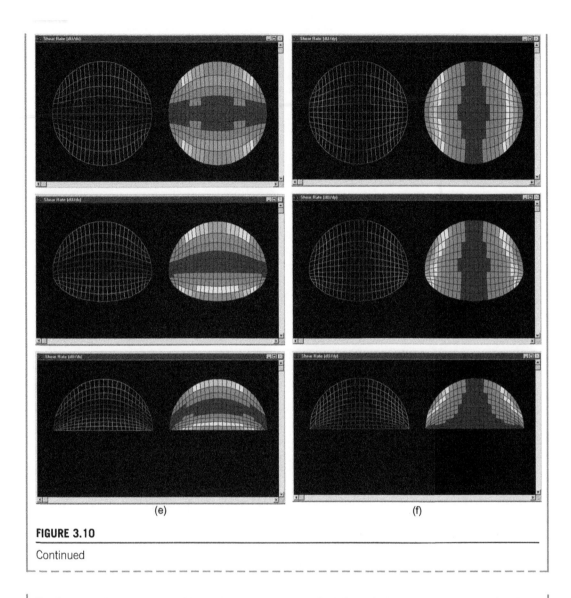

(e) (f)

FIGURE 3.10

Continued

flow is assumed here only to provide results that can be generalized dimensionlessly and therefore may be of greater utility (refer to the conclusion in the earlier example).

To demonstrate the wealth of physical quantities that can be predicted, we have duplicated typical high-level summaries; detailed area distributions of all quantities are also available. The assumed pressure gradient of "1 psi/ft" was taken for convenience and leads to flow rates that are large. However, because the flow is Newtonian, a thousand-fold reduction in pressure gradient will lead to a thousand-fold decrease in flow rate. Shear rates and viscous stresses scale similarly. This ability to rescale results makes our tabulated quantities useful in obtaining preliminary engineering estimates. Following are example results of six time steps selected for display. Detailed numerical results, for example, showing "typical" shear rates and viscous stresses whose

Example 3.3   105

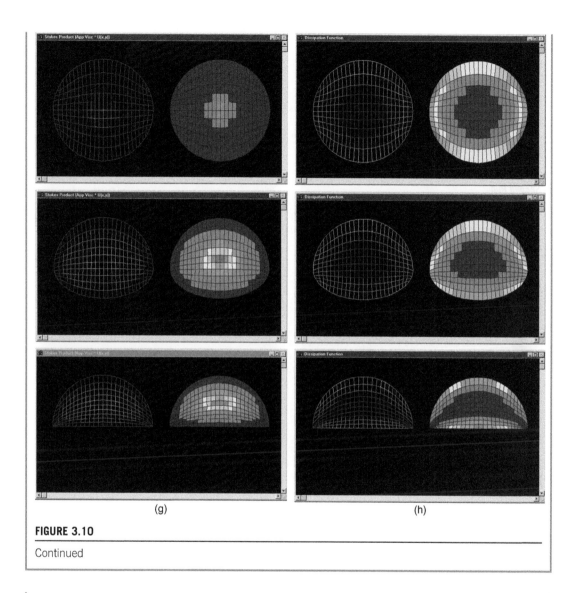

FIGURE 3.10

Continued

magnitudes must be rescaled in accordance with the above paragraph are given first. Then "snapshots" of the axial velocity field are given, in the same time sequence.

First Run, Initial Full Circle, without Hydrate Plug

```
NEWTONIAN FLOW OPTION SELECTED.
Newtonian flow, constant viscosity = 1.00000 cp
Axial pressure gradient assumed as .1000E +01 psi/ft.
Total volume flow rate = .7503E+08 gal/min
Cross-sectional area = .1129E+03 sq inch
```

TABULATION OF CALCULATED AVERAGE QUANTITIES:
Area weighted means for absolute values taken
over entire pipe (x,y) cross-sectional area
0 Axial flow velocity = .2266E +07 in/sec
0 Apparent viscosity = .1465E -06 lbf sec/sq in
0 Viscous stress, AppVis x dU/dx, = .1029E+00 psi
0 Viscous stress, AppVis x dU/dy, = .1230E+00 psi
0 Dissipation function = .2415E +06 lbf/(sec sq in)
0 Shear rate dU/dx = .7022E +06 1/sec
0 Shear rate dU/dy = .8394E +06 1/sec
0 Stokes product = .3321E +00 lbf/in

Second Run

NEWTONIAN FLOW OPTION SELECTED.
Newtonian flow, constant viscosity = 1.00000 cp
Axial pressure gradient assumed as .1000E +01 psi/ft.
Total volume flow rate = .6925E +08 gal/min
Cross-sectional area = .1088E +03 sq inch

TABULATION OF CALCULATED AVERAGE QUANTITIES:
Area weighted means for absolute values taken
over entire pipe (x,y) cross-sectional area
0 Axial flow velocity = .2159E +07 in/sec
0 Apparent viscosity = .1465E -06 lbf sec/sq in
0 Viscous stress, AppVis x dU/dx, = .1050E+00 psi
0 Viscous stress, AppVis x dU/dy, = .1176E+00 psi
0 Dissipation function = .2350E+06 lbf/(sec sq in)
0 Shear rate dU/dx = .7168E+06 1/sec
0 Shear rate dU/dy = .8026E+06 1/sec
0 Stokes product = .3163E+00 lbf/in

Third Run

NEWTONIAN FLOW OPTION SELECTED.
Newtonian flow, constant viscosity = 1.00000 cp
Axial pressure gradient assumed as .1000E +01 psi/ft.
Total volume flow rate = .6032E +08 gal/min
Cross-sectional area = .1047E +03 sq inch

TABULATION OF CALCULATED AVERAGE QUANTITIES:
Area weighted means for absolute values taken
over entire pipe (x,y) cross-sectional area
0 Axial flow velocity = .1974E +07 in/sec
0 Apparent viscosity = .1465E -06 lbf sec/sq in
0 Viscous stress, AppVis x dU/dx, = .1021E +00 psi
0 Viscous stress, AppVis x dU/dy, = .1066E +00 psi
0 Dissipation function = .2102E +06 lbf/(sec sq in)
0 Shear rate dU/dx = .6969E +06 1/sec
0 Shear rate dU/dy = .7275E +06 1/sec
0 Stokes product = .2893E +00 lbf/in

Fourth Run

NEWTONIAN FLOW OPTION SELECTED.
Newtonian flow, constant viscosity = 1.00000 cp
Axial pressure gradient assumed as .1000E +01 psi/ft.
Total volume flow rate = .4253E +08 gal/min
Cross-sectional area = .9642E +02 sq inch

Example 3.3   **107**

```
TABULATION OF CALCULATED AVERAGE QUANTITIES:
Area weighted means for absolute values taken
over entire pipe (x,y) cross-sectional area
O Axial flow velocity = .1538E +07 in/sec
O Apparent viscosity = .1465E-06 lbf sec/sq in
O Viscous stress, AppVis x dU/dx, = .9147E-01 psi
O Viscous stress, AppVis x dU/dy, = .8822E-01 psi
O Dissipation function = .1638E +06 lbf/(sec sq in)
O Shear rate dU/dx = .6243E +06 1/sec
O Shear rate dU/dy = .6021E +06 1/sec
O Stokes product = .2254E +00 lbf/in
```

Fifth Run

```
NEWTONIAN FLOW OPTION SELECTED.
Newtonian flow, constant viscosity = 1.00000 cp
Axial pressure gradient assumed as .1000E +01 psi/ft.
Total volume flow rate = .3417E +08 gal/min
Cross-sectional area = .9229E +02 sq inch

TABULATION OF CALCULATED AVERAGE QUANTITIES:
Area weighted means for absolute values taken
over entire pipe (x,y) cross-sectional area
O Axial flow velocity = .1300E +07 in/sec
O Apparent viscosity = .1465E-06 lbf sec/sq in
O Viscous stress, AppVis x dU/dx, = .8285E-01 psi
O Viscous stress, AppVis x dU/dy, = .7919E-01 psi
O Dissipation function = .1363E +06 lbf/(sec sq in)
O Shear rate dU/dx = .5654E +06 1/sec
O Shear rate dU/dy = .5405E +06 1/sec
O Stokes product = .1905E +00 lbf/in
```

Sixth, Final Run, with Large Blockage

```
NEWTONIAN FLOW OPTION SELECTED.
Newtonian flow, constant viscosity = 1.00000 cp
Axial pressure gradient assumed as .1000E +01 psi/ft.
Total volume flow rate = .2711E +08 gal/min
Cross-sectional area = .8816E +02 sq inch

TABULATION OF CALCULATED AVERAGE QUANTITIES:
Area weighted means for absolute values taken
over entire pipe (x,y) cross-sectional area
O Axial flow velocity = .1070E +07 in/sec
O Apparent viscosity = .1465E-06 lbf sec/sq in
O Viscous stress, AppVis x dU/dx, = .7323E-01 psi
O Viscous stress, AppVis x dU/dy, = .7136E-01 psi
O Dissipation function = .1115E +06 lbf/(sec sq in)
O Shear rate dU/dx = .4997E +06 1/sec
O Shear rate dU/dy = .4870E +06 1/sec
O Stokes product = .1568E +00 lbf/in
```

In Figure 3.11(a–f), sequential "snapshots" of the axial velocity field associated with a growing plug are shown. The reader should refer to the foregoing listings for the corresponding duct areas, volumetric flow rates, average shear rates and stresses, and so on. How is "scalability" applied? Consider, for example, that "1 psi/ft" implies a shear rate component of "0.4997E + 06 1/sec" in the last printout. A more practical "0.001 psi/ft" would be associated with a shear rate of "0.4997E + 03 1/sec."

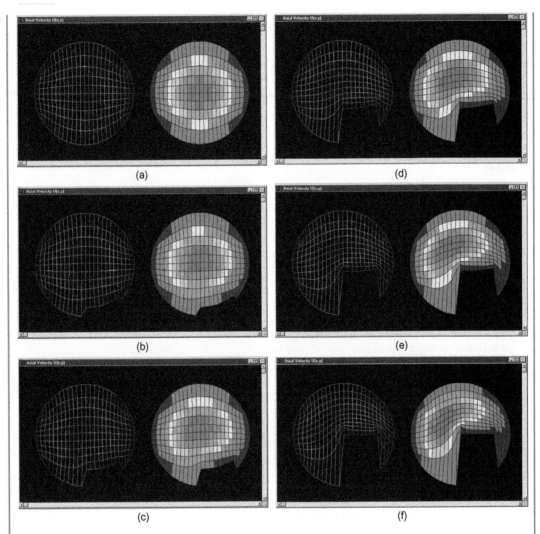

FIGURE 3.11

Velocity field, hydrate plug formation.

It is also interesting to compare the first and final runs. Initially, the full circle has an area of 112.9 in.^{2} and a volumetric flow rate of 0.7503E + 08 gpm. In the last simulation, these numbers reduce to 88.16 and 0.2711E + 08. Thus, a 22 percent reduction in flow area is responsible for a 64 percent decrease in flow rate! It is clear that even "minor" flowline blockages are not tolerable. Following these velocity diagrams, some discussion of the stress fields associated with the worst-case blockage is given.

Example 3.3   **109**

SIMULATION 3

Hydrate Plug with Newtonian Flow in Circular Duct (Viscous Stress Field)

In this example, we continue with Simulation 2, but focus on the largest blockage obtained in the final "snapshot." In particular, we consider the likelihood that the plug-like structure will remain in the form shown, given the erosive environment imparted by viscous shear stresses. To facilitate our discussion, we refer to Figure 3.12, which defines boundary points A, B, C, D, and E, and also interior point F. Figure 3.13(a) displays the "Stokes product," proportional to the product of apparent viscosity and velocity, which measures how well individual particles are convected with the flow. The maximum is located at F, where "in-stream" debris are likely to be found.

Figures 3.13(b) and 3.13(c) display both rectangular components of the viscous stress. The stresses $N(\Gamma)$ $\partial u/\partial x$ and $N(\Gamma)$ $\partial u/\partial y$ are strong, respectively, along BC and AB. Figure 3.13(d) shows the spatial distribution of the "dissipation function," which measures local heat generation due to internal friction, likely to be insignificant. However, the same function is also an indicator of total stress, which acts to erode surfaces that can yield. This figure suggests that B is most likely to erode. At the same time, stresses about our "hydrate plug" are lowest at D, suggesting that additional local growth is possible.

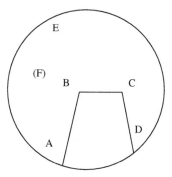

FIGURE 3.12

Generic plug diagram.

SIMULATION 4

Hydrate Plug with Power Law Flow in Circular Duct

In this example, we study the flow of a non-Newtonian Power law fluid past the worst-case blockage in Simulation 3. In particular, we examine the "total volumetric flow rate versus axial pressure gradient," or "Q versus dp/dz" signature of the flow. Before proceeding, it is instructive to reconsider the *exact* solution for Power law flow in a circular pipe:

$$Q/(\pi R^3) = [R\Delta p/(2KL)]^{1/n} n/(3n+1) \tag{3.67}$$

Results for "Q versus dp/dz" are plotted in Figure 3.14 for different values of "n," assuming a 6-in.-radius pipe and a fixed "K" value that would correspond to 100,000 cp if n = 1. In the Newtonian flow limit of n = 1, linearity is clearly seen; however, this exact solution shows that pronounced curvature is obtained as "n" decreases from unity. For any fixed value of dp/dz, it is also seen that Q is strongly dependent on the Power law index (Figure 3.15).

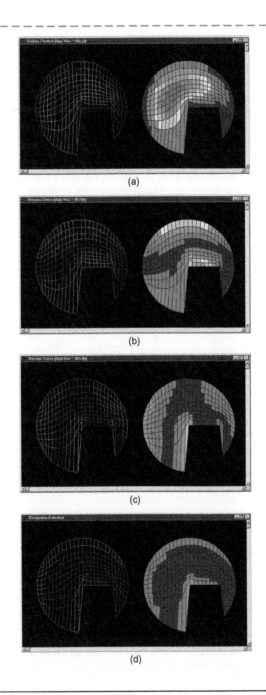

(a)

(b)

(c)

(d)

FIGURE 3.13

(a) Stokes product. (b) Viscous stress, $N(\Gamma)\ \partial u/\partial x$. (c) Viscous stress, $N(\Gamma)\ \partial u/\partial y$. (d) Dissipation function.

Example 3.3   111

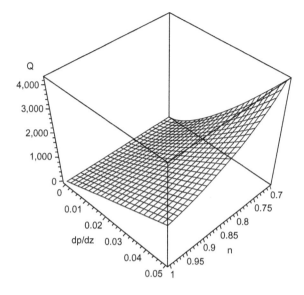

FIGURE 3.14

"Q versus dp/dz" for various "n."

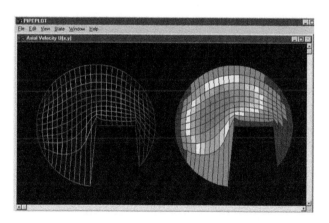

FIGURE 3.15

Typical Power law velocity profile.

We are interested in the corresponding results for Power law flow past the large blockage in the previous simulation. A number of runs were performed, holding fluid properties and geometry fixed, while "dp/dz" was varied. The particular values were selected because they gave "practical" volumetric flow rates. When dp/dz = 0.01 psi/ft, a flow rate of 651 gpm is obtained; at 0.10 psi/ft, the volumetric flow rate is not "6,510" but 11,570 gpm, clearly demonstrating the effects of nonlinearity. Values for dp/dz are shown in bold font in the tabulated results reproduced here, and "Q versus dp/dz" is plotted in Figure 3.16.

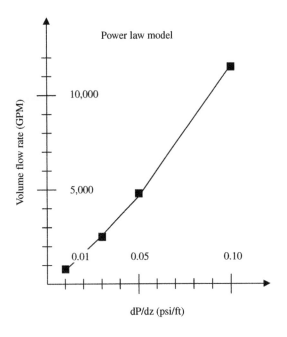

FIGURE 3.16

"Q versus dp/dz" nonlinear behavior.

First Run
    POWER LAW FLOW OPTION SELECTED.
    Power law fluid assumed, with exponent "n" equal
    to .8000E +00 and consistency factor of .1000E-03
    lbf sec^n/sq in.

    Axial pressure gradient assumed as .1000E-01 psi/ft.
    Total volume flow rate = .6508E +03 gal/min
    Cross-sectional area = .8816E +02 sq inch

    TABULATION OF CALCULATED AVERAGE QUANTITIES:
    Area weighted means for absolute values taken
    over entire pipe (x,y) cross-sectional area

    O Axial flow velocity = .2565E +02 in/sec
    O Apparent viscosity = .5867E-04 lbf sec/sq in
    O Viscous stress, AppVis x dU/dx, = .6413E-03 psi
    O Viscous stress, AppVis x dU/dy, = .6308E-03 psi
    O Dissipation function = .2344E-01 lbf/(sec sq in)
    O Shear rate dU/dx = .1191E +02 1/sec
    O Shear rate dU/dy = .1162E +02 1/sec
    O Stokes product = .1604E-02 lbf/in
Second Run
    POWER LAW FLOW OPTION SELECTED.
    Power law fluid assumed, with exponent "n" equal
    to .8000E +00 and consistency factor of .1000E-03

Example 3.3   113

lbf sec^n/sq in.
Axial pressure gradient assumed as .3000E-01 psi/ft.
Total volume flow rate = .2569E +04 gal/min
Cross-sectional area = .8816E +02 sq inch
TABULATION OF CALCULATED AVERAGE QUANTITIES:
Area weighted means for absolute values taken
over entire pipe (x,y) cross-sectional area
0 Axial flow velocity = .1013E +03 in/sec
0 Apparent viscosity = .4458E-04 lbf sec/sq in
0 Viscous stress, AppVis x dU/dx, = .1924E-02 psi
0 Viscous stress, AppVis x dU/dy, = .1892E-02 psi
0 Dissipation function = .2776E +00 lbf/(sec sq in)
0 Shear rate dU/dx = .4701E +02 1/sec
0 Shear rate dU/dy = .4587E +02 1/sec
0 Stokes product = .4813E-02 lbf/in

Third Run

POWER LAW FLOW OPTION SELECTED.
Power law fluid assumed, with exponent "n" equal
to .8000E +00 and consistency factor of .1000E-03
lbf sec^n/sq in.

Axial pressure gradient assumed as .5000E-01 psi/ft.
Total volume flow rate = .4866E +04 gal/min
Cross-sectional area = .8816E +02 sq inch

TABULATION OF CALCULATED AVERAGE QUANTITIES:
Area weighted means for absolute values taken
over entire pipe (x,y) cross-sectional area

0 Axial flow velocity = .1918E +03 in/sec
0 Apparent viscosity = .3923E-04 lbf sec/sq in
0 Viscous stress, AppVis x dU/dx, = .3206E-02 psi
0 Viscous stress, AppVis x dU/dy, = .3154E-02 psi
0 Dissipation function = .8761E +00 lbf/(sec sq in)
0 Shear rate dU/dx = .8901E +02 1/sec
0 Shear rate dU/dy = .8686E +02 1/sec
0 Stokes product = .8022E-02 lbf/in

Fourth Run

POWER LAW FLOW OPTION SELECTED.
Power law fluid assumed, with exponent "n" equal
to .8000E +00 and consistency factor of .1000E-03
lbf sec^n/sq in.

Axial pressure gradient assumed as .1000E +00 psi/ft.
Total volume flow rate = .1157E +05 gal/min
Cross-sectional area = .8816E +02 sq inch
TABULATION OF CALCULATED AVERAGE QUANTITIES:
Area weighted means for absolute values taken
over entire pipe (x,y) cross-sectional area

0 Axial flow velocity = .4561E +03 in/sec
0 Apparent viscosity = .3299E-04 lbf sec/sq in
0 Viscous stress, AppVis x dU/dx, = .6413E-02 psi
0 Viscous stress, AppVis x dU/dy, = .6308E-02 psi

```
0 Dissipation function = .4167E +01 lbf/(sec sq in)
0 Shear rate dU/dx = .2117E +03 1/sec
0 Shear rate dU/dy = .2066E +03 1/sec
0 Stokes product = .1604E-01 lbf/in
```

SIMULATION 5

Hydrate Plug, Herschel-Bulkley Flow in Circular Duct

In this set of runs, the "large blockage" example in Simulation 4 is reconsidered, with identical parameters, except that a nonzero yield stress of 0.005 psi is allowed. Thus, our "Power law" fluid model becomes a "Herschel-Bulkley" fluid. Whereas smooth velocity distributions are typical of Power law flows (e.g., Figure 3.15), the velocity field in flows with nonzero yield stress may contain "plugs" that move as solid bodies. For this simulation set, the plug flow velocity profiles obtained are typified by Figure 3.17.

At 0.01 psi/ft, our flow rate is now obtained as 95.1 gpm, and at 0.10 psi/ft, we have 1,001 gpm. These flow rates are an order of magnitude below those calculated previously; interestingly, the "Q versus dp/dz" response in this example is almost linear, although this is not generally true for Herschel-Bulkley fluids. As before, we provide "typical numbers" in the tabulated results that follow and also plot "Q versus dp/dz" for what is an "exceptional" data set in Figure 3.18.

First Run

```
HERSCHEL-BULKLEY FLOW OPTION SELECTED.
Power law curve assumed with exponent "n" equal
to .8000E +00 and consistency factor "k" of .1000E-03
lbf sec^n/sq in.

Yield stress of .5000E-02 psi taken throughout.
Axial pressure gradient assumed as .1000E-01 psi/ft.
Total volume flow rate = .9513E +02 gal/min
Cross-sectional area = .8816E +02 sq inch

Apparent viscosity and Stokes product set to
zero in plug regime for tabulation and display.
```

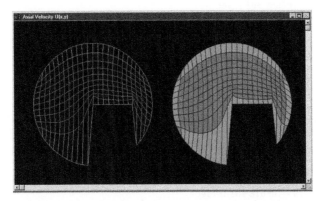

FIGURE 3.17

Plug flow in Herschel-Bulkley fluid.

Example 3.3   **115**

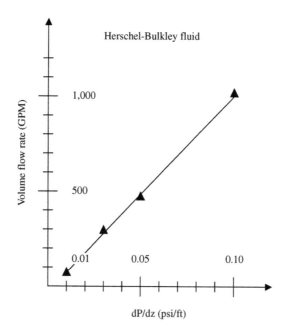

FIGURE 3.18

Near-linear behavior for "exceptional" data set.

```
TABULATION OF CALCULATED AVERAGE QUANTITIES:
Area weighted means for absolute values taken
over entire pipe (x,y) cross-sectional area

O Axial flow velocity = .3932E +01 in/sec
O Viscous stress, AppVis x dU/dx, = .2042E-03 psi
O Viscous stress, AppVis x dU/dy, = .1984E-03 psi
O Dissipation function = .1446E-02 lbf/(sec sq in)
O Shear rate dU/dx = .1180E +01 1/sec
O Shear rate dU/dy = .1070E +01 1/sec
```
Second Run
```
HERSCHEL-BULKLEY FLOW OPTION SELECTED.
Power law curve assumed with exponent "n" equal
to .8000E +00 and consistency factor "k" of .1000E-03
lbf sec^n/sq in.

Yield stress of .5000E-02 psi taken throughout.
Axial pressure gradient assumed as .3000E-01 psi/ft.
Total volume flow rate = .2854E +03 gal/min
Cross-sectional area = .8816E +02 sq inch
Apparent viscosity and Stokes product set to
zero in plug regime for tabulation and display.

TABULATION OF CALCULATED AVERAGE QUANTITIES:
Area weighted means for absolute values taken
over entire pipe (x,y) cross-sectional area
```

O Axial flow velocity = .1180E +02 in/sec
O Viscous stress, AppVis x dU/dx, = .6126E-03 psi
O Viscous stress, AppVis x dU/dy, = .5951E-03 psi
O Dissipation function = .1302E-01 lbf/(sec sq in)
O Shear rate dU/dx = .3539E +01 1/sec
O Shear rate dU/dy = .3211E +01 1/sec

Third Run

HERSCHEL-BULKLEY FLOW OPTION SELECTED.
Power law curve assumed with exponent "n" equal
to .8000E +00 and consistency factor "k" of .1000E-03
lbf sec^n/sq in.

Yield stress of .5000E-02 psi taken throughout.
Axial pressure gradient assumed as .5000E-01 psi/ft.
Total volume flow rate = .4757E +03 gal/min
Cross-sectional area = .8816E +02 sq inch
Apparent viscosity and Stokes product set to
zero in plug regime for tabulation and display.

TABULATION OF CALCULATED AVERAGE QUANTITIES:
Area weighted means for absolute values taken
over entire pipe (x,y) cross-sectional area

O Axial flow velocity = .1966E +02 in/sec
O Viscous stress, AppVis x dU/dx, = .1021E-02 psi
O Viscous stress, AppVis x dU/dy, = .9918E-03 psi
O Dissipation function = .3616E-01 lbf/(sec sq in)
O Shear rate dU/dx = .5899E +01 1/sec
O Shear rate dU/dy = .5351E +01 1/sec

Fourth Run

HERSCHEL-BULKLEY FLOW OPTION SELECTED.
Power law curve assumed with exponent "n" equal
to .8000E +00 and consistency factor "k" of .1000E-03
lbf sec^n/sq in.

Yield stress of .5000E-02 psi taken throughout.
Axial pressure gradient assumed as .1000E +00 psi/ft.
Total volume flow rate = .1001E +04 gal/min
Cross-sectional area = .8816E +02 sq inch

Apparent viscosity and Stokes product set to
zero in plug regime for tabulation and display.

TABULATION OF CALCULATED AVERAGE QUANTITIES:
Area weighted means for absolute values taken
over entire pipe (x,y) cross-sectional area

O Axial flow velocity = .4085E +02 in/sec
O Viscous stress, AppVis x dU/dx, = .2478E-02 psi
O Viscous stress, AppVis x dU/dy, = .2463E-02 psi
O Dissipation function = .2386E +00 lbf/(sec sq in)
O Shear rate dU/dx = .1606E +02 1/sec
O Shear rate dU/dy = .1637E +02 1/sec

Example 3.3   **117**

SIMULATION 6

Eroding a Clogged Bed

Here, we start with a clogged pipe annulus where the inner pipe rests on the bottom, with sand almost filled to the top. We postulate a simple erosion model, where light particles are washed away at speeds greater than a given critical velocity. In the runs shown below, this value is always exceeded, so that the sand bed will always erode. In this final simulation set, the hole completely opens up, providing a successful conclusion to this section!

To provide general results, we again consider a Newtonian flow, so that the specific results in the tabulations can be rescaled and recast more generally in the graph shown in Figure 3.20. While "Q versus dp/dz" is linear in Newtonian fluids, note that "Q versus N%" is not (see Figure 3.19). For that matter, even when a flow is Newtonian, the variation of Q versus any geometric parameter is typically nonlinear and computational modeling will be required.

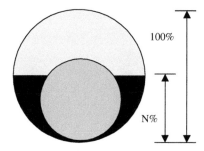

FIGURE 3.19

Clogged pipe simulation setup.

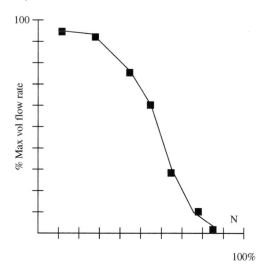

FIGURE 3.20

Generalized flow rate versus dimensionless "fill-up."

In the following results, a unit cp Newtonian fluid is assumed, and a pressure gradient of 0.001 psi/ft is fixed throughout. A 6.4-in. diameter is taken for the outer circle, with "y = 0" referring to its center elevation; a "yheight" of −3.2 in. implies "no clogging," while +2.0 is almost completely clogged. An inner 4.0-in. O.D. pipe rests at the very bottom of the annulus.

First Run

```
Enter YHEIGHT: -3.2
Total volume flow rate = .9340E +03 gal/min
Cross-sectional area = .2041E +02 sq inch

Area weighted means for absolute values taken
over entire pipe (x,y) cross-sectional area
0 Axial flow velocity = .1026E +03 in/sec
0 Apparent viscosity = .1465E-06 lbf sec/sq in
0 Viscous stress, AppVis x dU/dx, = .1951E-04 psi
0 Viscous stress, AppVis x dU/dy, = .2123E-04 psi
0 Dissipation function = .1226E-01 lbf/(sec sq in)
0 Shear rate dU/dx = .1331E +03 1/sec
0 Shear rate dU/dy = .1449E +03 1/sec
0 Stokes product = .1503E-04 lbf/in
```

Second Run

```
Enter YHEIGHT: -2.2
Total volume flow rate = .9383E +03 gal/min
Cross-sectional area = .1966E +02 sq inch

Area weighted means for absolute values taken
over entire pipe (x,y) cross-sectional area
0 Axial flow velocity = .1351E +03 in/sec
0 Apparent viscosity = .1465E-06 lbf sec/sq in
0 Viscous stress, AppVis x dU/dx, = .2453E-04 psi
0 Viscous stress, AppVis x dU/dy, = .2755E-04 psi
0 Dissipation function = .1635E-01 lbf/(sec sq in)
0 Shear rate dU/dx = .1674E +03 1/sec
0 Shear rate dU/dy = .1880E +03 1/sec
0 Stokes product = .1979E-04 lbf/in
```

Third Run

```
Enter YHEIGHT: -1.2
Total volume flow rate = .9157E +03 gal/min
Cross-sectional area = .1805E +02 sq inch
Area weighted means for absolute values taken
over entire pipe (x,y) cross-sectional area
0 Axial flow velocity = .1586E +03 in/sec
0 Apparent viscosity = .1465E-06 lbf sec/sq in
0 Viscous stress, AppVis x dU/dx, = .2506E-04 psi
0 Viscous stress, AppVis x dU/dy, = .3397E-04 psi
0 Dissipation function = .1947E-01 lbf/(sec sq in)
0 Shear rate dU/dx = .1710E +03 1/sec
0 Shear rate dU/dy = .2318E +03 1/sec
0 Stokes product = .2324E-04 lbf/in
```

Fourth Run

```
Enter YHEIGHT: 0.
Total volume flow rate = .7837E +03 gal/min
Cross-sectional area = .1492E +02 sq inch
Area weighted means for absolute values taken
```

Example 3.3   **119**

over entire pipe (x,y) cross-sectional area
O Axial flow velocity = .1769E +03 in/sec
O Apparent viscosity = .1465E-06 lbf sec/sq in
O Viscous stress, AppVis x dU/dx, = .1963E-04 psi
O Viscous stress, AppVis x dU/dy, = .4324E-04 psi
O Dissipation function = .2234E-01 lbf/(sec sq in)
O Shear rate dU/dx = .1340E +03 1/sec
O Shear rate dU/dy = .2951E +03 1/sec
O Stokes product = .2592E-04 lbf/in

Fifth Run

Enter YHEIGHT: 0.6
Total volume flow rate = .6089E +03 gal/min
Cross-sectional area = .1253E +02 sq inch

Area weighted means for absolute values taken
over entire pipe (x,y) cross-sectional area
O Axial flow velocity = .1714E +03 in/sec
O Apparent viscosity = .1465E-06 lbf sec/sq in
O Viscous stress, AppVis x dU/dx, = .1259E-04 psi
O Viscous stress, AppVis x dU/dy, = .4737E-04 psi
O Dissipation function = .2291E-01 lbf/(sec sq in)
O Shear rate dU/dx = .8593E +02 1/sec
O Shear rate dU/dy = .3233E +03 1/sec
O Stokes product = .2511E-04 lbf/in

Sixth Run

Enter YHEIGHT: 1.2
Total volume flow rate = .2823E +03 gal/min
Cross-sectional area = .8952E +01 sq inch

Area weighted means for absolute values taken
over entire pipe (x,y) cross-sectional area
O Axial flow velocity = .1133E +03 in/sec
O Apparent viscosity = .1465E-06 lbf sec/sq in
O Viscous stress, AppVis x dU/dx, = .8603E-05 psi
O Viscous stress, AppVis x dU/dy, = .3692E-04 psi
O Dissipation function = .1379E-01 lbf/(sec sq in)
O Shear rate dU/dx = .5871E +02 1/sec
O Shear rate dU/dy = .2520E +03 1/sec
O Stokes product = .1660E-04 lbf/in

Seventh Run

Enter YHEIGHT: 2.0
Total volume flow rate = .5476E +02 gal/min
Cross-sectional area = .4458E +01 sq inch

Area weighted means for absolute values taken
over entire pipe (x,y) cross-sectional area
O Axial flow velocity = .4484E +02 in/sec
O Apparent viscosity = .1465E-06 lbf sec/sq in
O Viscous stress, AppVis x dU/dx, = .4532E-05 psi
O Viscous stress, AppVis x dU/dy, = .2281E-04 psi
O Dissipation function = .5185E-02 lbf/(sec sq in)
O Shear rate dU/dx = .3093E +02 1/sec
O Shear rate dU/dy = .1557E +03 1/sec
O Stokes product = .6570E-05 lbf/in

Eighth Run
```
    Enter YHEIGHT: 2.5
    Total volume flow rate = .9648E +01 gal/min
    Cross-sectional area = .2126E +01 sq inch

    Area weighted means for absolute values taken
    over entire pipe (x,y) cross-sectional area
    0 Axial flow velocity = .1624E +02 in/sec
    0 Apparent viscosity = .1465E-06 lbf sec/sq in
    0 Viscous stress, AppVis x dU/dx, = .2321E-05 psi
    0 Viscous stress, AppVis x dU/dy, = .1360E-04 psi
    0 Dissipation function = .1832E-02 lbf/(sec sq in)
    0 Shear rate dU/dx = .1584E +02 1/sec
    0 Shear rate dU/dy = .9282E +02 1/sec
    0 Stokes product = .2380E-05 lbf/in
```

Velocity field "snapshots" at different stages of the unclogging process are given in Figure 3.21(a–f). Although we have described the problem in terms of debris removal for eccentric annuli in horizontal drilling, it is clear that the computations are also relevant to wax removal in a simple bundled pipeline, where wax has formed at the top, when heat has been removed temporarily (the plots shown should then be turned upside down).

The basic ideas on solids deposition and integrated non-Newtonian duct flow modeling have been developed in this chapter, and examples have been given that clearly demonstrate the dangers of even partial blockage. In summary, minor blockage can significantly decrease flow rate in a constant pressure gradient scenario. This also implies that minor blockages will require high start-up pressures when a pipeline system is recovering from stoppage. Here the problem can be severe, since temporary shutdowns generally allow blockages to solidify and adhere more securely. The "self-cleaning" ability of a flow is degraded under the circumstances.

EXAMPLE 3.4

Finite Difference Details for Annular Flow Problems

Reservoir engineers and structural dynamicists, for example, routinely use advanced finite difference and finite element methods in design calculations. But drillers have traditionally relied upon simpler handbook formulas and tables that are convenient at the rig site. Simulation methods are powerful, to be sure, but they have their limitations. This section explains the pitfalls and the philosophy one must adopt in order to bring state-of-the-art techniques to the field. We emphasize that *numerical methods do not always yield exact answers*. But more often than not, they produce excellent *trend information* that is useful in practical applications.

Concentric Newtonian flow

For our purposes, consider first the steady, concentric annular flow of a Newtonian fluid (see, for example, Bird et al. (2002)). The governing equations are

$$d^2u(r)/dr^2 + r^{-1} \, du/dr = (1/\mu)\partial p/\partial z \tag{3.68a}$$

$$u(R_i) = u(R_o) = 0 \tag{3.68b}$$

Example 3.4   **121**

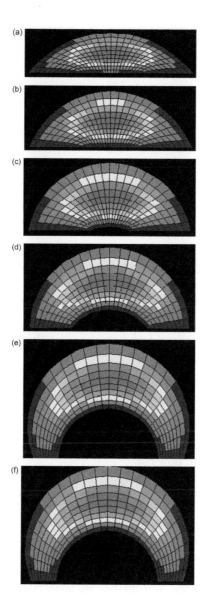

FIGURE 3.21

(a) Clogged annulus, "yheight" = 2.0 inches. (b) Clogged annulus, "yheight" = 1.2 inches. (c) Clogged annulus, "yheight" = 0.6 inches. (d) Clogged annulus, "yheight" = 0.0 inches. (e) Clogged annulus, "yheight" = −1.2 inches. (f) *Unclogged* annulus, "yheight" = −2.2 inches.

Table 3.1 Volumetric Flow Rate versus Mesh Number

# Meshes	GPM	% Error
2	783	25
3	929	11
4	980	6
5	1003	4
10	1035	1
20	1042	0
30	1044	0
100	1045	0

In Equations 3.68a and 3.68b, u(r) is the annular speed satisfying no-slip conditions at the inner and outer radii, R_i and R_o. The viscosity μ and the applied pressure gradient dp/dz are known constants. The exact solution was given earlier.

Let us examine a simple numerical solution. A "second-order accurate" scheme is derived by "central differencing" Equation 3.68a as follows,

$$(u_{j-1} - 2u_j + u_{j+1})/(\Delta r)^2 + (u_{j+1} - u_{j-1})/2r_j\Delta r = (1/\mu)\, \partial p/\partial z \qquad (3.69a)$$

where u_j refers to u(r) at the jth node at the r_j location, j being an ordering index. Equation 3.69a can be evaluated at any number of interior nodes for the mesh length Δr. The resulting "implicit" difference equations, when augmented by

$$u_1 = u_{jmax} = 0 \qquad (3.69b)$$

using Equation 3.68b, form a tridiagonal system of j_{max} unknowns that lends itself to simple solution for u_j and its total volumetric flow rate.

For our first run, we assumed $R_i = 4$ in., $R_o = 5$ in., $\partial p/\partial z = -0.0005$ psi/in., and $\mu = 2$ cp. Computed flow rates as functions of mesh density are given in Table 3.1. Note how the "100 mesh" solution is almost exact; but the "10 mesh" solution for flow rate, which is ten times faster to compute, is satisfactory for engineering purposes.

Now let us double the viscosity μ and recompute the solution. The gpms so obtained decrease exactly by a factor of two, and the dependence on viscosity is certainly brought out very clearly. However, the *trend information* relating changes in gpm to those in μ is accurately captured, even for coarse meshes. Sometimes, then, fine meshes are unnecessary. Similar comments apply to the pressure gradient dp/dz.

It is clear that the exact value of u(r) is mesh dependent; the finer the mesh, the better the answer. In some applications, it may be essential to find, through trial and error, a mesh distribution that leads to the exact solution or that is consistent with real data in some engineering sense. From that point on, "what if" analyses may be performed accurately with greater confidence. This rationale is used in reservoir engineering, where history matching with production data plays a crucial role in estimating reserves.

Example 3.4   **123**

For other applications, the exact numbers may not be as important as qualitative trends of different physical parameters. For example, how does hole eccentricity affect volumetric flow rate for a prescribed pressure gradient? For a given annular geometry, how does a decrease in the Power law exponent affect velocity profile curvature?

In structural engineering, it is well known that *uncalibrated* finite element analyses can accurately pinpoint *where* cracks are likely to form, even though the computed stresses may not be correct. For such qualitative objectives, the results of a numerical analysis may be acceptable "as is" provided the calculated numbers are not literally interpreted. Agreement with exact solutions, of course, is important, but often it is the very lack of such analytical solutions itself that motivates numerical alternatives. Thus, while consistency with exact solutions is desirable, in practice it is through the use of *comparative solutions* that computational methods offer their greatest value.

For annular flows and pipe flows in ducts having general cross-sectional geometries, this philosophy is appropriate because there are no analytical solutions or detailed laboratory measurements with which to establish standards for comparison. One should be satisfied as long as the solutions agree roughly with field data. The real objective, remember, aims at establishing *trends* with respect to *changes* in parameters like fluid rheology, flow rate, and hole eccentricity. We will show through extensive computations and correlation with empirical data that the models developed with our difference methods are correct and useful in this engineering sense. The ultimate acid test lies in validations with field applications, and these are addressed in Chapter 5.

We emphasize that steady eccentric flows are by no means as simple as the above example might suggest. In Equation 3.68a, the unknown speed u(r) depends on a single variable "r" only. For general annular flow problems, the velocity depends on two cross-sectional coordinates x and y, leading to a partial differential equation that is typically nonlinear for oilfield rheologies. The "two-point" boundary conditions in Equation 3.68b are replaced by no-slip velocity conditions enforced along two general arbitrary closed curves representing the borehole and pipe contours.

To implement these no-slip conditions accurately, "boundary-conforming meshes" must be used that provide high resolution in tight spaces. To be numerically efficient, these meshes must be *variable* with respect to all coordinate directions. The difference equations solved on such host meshes must be solved *iteratively*; for unlike Equations 3.69a and 3.69b, which apply to Newtonian flows with constant viscosities, the Power law, Bingham plastic, and Herschel-Bulkley fluids considered in this book satisfy nonlinear equations with problem-dependent apparent viscosities. The algorithms must be *fast, stable, and robust*; they must produce solutions without straining computing resources. Finally, computed solutions must be physically correct—this is the final arbiter that challenges all numerical simulations.

Eccentric flow details

The solution just given is straightforward because "concentric flow" implies *ordinary* differential equations, while "Newtonian fluid" means *constant* viscosities μ. For eccentric non-Newtonian flows, *partial* differential equations must be solved with *variable* viscosities N. We consider steady flows here. Moreover, it is impossible to select a simple grid as we did, say, in setting Δr constant; a curvilinear grid must be created numerically from the equations for general mappings. Once the mappings are available, using the procedure for Equation 3.58 described earlier, the axial momentum equation for u must be transformed to the new coordinates and solved iteratively. In general, a transformed equation with $B^{(1)}$ and $B^{(2)}$ variable coefficients takes the form

$$(\alpha u_{\xi\xi} - 2\beta u_{\xi\eta} + \gamma u_{\eta\eta} + B^{(1)}u_{\eta} + B^{(2)}u_{\xi})/J^2 = (1/N)\,\partial P/\partial z \qquad (3.70a)$$

If central differences are used for ξ and η derivatives, and the usual four-point molecule is used for the mixed derivative, then assuming constant grids $\Delta\xi$ and $\Delta\eta$, we can write the resulting equation in the form

$$(\gamma/\Delta\eta^2 - B^{(1)}/(2\Delta\eta))u_{i,j-1} - 2(\alpha/\Delta\xi^2 + \gamma/\Delta\eta^2)u_{i,j} + (\gamma/\Delta\eta^2 + B^{(1)}/(2\Delta\eta))u_{i,j+1}$$
$$= -(\alpha/\Delta\xi^2)(u_{i-1,j} + u_{i+1,j}) - B^{(2)}/(2\Delta\xi)(u_{i+1,j} - u_{i-1,j}) + (J^2/N)\partial p/\partial z \qquad (3.70b)$$
$$+ 2\beta(u_{i+1,j+1} - u_{i-1,j+1} - u_{i+1,j-1} + u_{i-1,j-1})/(4\Delta\xi\Delta\eta)$$

The first line takes the form $(\ )\,u_{i,j-1} + (\ )\,u_{i,j} + (\ )\,u_{i,j} + 1 = (\ )$, where we note that the parentheses contain different expressions. Equation 3.70b is the recursion relation used for iterative solutions of the steady flow formulation.

The procedure is straightforward. The rectangular computational domain is defined by the indexes $i = 1, 2, 3, \ldots$, imax and $j = 1, 2, 3, \ldots$, jmax, and we initialize the solution to a stored approximation or simply "0" if none are available. First we consider the line $i = 2$. Equation 3.70b is written for $j = 2, 3, \ldots$, jmax -1, yielding jmax -2 equations. To obtain solutions, no-slip boundary conditions are used to define the "$j = 1$" and "$j = $ jmax" equations. The resulting system is linear and can be solved using a tridiagonal equation solver. This process is repeated for $i = 3$, $i = 4, \ldots$, imax -1 until the entire i–j plane has been solved. At this point, u values along $i = 1$ and $i = $ imax are updated using boundary conditions. Then the apparent viscosity function $N(i,j)$, which depends on the rheological model assumed, is updated using the latest available values for u. The equation-solving process just described must be repeated until convergence is achieved.

Once the velocity field is available, physical quantities like apparent viscosity, shear rate, and viscous stress are obtained from their mathematical definitions in terms of velocity derivatives—for example, $u_x = (y_\eta u_\xi - y_\xi u_\eta)/J$ and $u_y = (x_\xi u_\eta - x_\eta u_\xi)/J$, where $J = (x_\xi y_\eta - x_\eta y_\xi)$. All physical quantities are then displayed in color and overlaid on the projection of the annulus in the computational plane. It is important to note that while $\Delta\xi$ and $\Delta\eta$ are constant in computational (ξ, η) space, they efficiently represent variable grids (with high densities in tight annular spaces) in the physical plane. Constant values of $\Delta\xi$ and $\Delta\eta$ allow further speed increases because more complicated difference formulas need not be used.

Equation 3.70a applies to boreholes with straight axes only. When this axis is curved, for example, the radius of curvature, R, enters the formulation and introduces centrifugal effects that modify the effective pressure gradient. These effects are important in the drilling of deviated and horizontal wells. Such effects are studied in Model 5.5 in Chapter 5. There we demonstrate that the pressure gradient $(1/N(\Gamma))\,\partial P/\partial z$ in Equation 3.70a is to be replaced by $(1/N)\,\partial P/\partial z - (1/R)\,\partial u/\partial x + (1/R^2)\,u$ so that

$$(\alpha u_{\xi\xi} - 2\beta u_{\xi\eta} + \gamma u_{\eta\eta} + B^{(1)}u_{\eta} + B^{(2)}u_{\xi})/J^2 = (1/N)\partial P/\partial z - (1/R)\partial u/\partial x + (1/R^2)u \qquad (3.70c)$$

where "x" is perpendicular to the borehole axis. The iterative process described above now applies to the modified equation, where $\partial u/\partial x$ is evaluated using the formula $u_x = (y_\eta u_\xi - y_\xi u_\eta)/J$.

Example 3.4   **125**

We emphasize that an "implicit" iterative scheme has been used, since coupled algebraic equations are involved. "Point-by-point" iterative methods, known as "explicit" methods, are simpler to program but may be numerically unstable. For further background development, the reader is referred to Press et al. (1992). Chapter 7 of Chin (2002) provides simple examples together with Fortran source code illustrating key differences between explicit and implicit schemes. For presentation purposes, the host model considered there is Laplace's equation for pressure taken in a simple rectangular domain.

Example 3.2   128

Steady, Two-Dimensional, Non-Newtonian, Single-Phase, Eccentric Annular Flow

This chapter describes detailed applications for steady, two-dimensional, non-Newtonian, single-phase, eccentric annular flow. We discuss general issues (e.g., Newtonian versus non-Newtonian effects, properties of "pressure gradient versus volumetric flow rate" curves for different fluid rheologies, the role of influx and outflux in affecting these curves, modeling of washouts)—topics that are amply illustrated with computation.

A particularly important application, that of swab-surge in drilling, is treated with respect to the new modeling capabilities offered in this book: high eccentricity, continuous mud circulation, pipe rotation, axial pipe movement, and so on (the effects of yield stress, particularly on plug zone determination in eccentric annuli, with and without pipe movement, are deferred to Chapter 5). We also take this opportunity to introduce the use of transient solvers in steady swab-surge calculations and to develop more general definitions for equivalent density calculations.

Some comments on swab-surge analysis are relevant to usage of commercial software. The subject itself is as old as drilling, but, unfortunately, little progress has been offered during the past decades. The usual concentric flow models are available, mostly limited to nonyield fluids; recent publications address yield stress effects but are restricted to slot flow models without any pipe movement. High eccentricity, general rheologies, pipe axial movement, and rotational capabilities, introduced here, are completely new. In recent years, certain oil service companies have offered "advanced" compressible flow models, claiming to reproduce field results with high accuracy. A cursory examination of the math reveals surprises—the equations contain a single "z" coordinate only, so that cross-sectional effects cannot possibly be modeled. In other words, eccentricity and fluid rheology influences are absent. Users should exercise caution in applying such models and question assumptions as needed.

EXAMPLE 4.1

Newtonian Flow Eccentric Annulus Applications

We introduce our steady, nonrotating, two-dimensional, single-phase, eccentric annular flow capabilities with the Windows user interface shown in Figure 4.1. Geometric properties are defined at the left—that is, the inner and outer circle center coordinates and radii, borehole axis curvature,

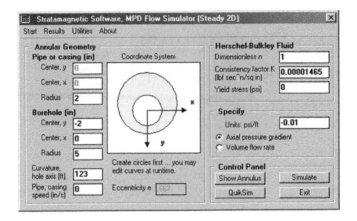

FIGURE 4.1

Steady, nonrotating flow interface.

and constant (positive, zero, or negative) drillpipe or casing speed. Fluid rheology is defined by Herschel-Bulkley parameters at the upper right, which also encompass Newtonian, Power law, and Bingham plastic flows. The entries shown apply to a Newtonian fluid, for which the Power n is unity, the yield stress vanishes, and the consistency factor K corresponds to a 100-cp viscosity (note that 1 cp = 0.0000001465 lbf sec/in.^{2}).

The algorithm solves the steady axial momentum equation written to boundary-conforming curvilinear coordinates when the pressure gradient is specified (and volumetric flow rate is to be determined) or when flow rate is given (and pressure gradient is the objective). Note the "plain English" design in Figure 4.1. Pre- and postprocessing analyses, grid generation setup, host equation development and numerical solution, plus color graphical displays (to be discussed) are completely automated. No expertise on the part of the user in numerical analysis or fluid mechanics is anticipated.

Convenient utilities are built into the user interface. For example, the passive (shaded) text box at the bottom center calculates eccentricities as circle properties are changed. Also, clicking on "Show Annulus" automatically displays annular layout together with a hypothetical 25 × 11 curvilinear grid, which, as shown in Figure 4.2, may be refined or changed at run time.

Extremely fast "no frills" simulation results are available by clicking "QuikSim." For the inputs in Figure 4.1, the axial velocity solution on a fine 61 × 41 mesh, as given in Figure 4.3, appears typically in two to three seconds, together with its convergence history, calculated flow rate, and computed cross-sectional area. The iterative relaxation method used for steady-state flow analysis is very stable and fast, and will, most of the time, provide accurate solutions in seconds. Sometimes, of course, solutions are not possible. For instance, when yield stress fluids are considered under very low-pressure gradient conditions, sought solutions will not be found because they do not physically exist.

In that case, iteration errors do not monotonically decrease to zero, and "red zones" depicting high velocities at the wide part of the annulus will not be found (refer to the inset of Figure 4.3 for an example of an unconverged simulation)—however, solution divergence is easily corrected by using a stronger pressure gradient. Physically, the stronger value found computationally is the one needed to move the fluid. The cross-sectional area is not computed from $\pi(R_{outer}^2 - R_{inner}^2)$,

Example 4.1   **129**

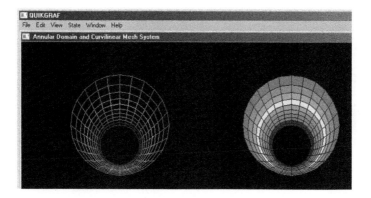

FIGURE 4.2

Coarse curvilinear grid, fine by conventional standards.

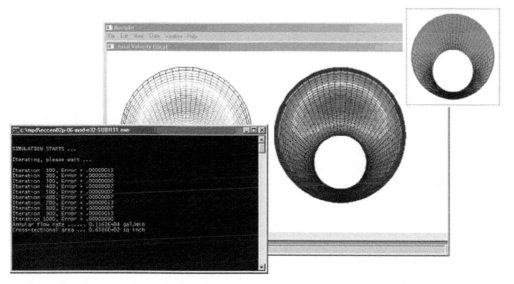

FIGURE 4.3

Fast simulations (unconvergent result, top right).

incidentally, although its value will be extremely close. Instead, totals are calculated by summing quadrilateral areas. This approach applies generally when our circles have been edited to model washouts, cuttings beds, and stabilizers.

Concentric annulus Newtonian flow validations

How can we be assured that calculated results are correct? After all, exact eccentric annular solutions even for simple Newtonian fluids do not exist. Concentric validations are reassuring. For the geometry shown in Figure 4.4, clicking "QuikSim" yields a flow rate of 736.2 gpm. The auxiliary

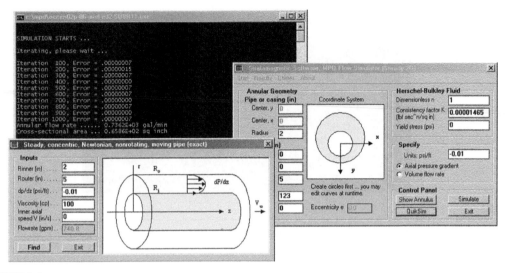

FIGURE 4.4

Concentric annulus comparison.

calculator shown, based on an exact Newtonian flow solution discussed later and available from the "Utilities" menu, gives an exact value of 740.8 gpm for an error of less than 1 percent.

The results just quoted assume stationary drillpipe or casing. However, both software screens in Figure 4.4 support constant inner pipe speeds that may be positive, zero, or negative. For the sign convention used in our mathematical model, a negative pressure gradient (indicating pressure decrease along the flow direction) yields a positive gpm flow rate. A positive pipe or casing speed, defining movement in the flow direction, will increase flow rate. For instance, when "50 in./sec" is entered into the screen at the lower left of Figure 4.4, our previous 740.8 gpm increases to an exact 1,045 gpm. When the same "50" is assumed for the eccentric finite difference model, the result, requiring identical computation time as before, is 1,047 gpm, again offering extremely high accuracy. If, alternatively, "−50 in./sec" is used to model pipe movement in a direction opposite to the main flow, the exact and approximate flow rates are, respectively, 436.5 gpm and 425.0 gpm, with an error of 2.7 percent.

More powerful modeling options, available through the "Simulate" button, permit mesh refinement and redefinition for problems where higher accuracy is required; they will be discussed later. The foregoing results for *concentric* annuli are reassuring and indicate, at least for the examples considered, that calculated velocities and flow rates are accurate. Of course, the numerical model hosted by Figure 4.1 is powerful because pipe movement is also easily considered for highly eccentric annuli. Figure 4.3 for the *eccentric* parameters of Figure 4.1 give a flow rate of 1,162 gpm for stationary pipe (higher than the 736.2 gpm found for concentric flow above). If "50 in./sec" is assumed, the result is 1,448 gpm, whereas the assumption "−50 in./sec" yields a reduced 975.9 gpm.

Now, a note on graphics. In Figure 4.3, the red zone at the wide part of the annulus indicates that maximum speeds are found there; if we had assumed a speed of 500 in./sec, the red zone would move toward and merge with the pipe boundary because both high speeds are comparable, as shown in Figure 4.5.

Example 4.2   **131**

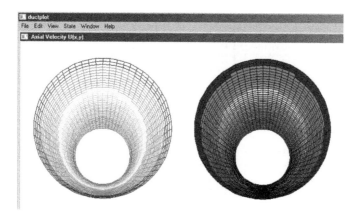

FIGURE 4.5

Fast pipe movement in the direction of the flow.

Velocity displays such as that in Figure 4.3 are important physically. While it is obvious that the fastest flow should be found at the widest location, the addition of steady pipe rotation, for instance, moves this maximum azimuthally and, in the transient case, propagates the entire zone in the azimuthal direction, a fact that may be useful in cuttings transport studies. Finally, we return to Figure 4.3, where we found a flow rate of 1,162 gpm. Here, "QuikSim" assumed a large borehole radius curvature of 123 ft. As an extreme case, we reduce this to 1.23 ft to find a rate decrease to 1,149 gpm. It is well known that decreasing radius of curvature, for a fixed pressure gradient, decreases flow rate because of centrifugal effects; for the Newtonian fluid acting in this annulus, calculated results indicate that the effects are minimal.

In the preceding discussions, we specified a (constant negative) pressure gradient dp/dz and calculated positive total volumetric flow rate. In many managed pressure drilling applications, it is often the pressure gradient that is desired when flow rate is specified. Then the pressure at the drillbit is known from "$P_{surface} - dp/dz \times L$," where $P_{surface}$ is the atmospheric or surface choke pressure and L is the borehole length. The simulator in Figure 4.3 supports this important calculation mode. Recall that the pressure gradient −0.01 psi/ft in Figure 4.3 gave 1,162 gpm. If "Volumetric flow rate" is instead selected in the "Specify" window and "1162" is entered in the input box, clicking "QuikSim" launches a sequence of automated inverse calculations. Here, iterations on dp/dz are performed using a half-step routine in which guesses are successively refined starting with a value applicable to drilling and cementing applications. After one minute of computing time, in which the complete boundary value problem is solved 13 times, the required value of −0.009961 psi/ft is obtained together with a color velocity plot (see Figure 4.6).

EXAMPLE 4.2

Power Law Flow in Eccentric Annuli

In Example 4.1, we focused on Newtonian flows because an exact solution for concentric annuli allowing pipe movement was available for validation purposes. We introduced our "QuikSim"

```
SIMULATION STARTS ...
Iterating on pressure gradient to match flow rate ...

Iteration  100, Error = .00000000
Iteration  200, Error = .00000011
Iteration  300, Error = .00000000
Iteration  400, Error = .00000000
Iteration  500, Error = .00000000
Iteration  600, Error = .00000000
Iteration  700, Error = .00000011
Iteration  800, Error = .00000000
Iteration  900, Error = .00000000
Iteration 1000, Error = .00000000

O  Axial pressure gradient of -.1000E+00 psi/ft
    yields volume flow rate of 0.1162E+05 gal/min.
    Iterations continuing ...

Flow rate target error is, .8997E+03 %

Iteration  100, Error = .00000000
Iteration  200, Error = .00000011
Iteration  300, Error = .00000000
Iteration  400, Error = .00000000
Iteration  500, Error = .00000000
Iteration  600, Error = .00000000
Iteration  700, Error = .00000011
Iteration  800, Error = .00000000
Iteration  900, Error = .00000000
Iteration 1000, Error = .00000000

O  Axial pressure gradient of -.5000E-01 psi/ft
    yields volume flow rate of 0.5808E+04 gal/min.
    Iterations continuing ...

Flow rate target error is, .3998E+03 %
  .
  .
  .
Iteration  100, Error = .00000000
Iteration  200, Error = .00000000
Iteration  300, Error = .00000007
Iteration  400, Error = .00000007
Iteration  500, Error = .00000007
Iteration  600, Error = .00000000
Iteration  700, Error = .00000007
Iteration  800, Error = .00000007
Iteration  900, Error = .00000007
Iteration 1000, Error = .00000000

O  Axial pressure gradient of -.9961E-02 psi/ft
    yields volume flow rate of 0.1157E+04 gal/min.
    Iterations continuing ...

Pressure gradient found iteratively, -.9961E-02 psi/ft,
to yield 0.1157E+04 gal/min vs target 0.1162E+04 gal/min.
Note:  Iterations terminate within 1% of target rate.
Refine result by manually changing pressure gradient.
Annular flow rate ...... 0.1157E+04 gal/min
Cross-sectional area ... 0.6586E+02 sq inch
```

FIGURE 4.6

Iterative calculation for dp/dz, with flow rate specified.

option that allows users to obtain fast "no frills" (but very accurate) solutions. Again, modeling parameters include borehole curvature and pipe movement in eccentric boreholes. Here we extend our study to nonlinear Power law flows; yield stress effects, which involve some subtlety, will be treated separately. We additionally explore more comprehensive options under the "Simulate" button and provide more details under the "Results" menu.

We first introduce a baseline geometry and its QuikSim solution in Figure 4.7. This is a concentric annulus, and, for the parameters shown, the computed flow rate from the finite difference analysis is 1,494 gpm. An exact analytical solution for Herschel-Bulkley fluids with yield stress is available for concentric annuli and accessible from the "Utilities" menu; however, a stationary pipe is required (this is discussed elsewhere in this book). The exact solution (with zero yield stress here) gives 1,518 gpm, so our solution incurs an error of less than 2 percent.

Having established the accuracy of our non-Newtonian method, we explore the effects of borehole anomalies, in particular the consequences of real-world eccentricities. For example, how do

Example 4.2   **133**

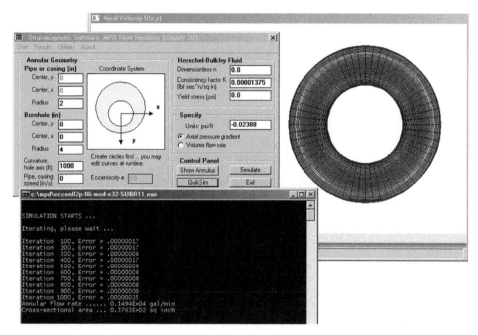

FIGURE 4.7

Power law flow in concentric annulus.

cuttings beds (which reduce flow area) and washouts (which increase area) affect flow properties? When flow rate is specified, what are the effects on pressure drops for managed pressure drilling?

To address these questions, we click "Simulate," which offers more comprehensive modeling options. These provide greater meshing flexibility and convenience. We emphasize that in numerical analysis, different meshes lead to solutions of varying accuracy. But very often, coarser systems are used to perform numerous fast runs for comparative purposes. In the QuikSim mode, a fine 61×41 mesh is hardcoded for high accuracy. Here we will use a 25×11 grid to demonstrate mesh sensitivities in gpm prediction, but mainly, as will be evident, for presentation clarity and space limitations. When run in pure concentric mode, the 1,494 gpm obtained previously is now replaced by 1,388 gpm for a 7.6 percent change. This new number is the basis for several comparisons. We first assess the effect of cuttings beds. Clicking "Simulate" launches a DOS screen in which $x-y$ conventions and coordinates are displayed:

```
Pipe radius .2000E +01, centered at X = 0.000E +00, Y = 0.000E +00.
Hole radius .4000E +01, centered at X = 0.000E +00, Y = 0.000E +00.
All distances and coordinates in inches.
```

```
                          POSITIONS (INCHES):
Node:      Xinner          Yinner          Xouter          Youter
  1     0.2000E +01     0.0000E +00     0.4000E +01     0.0000E +00
  2     0.1932E +01    − 0.5176E +00     0.3864E +01    − 0.1035E +01
  3     0.1732E +01    − 0.1000E +01     0.3464E +01    − 0.2000E +01
```

4	0.1414E +01	− 0.1414E +01	0.2828E +01	− 0.2828E +01
.				
.				
.				
12	− 0.1932E +01	− 0.5176E +00	− 0.3864E +01	− 0.1035E +01
13	− 0.2000E +01	− 0.3020E-06	− 0.4000E +01	− 0.6040E-06
14	− 0.1932E +01	0.5176E +00	− 0.3864E +01	0.1035E +01
15	− 0.1732E +01	0.1000E +01	− 0.3464E +01	0.2000E +01
16	− 0.1414E +01	0.1414E +01	− 0.2828E +01	0.2828E +01
17	− 0.1000E +01	0.1732E +01	− 0.2000E +01	0.3464E +01
18	− 0.5176E +00	0.1932E +01	− 0.1035E +01	0.3864E +01

The user is reminded that

```
You may modify (x,y) coordinates point-by-point to
include cuttings bed, borehole swelling and erosion,
and also, noncircular drill collar effects ...

Points are individually queried in clockwise manner
starting from bottom of pipe/annulus at P .... again:

X/Y orientation:
o-----> Y
|
|
P
|
V X
```

Then the option to modify borehole wall shape and inner circular contour is offered. In this first example, only the former is changed. For instance, we have simple queries, as shown below, with responses given in bold font.

```
Modify borehole wall shape?   Y/N:   y

Point 1:   X =   4.0000,   Y =   0.0000
Modify above coordinates?   Y/N:   y
o   Enter new X value:   2.828
o   Enter new Y value:   0.

Point   2:   X =   3.8637,   Y =   − 1.0353
Modify above coordinates? Y/N: y
o   Enter new X value:   2.828
o   Enter new Y value:   − 1.0353
   .
   .
   .
```

Example 4.2   **135**

A list of 24 points is presented, and we alter 5 points. From the "Results – Text output" menu, the run summary lists the original coordinates as

POSITIONS (INCHES):

Node:	Xinner	Yinner	Xouter	Youter
1	0.2000E +01	0.0000E +00	0.4000E +01	0.0000E +00
2	0.1932E +01	− 0.5176E +00	0.3864E +01	− 0.1035E +01
3	0.1732E +01	− 0.1000E +01	0.3464E +01	− 0.2000E +01
4	0.1414E +01	− 0.1414E +01	0.2828E +01	− 0.2828E +01
5	0.1000E +01	− 0.1732E +01	0.2000E +01	− 0.3464E +01
6	0.5176E +00	− 0.1932E +01	0.1035E +01	− 0.3864E +01
7	0.1510E − 06	− 0.2000E +01	0.3020E -06	− 0.4000E +01
8	− 0.5176E +00	− 0.1932E +01	− 0.1035E +01	− 0.3864E +01
9	− 0.1000E +01	− 0.1732E +01	− 0.2000E +01	− 0.3464E +01
10	− 0.1414E +01	− 0.1414E +01	− 0.2828E +01	− 0.2828E +01
11	− 0.1732E +01	− 0.1000E +01	− 0.3464E +01	− 0.2000E +01
12	− 0.1932E +01	− 0.5176E +00	− 0.3864E +01	− 0.1035E +01
13	− 0.2000E +01	− 0.3020E -06	− 0.4000E +01	− 0.6040E -06
14	− 0.1932E +01	0.5176E +00	− 0.3864E +01	0.1035E +01
15	− 0.1732E +01	0.1000E +01	− 0.3464E +01	0.2000E +01
16	− 0.1414E +01	0.1414E +01	− 0.2828E +01	0.2828E +01
17	− 0.1000E +01	0.1732E +01	− 0.2000E +01	0.3464E +01
18	− 0.5176E +00	0.1932E +01	− 0.1035E +01	0.3864E +01
19	0.2385E -07	0.2000E +01	0.4770E -07	0.4000E +01
20	0.5176E +00	0.1932E +01	0.1035E +01	0.3864E +01
21	0.1000E +01	0.1732E +01	0.2000E +01	0.3464E +01
22	0.1414E +01	0.1414E +01	0.2828E +01	0.2828E +01
23	0.1732E +01	0.1000E +01	0.3464E +01	0.2000E +01
24	0.1932E +01	0.5176E +00	0.3864E +01	0.1035E +01

Note that the starting circles need not be concentric; any eccentricity is permissible. The modified points are also listed; in particular, we show only those lines containing the (bold) cuttings bed we introduced.

FINAL (POSSIBLY MODIFIED) PIPE/HOLE COORDINATES:

POSITIONS (INCHES):

Node:	Xinner	Yinner	Xouter	Youter
1	0.2000E +01	0.0000E +00	**0.2828E +01**	0.0000E +00
2	0.1932E +01	− 0.5176E +00	**0.2828E +01**	− 0.1035E +01
3	0.1732E +01	− 0.1000E +01	**0.2828E +01**	− 0.2000E +01
4	0.1414E +01	− 0.1414E +01	**0.2828E +01**	− 0.2828E +01
.				
.				
.				

```
COMPUTED MESH SYSTEM:
                            11    11    11
                      11    10     9    10    11
                      10     9     8     9    10
            11          9     7   7   7      9        11
              10        7 6   6   5   6   6 7     10
                8       5   4   3   4   5      8
         1110          6   3 2 2   1   2 2 3    6         1011
            9 7       4   1              1   4      7 9
              6 5 3   1                  1   3 5 6
         1110 9       2 1                  1 2          81011
              7 6 4 2                        2 4 5 7

       11 9 8 7 5 3 1                        1 3 5 7 8 91011

              7 6 4 2                        2 4 5 7
         1110 9       2 1                  1 2             91011
              6 3   1                     1   3 5 6
            9 7       4   1              1   4      7 9
         1110           7   5 3 2   2   2 2 5   6 7     1011
                9       8 7   6   6   6   6 8    8 9
              10  10 9   9 8   9   7 9   910 10
              11     11    11    11    11    11    11
```

FIGURE 4.8

Curvilinear grid (the reader should "connect the dots").

22	0.1414E+01	0.1414E+01	0.2828E+01	0.2828E+01
23	0.1732E+01	0.1000E+01	0.2828E+01	0.2000E+01
24	0.1932E+01	0.5176E+00	0.2828E+01	0.1035E+01

Our interactive screens and text output summary provide more numerical detail than is possible with color plots. For instance, quantitative information about the curvilinear grid generated is offered, as shown in Figure 4.8. Then simulation commences, and as the results show, stable and rapid convergence is achieved. Screen output shows that the annular flow rate is 1,086 gpm as opposed to 1,388 gpm for a 28 percent reduction. The new cross-sectional area is reduced to 32.88 in.^{2} from $\pi(4^2-2^2)$, or 37.70 in.^{2}. The complete solution, from grid generation to solution, requires only seconds on typical computers.

```
SIMULATION STARTS ...
Power law fluid assumed with exponent "n" equal
to .8000E+00 and consistency factor of .1375E-04
lbf sec^n/sq in.
A yield stress of .0000E+00 psi, is taken.
Axial pressure gradient assumed as -.2388E-01 psi/ft.

Iteration 100, Error = .00000020
Iteration 200, Error = .00000013
    .

    .
Iteration 800, Error = .00000000
Iteration 900, Error = .00000007
Iteration 1000, Error = .00000007
```

Also provided but not shown are detailed numerical tabulations for all physical properties at all coordinate points, in addition to the following numerical text displays overlaid on annular shape for all relevant physical properties (in each case, the first two significant digits are printed for convenience). These can be very informative. In Figure 4.9, for example, we find that maximum velocities at the top (i.e., 21 in./sec) are five times those at the bottom. These numbers may be useful in hole-cleaning applications.

Example 4.2   **137**

```
COMPUTED AXIAL VELOCITY (IN/SEC):
                                    0    0    0
                               0    9   15    9    0
                               9   15   19   15    9
                          0   15   21   21   21   15         0
                          9  2121  21   19   21  2121        9
                         19   19   17   12   17   19    19
                    0  9      21  12  6 6    0    6  612  21      9  0
                    1521    17   0              0   17   2115
                       211912   0                 0  121921
              0  915       6  0                      0  6       19  9  0
                 212015  6                             6151921

              01519201912  0                          01219201915  9  0

                  202015  6                           6151820
              0  915       5  0                       0  5       15  9  0
                    1810   0                         0  101718
                  1519   11   0                      0  11    1915
              0  9      15   8  5  1   1    1  2  8  1515        9  0
                       13   9 9   5   5    5   9 9  1513
                       8   5 8   4 5   4    5 4   8 5   8
                       0    0    0    0    0    0    0    0
```

FIGURE 4.9

Axial velocity, U.

```
                    L    L    L
               L    T    T    T    L
               T    T    T    T    T
          L    T    T    T    T    T         L
            T   TT   T    T    T   TT    T
            T    T    T    T    T    T    T
         L T       T   TTT   L   TTT   T      T L
         TT      T  L           L   T      TT
            TTT   L               L   TTT
       L T T       T L           L T       T T L
         T T T T                   T T T T

       L T T T T T L           L T T T T T T L

          T T T T                   T T T T
       L T T       T L           L T       T T L
            TT   L               L   TTT
          TT    T  L           L   T      TT
         L T       T  TTT   T   TTT   TT      T L
            T   TT   T    T    T   TT   TT
            T    TT   TT   T   TT   TT    T
          L    L    L    L    L    L    L
```

FIGURE 4.10

Laminar-turbulent (L-T) stability map.

No claim is made to model the enormous difficulties associated with turbulence, but a simple tool is provided for user convenience. At run time, queries are made for fluid specific gravity and critical Reynolds number, here taken as 1.5 and 2,100, respectively. Then a flow stability map like that in Figure 4.10 is provided as a flow analysis guide to the engineer. Average Reynolds numbers for the bottom half and for the entire annulus are also given.

In Newtonian flows, the viscosity is constant throughout the flow domain (assuming that there are no temperature or pressure dependencies). However, in non-Newtonian flows, the apparent viscosity varies within the cross section and will depend on pressure gradient or flow rate. It also depends on the size and shape of the vessel. Figure 4.11 shows the apparent viscosity distribution obtained for our cuttings bed example. Components of shear rate and viscous stress in the x and y directions, useful in hole-cleaning applications, are book-kept separately, since components parallel and perpendicular to the cuttings bed play different bed removal functions. Also, Figures 4.12 through 4.17 are numerical diagrams for both dissipation function and Stokes product.

In many engineering problems, averages provide important tools for correlation purposes. For example, the average viscous stress at the bottom of the annulus is a good indicator of cuttings

```
COMPUTED APPARENT VISCOSITY (LBF SEC/SQ IN):
                            40    40    40
                        40    43    46    43    40
                        43    46    53    46    43
                40    46    69  69  69    46      40
                43  6968  68  53  68  6869    43
                53    53  46  43  46  53    53
        4043      68  434141  38  414143  68      4340
        4769    46  38            38  46    6947
          685343  38            38  435368
    404347    4138                3841      534340
        70684641                41465370

    40475470534339                3943537054474340

        71664741                41475371
    404447    4139                3941      474440
          6143  40            40  435261
        4868    47  42          42  47    6848
    4044      59  544747  49  474554  5659      4440
          50  5256  69  80  69  5752  5550
          47  4548  5156  53  6451  4845  47
          43    42    44    47    44    42    43
```

FIGURE 4.11

Apparent viscosity, η.

```
PLOT OF STRESS "AppVisc x dU(x,y)/dx" VS (X,Y):
COMPUTED (ABSOLUTE VALUE OF) VISCOUS STRESSES (PSI):
                            15    15    15
                        14    13    10    13    14
                        11     9     5     9    11
                11       8     2   2   2     8       11
                 9     1 1   2   6   2   1 1       9
                     4     5   9  14   9   5     4
            8 6       1  121515  20  151512   1       6 8
              5 0       7  17        17   7     0 5
                1 3 7  14            14   7 3 1
        4 3 2       810              10 8       1 3 4
              0 0 2 4                4 2 1 0

        0 0 0 0 0 0 0                0 0 0 0 0 0 0

              1 0 2 4                4 2 0 1
        4 3 2       7 9              9 7       2 3 4
                0 5  11            11   5 1 0
                5 2     4  11      11   4     2 5
            7 6       2   2 7 9   8   9 9 2   0 2       6 7
                7     6 3   0   1   0   0 6   5 7
                9   11 9   7 4   5   2 7   911     9
              10      14    11    7    11    14    10
```

FIGURE 4.12

Viscous stress, $\eta\, \partial U/\partial x$.

```
PLOT OF STRESS "AppVisc x dU(x,y)/dy" VS (X,Y):
COMPUTED (ABSOLUTE VALUE OF) VISCOUS STRESSES (PSI):
                            4     0     4
                        8     3     0     3     8
                        6     2     0     2     6
                11       5     0   0   0     5       11
                 9     1 1   0   0   0   1 1       9
                     4     2   2   0   2   2     4
        1411       1   7 8 4   0   4 8 7   1       1114
              8 1       7  10        10   7     1 8
                1 512  14            14  12 5 1
        1513 9      1517              1715       51315
              2 2 915              15 9 5 2

    15 9 5 1 61419              1914 6 1 5 91315

              1 2 915              15 9 6 1
        1512 9      1415              1514       91215
                312  11            11  12 6 3
                6 0     8  6       6   8     0 6
        1210       2   5 6 2   2   2 6 5   4 2       1012
                1   2 3   2   1   2   4 2   0 1
                3   0 1   0 1   0   1 0   1 0     3
              5       0     0     0     0     0     5
```

FIGURE 4.13

Viscous stress, $\eta\, \partial U/\partial y$.

Example 4.2   **139**

```
COMPUTED DISSIPATION FUNCTION (LBF/(SEC X SQ IN)):
                            6    6    6
                       6    4    2    4    6
                       4    2    0    2    4
                  6    2    0    0    0    2       6
                  4    0 0  0    0    0  0 0    4
                  0    0    2    4    2    0     0
            6 4       0    4 7 7  10   7 7 4   0        4 6
            2 0    2  10         10   2       0 2
                0 0 4 10               10   4 0 0
      6 4 2       710                 10 7        0 4 6
          0 0 2 7                      7 2 0 0

    6 2 0 0 0 410                      10 4 0 0 0 2 4 6

          0 0 2 6                       6 2 0 0
      6 3 1       6 8                   8 6       1 3 6
              0 4 6                   6   4 0 0
          1 0    1  4                 4   1      0 1
      5 3       0    0 2 1   1   1 3 0   0 0     3 5
              1  0 0    0    0   0  0 0    0 1
              2   3 1  1 0   0   0 1  1 3    2
              3    4    2    1   2    4    3
```

FIGURE 4.14

Dissipation function.

```
PLOT OF SHEAR RATE "dU(x,y)/dx" VS (X,Y):
COMPUTED (ABSOLUTE VALUE OF) SHEAR RATES (1/SEC):
                           39    40    39
                      35   30    21    30   35
                      26   20    11    20   26
                 28   18    2    3    2   18       28
                 21    2 2  3   11    3   2 2     21
                    7    9  20  33   20   9      7
            2015      2 283640  51  403628  2        1520
              10 1   15  44         44  15      110
                 1 515  36           36  15 5 1
            10 7 5     2125              2521        2 710
              0 1 510                   10 5 3 0

          0 0 0 0 0 0 0                 0 0 0 0 0 0 0

              1 0 410                   10 4 1 1
            10 8 5    1823              2318      5 810
                 013 29               29  13 3 0
              10 3    10  27           27 10      310
            1814     4   41519  17  1921 4   0 4    1418
                14   11 5   0    1    0   011  914
                19  2618 14 8    9   314 1826 19
                24    33    25   15   25    33    24
```

FIGURE 4.15

Shear rate, $\partial U/\partial x$.

```
PLOT OF SHEAR RATE "dU(x,y)/dy" VS (X,Y):
COMPUTED (ABSOLUTE VALUE OF) SHEAR RATES (1/SEC):
                           10    0    10
                      20    8    0    8   20
                      15    5    0    5   15
                 28   10    0    0    0   10       28
                 21    1 1  0    0    0  1 1      21
                    7    5  5    0    5   5      7
            3526      2 152110   0  102115  2        2635
              18 2   15  25         25  15     218
                 2 928  36           36  28 9 2
            392920    3644               4436       102939
              2 32040                   402010 2

          402010 2113250               503211 210203040

              1 32039                   392011 1
            382819    3440               4034      192838
                 527 29               29  2711 5
              13 0    15  15           15 15      013
            3122     4   913 4   4   414 9   8 4    2231
                 3    4 6  3    1    3   7 4   1 3
                 8   0 2  1 1    0   2 1   2 0    8
                13    0    0    0    0    0   13
```

FIGURE 4.16

Shear rate, $\partial U/\partial y$.

```
PLOT OF "STOKES PRODUCT" OR "VELOCITY X
APPARENT VISCOSITY" VS (X,Y):
COMPUTED STOKES PRODUCT(LBF/IN):
                             0     0     0
                       0     4     7     4     0
                       4     7    10     7     4
                 0     7    14    14    14     7     0
                 4   1414   14    10    14  1414    4
                10    10     7     5     7    10    10
           0 4       14     5 2 2    0   2 2 5   14         4 0
                714        7    0        0    7     14 7
                1410 5     0              0   51014
          0 4 7        2 0                0 2         10 4 0
               1414 7 2                   2  71014

          0 7101410 5 0                   0 5101410 7 4 0

               1413 7 2                   2  71014
          0 4 7        2 0                0 2          7 4 0
                11 4     0                0   4  911
                713        5    0         0    5    13 7
           0 4          9     4 2 0    0 1 4    8 9          4 0
                        6    5 5   3    4    3  5 5   8 6
                    4    2 3   2 2    2    3 2   3 2    4
                    0     0     0     0     0     0     0
```

FIGURE 4.17

Stokes product, ηU.

```
TABULATION OF CALCULATED AVERAGE QUANTITIES, I:
Area weighted means of absolute values taken over
BOTTOM HALF of annular cross-section ...
O  Axial annular velocity  (inches/sec):   .8940E+02
O  Apparent viscosity (lbf sec / sq in):   .5089E-05
O  Viscous stress, AppVis x dU/dx (psi):   .4739E-03
O  Viscous stress, AppVis x dU/dy (psi):   .6356E-03
O  Dissipation fnction (lbf/(sec sqin)):   .2425E+00
O  Shear rate dU/dx (Recip sec, 1 /sec):   .1013E+03
O  Shear rate dU/dy (Recip sec, 1 /sec):   .1407E+03
O  Stokes product Vel x AppVis (lbf/in):   .4854E-03

TABULATION OF CALCULATED AVERAGE QUANTITIES, II:
Area weighted means of absolute values taken over
ENTIRE annular (x,y) cross-section ...
O  Axial annular velocity  (inches/sec):   .1082E+03
O  Apparent viscosity (lbf sec / sq in):   .5033E-05
O  Viscous stress, AppVis x dU/dx (psi):   .5967E-03
O  Viscous stress, AppVis x dU/dy (psi):   .6388E-03
O  Dissipation fnction (lbf/(sec sqin)):   .2941E+00
O  Shear rate dU/dx (Recip sec, 1 /sec):   .1327E+03
O  Shear rate dU/dy (Recip sec, 1 /sec):   .1436E+03
O  Stokes product Vel x AppVis (lbf/in):   .5886E-03
```

FIGURE 4.18

Average quantities for half and entire domains.

transport efficiency, because it is mechanical stress that removes debris. Averages for various physical quantities are computed for the entire annulus and for the bottom half, thus providing the engineer with "ballpark" numbers for potential correlation applications. In addition, simple line plots together with tabulated values are given for all physical quantities at the top of the hole and just next to the hole bottom, as shown in Figures 4.18 and 4.19.

Again, in this example we have shown how we can edit the borehole outer contour and easily introduce a flat cuttings bed. Other bed inclination effects are also easily incorporated. In Figure 4.20, the velocity distribution for the original concentric baseline annulus is shown; that associated with the cuttings bed modification is given at the center, while to the far right, we have introduced an asymmetric washout plus stabilizers.

The flows obtained for the previous three calculations are complicated, hardly similar, and qualitatively quite different. Despite the differences, we emphasize that all three required identical

Example 4.2   141

```
VERTICAL SYMMETRY PLANE ABOVE DRILL PIPE
Axial velocity distribution (in/sec):
    X                      0
                            _____    ___

   1.00    0.0000E+00      |
   1.27    0.9453E+02      |                *
   1.52    0.1616E+03      |                     *
   1.75    0.1985E+03      |                          *
   1.97    0.2116E+03      |                             *
   2.17    0.2112E+03      |                            *
   2.36    0.1989E+03      |                           *
   2.54    0.1702E+03      |                      *
   2.70    0.1253E+03      |               *
   2.86    0.6528E+02      |        *
   3.00    0.0000E+00      |


VERTICAL SYMMETRY PLANE BELOW DRILL PIPE
Axial velocity distribution (in/sec):
    X                      0
                            _____

   6.93    0.0000E+00      |
   7.02    0.1758E+02      |        *
   7.11    0.3267E+02      |              *
   7.20    0.4390E+02      |                   *
   7.29    0.5110E+02      |                       *
   7.38    0.5450E+02      |                         *
   7.47    0.5458E+02      |                          *
   7.56    0.5047E+02      |                        *
   7.65    0.4067E+02      |                  *
   7.74    0.2390E+02      |             *
   7.83    0.0000E+00      |


VERTICAL SYMMETRY PLANE ABOVE DRILL PIPE
Apparent viscosity distribution (lbf sec/sq in):
    X                      0
                            _____

   1.00    0.4028E-05      |           *
   1.27    0.4364E-05      |             *
   1.52    0.4699E-05      |               *
   1.75    0.5365E-05      |                    *
   1.97    0.6955E-05      |                          *
   2.17    0.6857E-05      |                          *
   2.36    0.5349E-05      |                  *
   2.54    0.4692E-05      |               *
   2.70    0.4309E-05      |             *
   2.86    0.4101E-05      |           *
   3.00    0.3893E-05      |          *


VERTICAL SYMMETRY PLANE BELOW DRILL PIPE
Apparent viscosity distribution (lbf sec/sq in):
    X                      0
                            _____    ___

   6.93    0.4672E-05      |          *
   7.02    0.4998E-05      |            *
   7.11    0.5324E-05      |             *
   7.20    0.5863E-05      |                *
   7.29    0.6808E-05      |                    *
   7.38    0.8066E-05      |                          *
   7.47    0.7193E-05      |                      *
   7.56    0.6010E-05      |                 *
   7.65    0.5389E-05      |             *
   7.74    0.5051E-05      |           *
   7.83    0.4714E-05      |          *


VERTICAL SYMMETRY PLANE ABOVE DRILL PIPE
Viscous stress, AppVis x dU/dx  (psi):
    X                                  0
                            _____    __

   1.00    0.1639E-02                  |                *
   1.27    0.1356E-02                  |             *
   1.52    0.1008E-02                  |          *
   1.75    0.5934E-03                  |     *
   1.97    0.2100E-03                  |*
   2.17   -0.2224E-03           *      |
   2.36   -0.6002E-03          *       |
   2.54   -0.1014E-02      *           |
   2.70   -0.1425E-02    *             |
   2.86   -0.1738E-02  *               |
   3.00   -0.2012E-02                  |
```

FIGURE 4.19

Line graphs and tabulations.

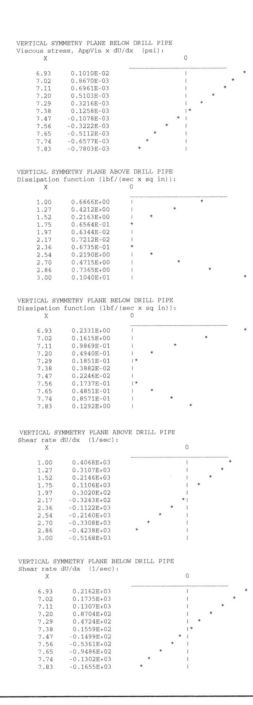

```
VERTICAL SYMMETRY PLANE BELOW DRILL PIPE
Viscous stress, AppVis x dU/dx  (psi):
    X                                              0
   6.93    0.1010E-02                       |              *
   7.02    0.8670E-03                       |           *
   7.11    0.6961E-03                       |         *
   7.20    0.5103E-03                       |       *
   7.29    0.3216E-03                       |    *
   7.38    0.1258E-03                       |*
   7.47   -0.1078E-03                    *  |
   7.56   -0.3222E-03              *        |
   7.65   -0.5112E-03          *            |
   7.74   -0.6577E-03       *               |
   7.83   -0.7803E-03     *                 |

VERTICAL SYMMETRY PLANE ABOVE DRILL PIPE
Dissipation function (lbf/(sec x sq in)):
    X                                              0
   1.00    0.6666E+00   |                       *
   1.27    0.4212E+00   |               *
   1.52    0.2163E+00   |       *
   1.75    0.6564E-01   *
   1.97    0.6344E-02   |
   2.17    0.7212E-02   |
   2.36    0.6735E-01   *
   2.54    0.2190E+00   |       *
   2.70    0.4715E+00   |              *
   2.86    0.7365E+00   |                    *
   3.00    0.1040E+01   |                            *

VERTICAL SYMMETRY PLANE BELOW DRILL PIPE
Dissipation function (lbf/(sec x sq in)):
    X                                              0
   6.93    0.2331E+00   |                          *
   7.02    0.1615E+00   |                    *
   7.11    0.9869E-01   |              *
   7.20    0.4940E-01   |      *
   7.29    0.1851E-01   |*
   7.38    0.3882E-02   |
   7.47    0.2246E-02   |
   7.56    0.1737E-01   |*
   7.65    0.4851E-01   |      *
   7.74    0.8571E-01   |            *
   7.83    0.1292E+00   |                  *

VERTICAL SYMMETRY PLANE ABOVE DRILL PIPE
Shear rate dU/dx  (1/sec):
    X                                              0
   1.00    0.4068E+03                       |            *
   1.27    0.3107E+03                       |           *
   1.52    0.2146E+03                       |        *
   1.75    0.1106E+03                       |   *
   1.97    0.3020E+02                       |
   2.17   -0.3243E+02                    * |
   2.36   -0.1122E+03                *      |
   2.54   -0.2160E+03            *          |
   2.70   -0.3308E+03        *              |
   2.86   -0.4238E+03     *                 |
   3.00   -0.5168E+03                       |

VERTICAL SYMMETRY PLANE BELOW DRILL PIPE
Shear rate dU/dx  (1/sec):
    X                                              0
   6.93    0.2162E+03                       |             *
   7.02    0.1735E+03                       |           *
   7.11    0.1307E+03                       |         *
   7.20    0.8704E+02                       |       *
   7.29    0.4724E+02                       |   *
   7.38    0.1559E+02                       |*
   7.47   -0.1499E+02                    * |
   7.56   -0.5361E+02               *       |
   7.65   -0.9486E+02           *           |
   7.74   -0.1302E+03       *               |
   7.83   -0.1655E+03     *                 |
```

FIGURE 4.19

(Continued).

Example 4.2   **143**

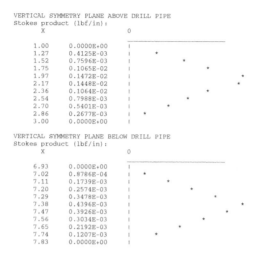

```
VERTICAL SYMMETRY PLANE ABOVE DRILL PIPE
Stokes product (lbf/in):
   X                            0
                               _____
  1.00      0.0000E+00         |
  1.27      0.4125E-03         |        *
  1.52      0.7596E-03         |            *
  1.75      0.1065E-02         |               *
  1.97      0.1472E-02         |                   *
  2.17      0.1448E-02         |                  *
  2.36      0.1064E-02         |               *
  2.54      0.7988E-03         |            *
  2.70      0.5401E-03         |        *
  2.86      0.2677E-03         |    *
  3.00      0.0000E+00         |

VERTICAL SYMMETRY PLANE BELOW DRILL PIPE
Stokes product (lbf/in):
   X                            0
                               _____
  6.93      0.0000E+00         |
  7.02      0.8786E-04         |  *
  7.11      0.1739E-03         |     *
  7.20      0.2574E-03         |        *
  7.29      0.3478E-03         |           *
  7.38      0.4396E-03         |              *
  7.47      0.3926E-03         |             *
  7.56      0.3034E-03         |          *
  7.65      0.2192E-03         |       *
  7.74      0.1207E-03         |    *
  7.83      0.0000E+00         |
```

FIGURE 4.19

(Continued).

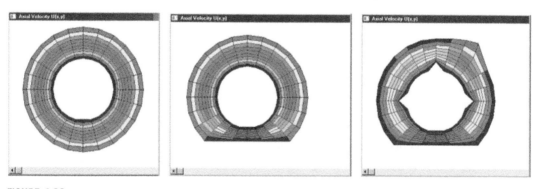

FIGURE 4.20

Pressure gradient, −0.02388 psi/ft throughout (flow rates, from left to right: 1,388 gpm, 1,086 gpm, and 1,040 gpm).

computation times, the main difference being the degree of labor required to enter coordinate changes. User interface improvements are planned. Again, the simulator allows us to quantify the effects of fluid nonlinearities and geometric anomalies, both of which are important to managed pressure drilling.

As a final calculation, we ask, "Which pressure gradient is required to obtain the original flow rate of 1,388 gpm (obtained for the concentric annulus) for the cuttings bed configuration at the center of Figure 4.20? This is easily answered by entering parameters as shown in Figure 4.21 and reentering the coordinate modifications used before.

For our coarse mesh, 12 iterations, each involving completely converged solutions of the velocity problem were required, taking about 30 seconds of computing time. In fact, the final screen output shows the following.

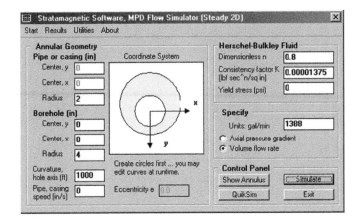

FIGURE 4.21

"Volumetric flow rate" specified mode.

```
Iterating on pressure gradient to match flow rate ...

o Axial pressure gradient of -.2930E-01 psi/ft
  yields volume flow rate of 0.1384E+04 gal/min.
  Iterations continuing ...

Pressure gradient found iteratively, -.2930E-01 psi/ft,
to yield 0.1384E+04 gal/min versus target 0.1388E+04 gal/min.
Note: Iterations terminate within 1% of target rate.
Refine result by manually changing pressure gradient.
```

Our results indicate that the blockage introduced by the cuttings bed worsened the pressure gradient from -0.02388 psi/ft to -0.02930 psi/ft, a consequence that may prove unacceptable for drilling safety (this represents a 23 percent increase in equivalent circulating density). We emphasize that our coarse mesh was used only to reduce the number of pages in this book. In real applications, the finer meshes supported by this simulator should be used. The focus of Examples 4.1 and 4.2 has been on problems where validations with exact solutions are available and, in particular, that convey a sense of the "numbers" describing physical properties in the annulus. For the remainder of this book, we turn to improved graphical displays, analysis tools, and menu options; detailed tabulated results are, of course, always available at the user's option.

EXAMPLE 4.3

Turbulence Modeling and Power Law Flow Analogy

The classic paper "Turbulent Flow of Non-Newtonian Systems" by Dodge and Metzner (1959) derived a general form of the logarithmic friction factor and Reynolds number correlation relationship using dimensional arguments. It is considered the standard for non-Newtonian turbulent pipe

Example 4.4   **145**

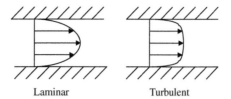

Laminar   Turbulent

FIGURE 4.22

Laminar versus turbulent velocity profiles in a circular pipe.

flow over the past fifty years and remains the most trusted for predictions of turbulent losses in Power law fluids. The work, for Power law fluids, has spawned numerous scholarly extensions to more complicated rheologies and geometries. It is not possible, given project constraints, to review these; however, some key ideas may be applicable to the subject of eccentric annular flow modeling.

Figure 4.22 for Newtonian flow, from Schlichting (1968), shows laminar and turbulent velocity profiles in a circular pipe. For the same flow rate, the latter profile is significantly flatter than the paraboloidal shape at the left. Dodge and Metzner (1959) observed that, in a Power law fluid, as n decreases toward zero, the laminar profile becomes progressively flatter and perfectly flat in the limit of zero n. Now, on passing from laminar to turbulent flow, the laminar velocity profile is flattened by turbulent momentum transfer from high-velocity to low-velocity areas. Hence, turbulence has the same effect on velocity profile, as does a decreasing value of n. In the case where n = 0, again with the laminar profile flat, there is no distinction between laminar and turbulent profile shape, so the two friction factor—Reynolds number relationships become identical.

Of course, for general eccentric annular flows, the Dodge-Metzner correlation cannot be used because it applies only to circular pipes. But the authors' comments motivate us to explore the possibility of modeling turbulent velocity profiles and pressure drops using very low values of the Power law exponent. In Figures 4.23 and 4.24, we show computed velocity profiles and volumetric flow rates for a highly eccentric annulus assuming, respectively, 0.1 and 0.03 values for n. Both velocity profiles display the required flatness. Here n and K may be related to effective turbulent eddy as opposed to laminar viscosities. Their values may be related to wall roughness or inlet disturbance levels, but might practically be viewed as history-matching parameters. Numerical computations in QuikSim mode require about five seconds, no more than any other eccentric flow simulations, and are extremely stable.

EXAMPLE 4.4

Pressure Gradient versus Flow Rate Curve Computation for Non-Newtonian Eccentric Annuli

The formulas for stress $\tau = \{1 - \exp(-\eta_0 \ d\gamma/dt/\tau_0)\}\{\tau_0 + K \ (d\gamma/dt)^n\}$ and apparent viscosity $\eta = \tau/(d\gamma/dt) = \{1 - \exp(-\eta_0 \ d\gamma/dt/\tau_0)\}\{\tau_0/(d\gamma/dt) + K \ (d\gamma/dt)^{n-1}\}$ are extended Herschel-Bulkley constitutive relationships that apply to intrinsic fluid properties only (these theoretical representations are explained in detail elsewhere in this book). The constants n, K, τ_0, and η_0 are determined from laboratory measurements using viscometers with simple geometries whose data can be

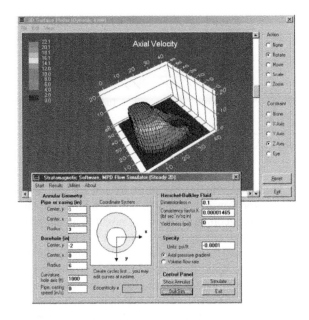

FIGURE 4.23

Low n = 0.1 simulation (302.7 gpm).

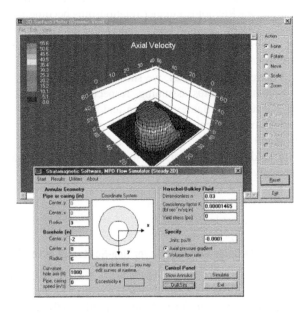

FIGURE 4.24

Very low n = 0.03 simulation (787.6 gpm).

Example 4.4   **147**

interpreted exactly. These relationships apply only to microscopic properties. These equations are used in flow simulators such as ours to determine pressure gradients, flow rates, and detailed macroscopic properties throughout the annular cross section when complicated geometries are specified.

In steady flow applications, the "pressure gradient versus volumetric flow rate" curve is one important analysis objective (refer to the menu in Figure 1.20). It describes macroscopic behavior and depends on annular geometry, drillpipe or casing axial speed, pressure gradient or flow rate, and rotational speed. Unlike, say, Newtonian flow in concentric annuli (e.g., Equation 5.1f), where flow rate varies linearly with pressure gradient and inversely with viscosity in the stationary case, the relationships for non-Newtonian flow are complicated by nonlinearity and a variable apparent viscosity dependent on geometry and rate.

Eccentricity effects

To demonstrate the subtleties of nonlinearity, we consider inner and outer radii of 3 and 6 in., respectively, and apply the steady solver in Figure 4.25 to a Power law fluid with $n = 0.415$ and $K = 0.0000568$ lbf $sec^n/in.^2$ (for which the yield stress is zero). The dp/dz versus flow rate curve in

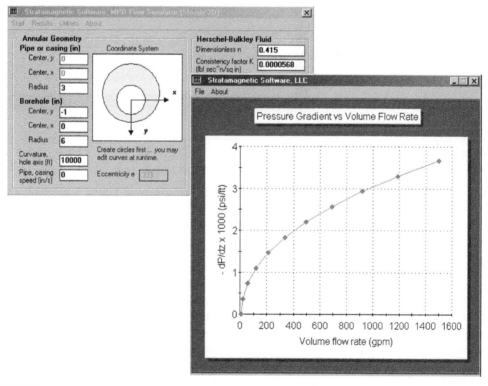

FIGURE 4.25

Flow rate versus dp/dz (0.333 eccentricity).

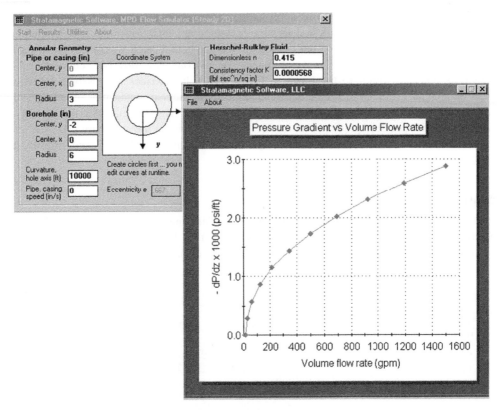

Flow rate versus dp/dz (0.667 eccentricity).

Figure 4.25 corresponds to an eccentricity of 0.333, while that shown in Figure 4.26 corresponds to one of 0.667. Note the differences found between the two results. Also note that our plotting utility will give both "dp/dz versus gpm" and "gpm versus dp/dz" results depending on user preference.

Effect of axial pipe movement

In the next examples, we illustrate the effects of axial pipe or casing movement for our 0.667 eccentricity geometry. We assume a high speed of $+100$ in./sec to emphasize key ideas. Even when the pressure gradient vanishes, there is net positive flow because the drillpipe is dragging fluid in the positive direction (for dp/dz = 0 psi/ft. and $U_{pipe} = +100$ in./sec, we have $+669.0$ gpm). The bottom part of Figure 4.27 shows that if pressure increases in the direction of flow, thus opposing motion, positive flow rate is nonetheless achieved due to dragging. In Figure 4.28, we consider a high negative speed of -100 in./sec. The line graph indicates that a strong pressure gradient is needed just

Example 4.4   **149**

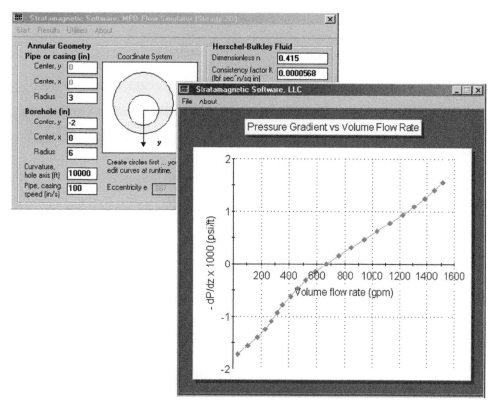

FIGURE 4.27

Positive drillpipe or casing speed.

to maintain a small positive flow rate because the drillpipe is dragging fluid to the left. In both figures, the pressure gradient versus flow rate relationship is almost linear. Solutions for very negative flow rate, which are physically possible, are not calculated or plotted here.

We might note that "dp/dz versus gpm" curve calculations are not as straightforward as they appear. Since pressure gradients found in practice may range anywhere from 0.0001 psi/ft to 0.1 psi/ft, performing complete annular flow computations at, say, 0.0001 psi/ft increments may lead to hour-long computing times. The strategy employed is simple. We assume a maximum flow rate of 1,500 gpm and use our steady solver in "flow rate specified" mode to determine the pressure gradient characteristic of the annular geometry and rheology. This number is then divided by a reasonable number, say 20, and then 20 "dp/dz specified" runs are performed to determine the corresponding gpms. Example calculated results appear in Figures 4.25, 4.26, 4.27, and 4.28. Our "dp/dz versus gpm" curve generation option is accessed directly from the menu in Figure 1.20 and is completely automated. No other software interfaces are called by the high-level menu.

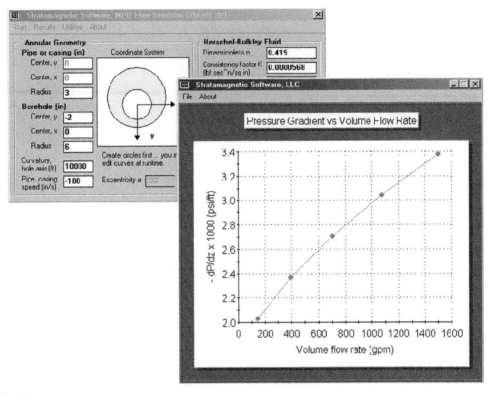

FIGURE 4.28

Negative drillpipe or casing speed.

EXAMPLE 4.5

Effects of Influx-Outflux along the Borehole Path for Non-Newtonian Eccentric Annuli without Rotation

Our steady non-Newtonian flow solver is exact two-dimensionally, providing accurate flow rates and field properties when annular geometries with impermeable walls are specified. In many drilling applications, fluid influxes and outfluxes will be found along the path of the borehole, so that the volumetric flow rates at a particular location will differ from that at another. A simple utility was developed to provide approximate solutions for total pressure drop when local fluid gains or losses can be estimated. This is accessed from the Utilities menu by calling "Influx (outflux) . . . total pressure drop." The action produces an "Influx (outflux) interval data" form, as shown in Figure 4.29, which is completed by the user, allowing up to ten different borehole intervals with different lengths and net flow rates.

The large variations shown in Figure 4.29 are assumed for illustrative purposes only so that the reader can follow the interpolation process (described below) by eye. Once "Saved" is selected in Figure 4.29 and "Total pressure drop" is clicked, the software algorithm automatically constructs the required "dp/dz versus gpm" curve, as explained in Example 4.4. Then the pressure drop

Example 4.6   **151**

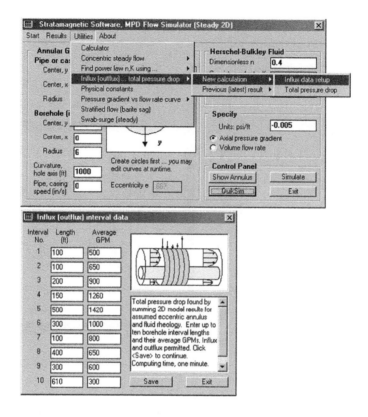

FIGURE 4.29

Creating influx (outflux) interval data.

calculations corresponding to those in the user's influx table are obtained by interpolation from the general curve and summed. Results are summarized as shown in Figure 4.30. This option can also be used for inverse applications. For instance, if the total drop in pressure for an interval can be estimated or is known from a logging measurement, repeated application of the method can be used to predict net fluid influx or outflux.

We emphasize that the modeling option in Figure 4.29 assumes a nonrotating drillpipe. Depending on rotational rate and annular eccentricity, the total pressure drop along the borehole path may be higher or lower than the value determined on a stationary basis. We discuss rotational effects in Chapter 7 and, in particular, how they can be computed using the "Transient 2D" simulator.

EXAMPLE 4.6

Steady-State Swab-Surge in Eccentric Annuli for Power Law Fluids with and without Circulation (No Rotation)

In this example, we discuss applications of our steady-state, non-Newtonian flow simulator to swab-surge analysis for eccentric annuli with and without mud circulation. This problem is

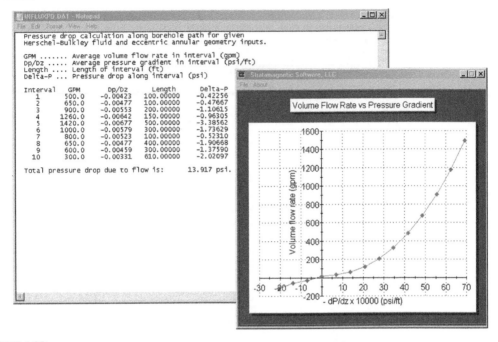

FIGURE 4.30

Total pressure drop computed.

important and complementary to new hardware capabilities in managed pressure drilling that allow continuous mud circulation while tripping in and out of the hole. We focus implicitly on long deviated and horizontal wells for which hole eccentricity is very important. Existing models are either concentric, which are inapplicable, or one-dimensional, in which case any details of the annular cross section are impossible to model. Therefore, our work describes completely new methods that support accurate prediction of pressure distributions in the hole.

Basic concepts

Our simulator predicts the constant pressure gradient $\partial P/\partial z$ needed to induce a specified volumetric flow rate Q for any Herschel-Bulkley fluid in an eccentric annulus. By convention, when Q is positive or "flowing to the right," the pressure P falls in the direction of increasing z. Analogously, when Q is negative or "flowing to the left," P increases with increasing z. Let us first consider flows *without* mud circulation. In the top diagram of Figure 4.31, the drillpipe and bit are shown moving toward the bottom of the hole and displacing fluid as it moves to the left. This fluid must then flow to the right as shown and will produce a positive Q. Now, the equation for pressure is simply $P = z\,\partial P/\partial z + \text{constant}$. If $z = 0$ represents the surface where $P = P_{\text{surf}}$ is fixed by the driller and $z = -L$ is the bit location with L being the borehole length, then the pressure at the bit is just $P_{\text{bit}} = -L\,\partial P/\partial z + P_{\text{surf}}$. Since $\partial P/\partial z < 0$, we have $P_{\text{bit}} \gg P_{\text{surf}}$, which formally shows that in a "surge" situation the bottomhole pressure greatly exceeds that at the surface.

Example 4.6   **153**

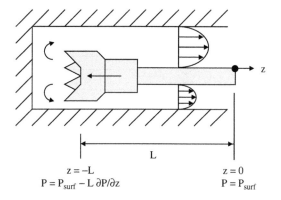

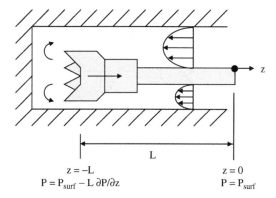

FIGURE 4.31

Coordinate system and conventions.

Next, consider the bottom diagram in Figure 4.31. Here we "swab" the drillstring, pulling it out of the hole. To fill the void left by the drillbit, the flow, Q, must travel toward the left, for which we have $\partial P/\partial z > 0$. Then $P_{bit} = -L\,\partial P/\partial z + P_{surf}$ implies that $P_{bit} \ll P_{surf}$, which formally shows that pressure is greatly reduced at the bit. Increased pressures at the bit are associated with formation invasion and the possibility of fracturing the rock, while decreased pressures may increase the likelihood of blowouts.

The main simulation objective is accurate prediction of P_{bit} as a function of annular geometry, fluid rheology, and (positive or negative) tripping speed in the presence of mud circulation at any pump rate. To produce meaningful results, the simulator must be able to model general eccentricities, arbitrary Herschel-Bulkley parameters, and nonzero drillpipe speeds for any pump rate, the same as the steady-state flow simulator described here will in an *exact* manner. There are several scenarios that must be considered in addressing this problem; they are outlined in Figure 4.32. Surge situations, as shown in parts (a) and (b) of the figure, are straightforward to model.

In part (a) of Figure 4.32, without mud flow, the net flow $Q > 0$ simply flows to the right. When mud is pumped down the drillstring, as shown in part (b), the flow rate Q is simply increased, as

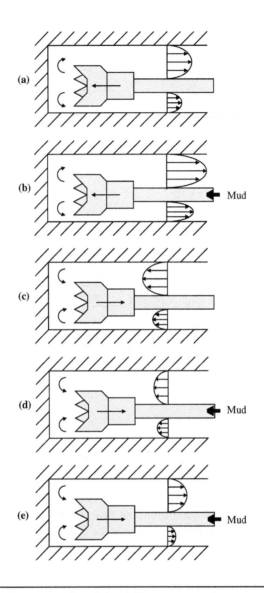

FIGURE 4.32

Five scenarios in continuous flow managed pressure drilling.

shown by the exaggerated velocity profile. Swab scenarios are slightly more subtle. In part (c), without mud flow, pulling the drillstring out of the hole induces a negative flow $Q < 0$ to the left. In part (d), mud is pumped down the drillstring at a low pump rate. If the rate is low enough, Q will still be negative. On the other hand, if the pump rate is high, as suggested in part (e), the net flow will come out of the hole, with $Q > 0$ now being positive. *In this limit, pulling the drillstring out of the hole is consistent with pressures at the bit that exceed those at the surface.*

Example 4.6   **155**

The foregoing five scenarios are obvious in retrospect, and we have summarized them only because they do not arise in more conventional studies where mud does not circulate. Note that the equation "$P_{bit} = -L\ \partial P/\partial z + P_{surf}$" is all that is necessary to calculate pressure at the bit. Again, L is the hole or drillstring length, P_{surf} is the known pressure at the surface choke, and $\partial P/\partial z$ represents output produced by the simulator.

Macroscopic rheological properties

Unlike Newtonian flows, where the viscosity is a constant once and for all (assuming no pressure or temperature dependencies), the apparent viscosity in a non-Newtonian flow varies throughout the cross section and depends on geometrical details plus flow rate or pressure gradient. This is not to say that viscosity is unimportant: It is a useful correlator for cuttings transport and hole-cleaning efficiency and may be significant in stuck pipe assessment. Apparent viscosity, we emphasize, is not a property intrinsic to the fluid; however, for Herschel-Bulkley fluids, "n," "K," and "τ_{yield}" are.

These "microscopic" properties are input into the simulator to create an all-important "pressure gradient versus flow rate curve" that describes "macroscopic" properties for the overall flow. This curve is important to swab-surge analysis: Once the combined flow rate due to surface pumping plus tripping is known, it gives the pressure gradient required for use in the equation "$P_{bit} = -L\ \partial P/\partial z + P_{surf}$." We will give examples of different curves obtained for different fluid types and annular geometries next. We will introduce the basic analysis concepts by way of software modules that have been developed to host our calculations.

Newtonian fluids

The three Herschel-Bulkley parameters noted above can be determined from viscometer measurements using any number of regression techniques available in the literature. (For zero-yield flows of Newtonian and Power law fluids, n and K can be determined using the built-in utilities shown in Figure 1.18.) Once these are available, they are entered into the top right text boxes of the simulator interface in Figure 4.33, where, for the present example, we have assumed the properties of water at 1 cp. For the concentric geometry indicated, clicking on "QuikSim" leads to a flow rate of 943.5 gpm.

Next, in Figure 4.34, we increase the eccentricity, ε, from 0.0 to 0.667 for the same input parameters and obtain the greatly increased flow rate of 1,521 gpm. (It is well known that increases in eccentricity generally lead to increases in flow rate under the same assumed pressure gradient.) Figures 4.33 and 4.34 represent the results of "single analysis mode" simulations when detailed results like those in Figures 1.12, 1.13, 1.14, and 1.15 are required. Much quicker results are obtained when the option in Figure 4.35 is selected. This option ignores the "pressure gradient specified" or "flow rate specified" prescriptions and leads, within a minute or two, to the results in Figure 4.36, here for our eccentric annulus. It is important to observe two features characteristic of Newtonian flows.

First, the "pressure gradient versus flow rate curve" passes through the origin; second, the curve is a straight line whose slope depends only on the geometry of the annulus. Once this slope is determined for a specific eccentric annulus at any given pressure gradient, either computationally or experimentally, the same applies to all pressure gradients. In this sense, Newtonian flows represent an exception to general nonlinear fluid rheologies, where every case must be treated on an individual

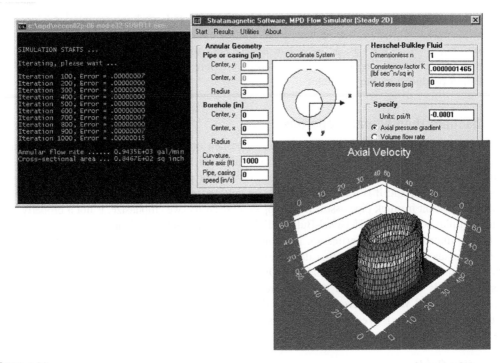

FIGURE 4.33

Newtonian concentric ($\varepsilon = 0.0$) flow.

basis. The straight line nature of the curve means that changes in flow rate lead to proportional changes in pressure gradient.

Finally, we note that for the "pressure gradient versus flow rate curve" option in Figure 4.35, we fixed the pipe or casing speed to zero for our calculations. In general, this can be a positive or negative constant, making the resulting curve useful in swab-surge applications when tripping at rapid speeds (compared to a nominal speed in the annulus). We will give example calculations later in this example.

Power law fluids

Next we reconsider the above concentric and eccentric geometries for zero-yield power fluids with $n = 0.415$ and $K = 0.0000944$ lbf $\sec^n$/in.^{2} (this unweighted mud was used in a recent laboratory study). The significant departure of n from unity implies large nonlinearities. This is reflected in the highly curved lines in Figures 4.37 and 4.38, showing that incremental changes in flow rate do not lead to proportional changes in pressure gradient—the exact changes are rate dependent. Also note the significant differences going from concentric (vertical well) to eccentric (deviated or horizontal well) applications. These results serve as a warning that models based on oversimplified geometric assumptions can lead to operational hazards.

Example 4.6   **157**

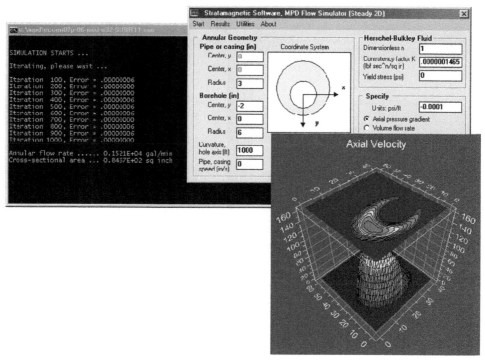

FIGURE 4.34

Newtonian eccentric ($\varepsilon = 0.667$) flow.

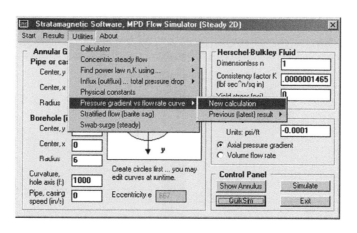

FIGURE 4.35

Newtonian dp/dz versus flow rate calculation ($\varepsilon = 0.667$).

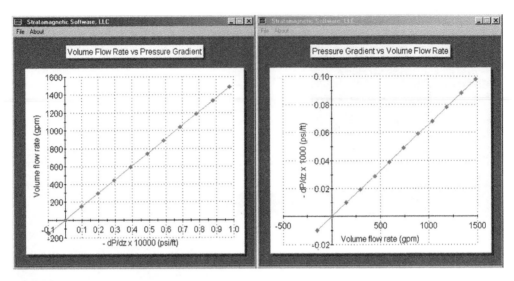

FIGURE 4.36

Newtonian dp/dz versus flow rate behavior ($\varepsilon = 0.667$).

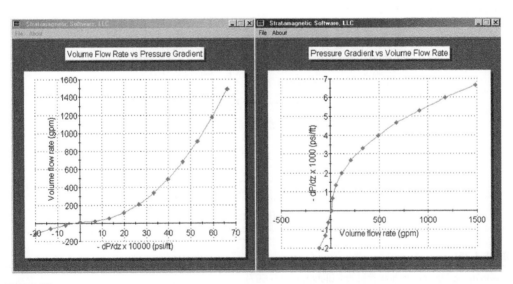

FIGURE 4.37

Power law concentric flow ($\varepsilon = 0.0$).

Example 4.6   **159**

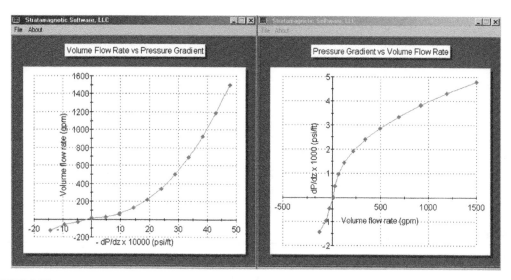

FIGURE 4.38

Power law eccentric flow ($\varepsilon = 0.667$).

FIGURE 4.39

Assumptions for surge run with pumps off.

Swab-surge examples

Now we consider an application for "tripping with pumps off" and "tripping with continuous circulation" that demonstrates the subtleties of flow nonlinearity. If we invoke the "Swab-surge (steady)" option from the main interface in Figure 1.22, we obtain the Swab-Surge Worksheet in Figure 4.39. (The embedded calculations conservatively assume that the drillbit completely blocks the annulus and that fluid does not pass through the nozzles.)

We at first turn off the mud pump while assuming a hole radius of 4 in. and a "tripping in" speed of 5,000 ft/hr. The Worksheet indicates that, following the convention of Figure 4.31, we have a positive induced flow rate of +217.6 gpm, while the drillpipe speed is negative with a value of −16.67 in./sec. (The drillbit is assumed to completely block the hole.) The Worksheet instructs the user to enter "217.6" and "−16.67," as we have in Figure 4.40 for the eccentric annulus and Power law fluid assumed. Clicking on "Show Annulus" produces the display in Figure 4.41. The required pressure gradient dp/dz is −0.006494 psi/ft (minus values indicate high surge pressures at the bit).

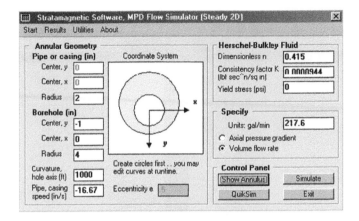

FIGURE 4.40

Additional assumptions for surge run with pumps off.

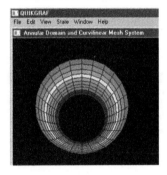

FIGURE 4.41

Eccentric annulus and curvilinear grid assumed (internal grid used in computations is finer).

Now consider an identical situation except that the pump is circulating at 500 gpm. The screens analogous to Figures 4.39 and 4.40 were given previously. Clicking on "QuikSim" (as before) shows that the required pressure gradient now becomes −0.01045 psi/ft. This pressure drop is steeper than before, as expected, because the flow rate is higher. It is interesting that the flow rate ratio between the two previous runs is 717.6/217.6, or 3.30. The ratio of pressure gradients, however, is 0.01045/0.006494, or 1.61 (Figures 4.42 and 4.43). In a Newtonian flow, the "3.30" and "1.61" numbers would have been identical. For non-Newtonian flows, they typically are not, and general conclusions cannot be given—results must be found by case-by-case computations. This example points to the danger of using Newtonian models even for crude estimates.

In the next calculation, we consider "tripping out" in a swab application with the pumps off. Instead of "+217.6" and "−16.67" as we had before, Figure 4.44 shows that the relevant numbers are reversed, with "−217.6" and "+16.67." When these replace their counterparts in Figure 4.40, "QuikSim" analysis correctly shows that the axial pressure gradient is now +0.006494 psi/ft instead of −0.006494 psi/ft. This positive sign, as discussed earlier, indicates lower pressures relative to

Example 4.6   **161**

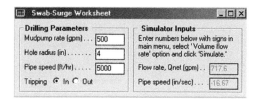

FIGURE 4.42

Assumptions for surge run with pumps on.

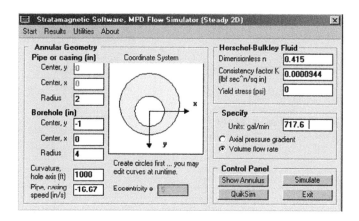

FIGURE 4.43

Additional assumptions for surge run with pumps on.

FIGURE 4.44

Assumptions for swab run with pumps off.

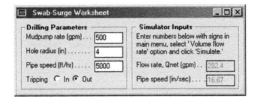

FIGURE 4.45

Assumptions for swab run with pumps on.

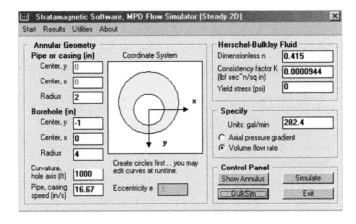

FIGURE 4.46

Additional assumptions for swab run with pumps on.

those at the surface. Now let us recall the equation "$P_{bit} = -L\ \partial P/\partial z + P_{surf}$" for pressure at the drillbit. Suppose that $P_{surf} = 14.7$ psi is open to the atmosphere. Then we can express bit pressure in psi if L is given in feet via $P_{bit} = 14.7 - 0.006494\ L$. In this example, P_{bit} vanishes if L = 2,264 ft, at which point the possibility of a blowout increases significantly.

What would be the effect if, as in Figure 4.42, we ran the mud pump at 500 gpm? The corresponding Swab-Surge Worksheet would appear as it does in Figure 4.45, showing a net flow rate of 282.4 gpm. The calculation suggested by Figure 4.46 gives a negative pressure gradient of -0.005811 psi/ft. This shows that our 500-gpm pump rate is enough to prevent overly low pressures when tripping out at 5,000 ft/hr. While we have focused on low pressures that may allow blowouts, it is obvious that a similar analysis allows us to select pump rates that will not fracture the formation when the frac gradient is known.

Neutral pressure gradient operation

Our simulator allows us to pose and solve still another problem of interest in swabbing operations. Suppose, as previously, that we wish to trip out at 5,000 ft/hr, or 16.67 in./sec. We found from a prior analysis that this action is responsible for a negative flow rate of -217.6 gpm, with the

Example 4.6   **163**

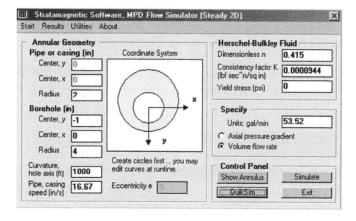

FIGURE 4.47

Surface mud pump rate needed for vanishing axial pressure gradient while tripping out.

FIGURE 4.48

Calculation providing zero axial pressure gradient.

left-bound annular fluid flow arising from the need to fill the borehole void left by the retreating drillstring. We ask ourselves which net flow rate would allow us to maintain a "neutral pressure gradient" of 0.00 psi/ft—that is, one that allows us to have a constant pressure along the annulus equal to the surface choke pressure. If we run the simulator with +16.67 in./sec and 0.00 psi/ft in "specify axial pressure gradient" mode, we obtain a net flow rate of 53.52 gpm. This 53.52 gpm is, of course, the flow rate obtained by simply dragging the drillstring along without an imposed pressure gradient. In other words, the pump must be operated at 217.6 + 53.52, or 271.1, gpm to create a simple dragging flow and to produce the required zero pressure gradient.

This "reverse thinking" can be verified directly. It is easily validated by the forward calculation shown in Figure 4.47. This calls for us to enter 53.52 in the volumetric flow rate screen of Figure 4.48. Clicking "QuikSim" leads to an extremely small −0.00001221 psi/ft, which allows us to impress surface choke pressure directly on the drillbit. Pressure is constant along the borehole. This predictive capability is a direct result of our ability to model drillpipe movement in a rigorous computational manner in very complicated borehole environments. We again note that the simulator was applied to a highly nonlinear Power law fluid with pipe movement in a very eccentric annulus.

EXAMPLE 4.7

Steady-State Swab-Surge in Concentric Annuli for Power Law Fluids with Drillpipe Rotation but Small Pipe Movement

The approach taken to modeling swab-surge effects in Example 4.6 is straightforward. Basically, the Swab-Surge Worksheet is used to compute a kinematic volumetric flow rate correction to the mud pump flow rate that accounts for changes in void space near the drillbit due to tripping out or in. The new flow rate is then used in the annular flow analysis together with the correctly signed drillpipe speed. We employ this approach throughout for swab-surge applications. When the drillpipe rotates, a closed-form analytical solution for the complete flow field is developed in Chapter 5 that allows general steady rotation at any rpm, provided the annulus is concentric and stationary in the axial direction. This latter assumption is satisfactory for slow tripping speeds, as they invariably should be in operations given safety considerations. The simpler simulator is accessed as shown in Figure 4.49.

Four run-time options are shown in the screen in the figure. The first two provide detailed results for single run sets (detailed examples are developed in Chapter 5). The third and fourth

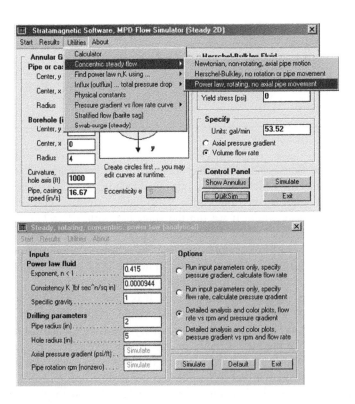

FIGURE 4.49

Concentric, rotating, Power law flow.

Example 4.8   **165**

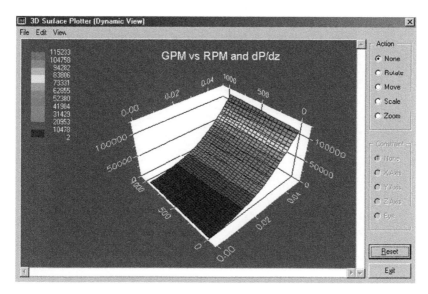

FIGURE 4.50

GPM versus RPM and dP/dz.

options provide fast calculations for "GPM versus RPM and dP/dz" and "dP/dz versus RPM and GPM," typically requiring about 15 seconds of computing time, with automated three-dimensional color plots that allow zooming and mouse rotation. Results shown in Figures 4.50 and 4.51 clearly illustrate the roles of rotation and pressure gradient that must be understood in managed pressure drilling applications.

EXAMPLE 4.8

Steady-State Swab-Surge in Eccentric Annuli for Herschel-Bulkley Fluids with Drillpipe Rotation and Axial Movement

In Example 4.7, we addressed pressure gradient computations for general flow rates and rotational speeds for Power law fluids in a concentric annulus under steady conditions without axial pipe movement. For such flows, the convective terms in the momentum equations vanish identically. The effect of rotation is restricted to shear thinning so that, for a given pressure gradient, increases in rotation rate will reduce apparent viscosity and increase volumetric flow. These effects are well known in the older literature and apply mainly to vertical wells.

Run A

In deviated and horizontal wells, annular eccentricity is the rule. While shear thinning remains important, a nonlinear convective term (whose magnitude is proportional to density and rotational

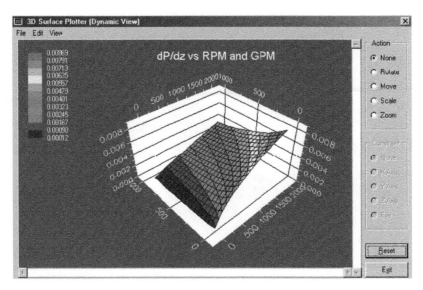

FIGURE 4.51

dP/dz versus RPM and GPM.

speed and is variable throughout the cross section) appears and modifies the local axial pressure gradient. For most practical geometries, this will reduce the flow relative to that found for the eccentric nonrotating problem for the same applied pressure gradient. Equivalently, for the same flow rate the pressure drop increases significantly. These properties are important in managed pressure drilling.

The direct computation of steady rotating flow in an eccentric annulus is often a numerical process that is unstable. Solutions have been published by various authors who have given few computational details related to convergence properties and computing times. Such schemes tend to destabilize at higher specific gravities and rotational speeds and, unfortunately, in the ranges typical of most drilling applications. Fortunately, steady rotating flow solutions can be computed by solving the transient formulation asymptotically for large times.

Figure 4.52 shows that we have set up flow simulations for a Power law fluid in an eccentric annulus with axial pipe movement but no rotation. The problem is integrated in time starting with quiescent conditions. Figure 4.53 shows computed volumetric flow rates reaching constant levels at 941.0 gpm after about one minute of computing time (this is interestingly, but fortuitously, also the physical time scale) with convergence to steady state achieved very stably. The maximum axial flow is found, as expected, at the wide side of the annulus.

Run B

Repeating the foregoing simulation to allow drillstring rotation is straightforward. For example, we simply change the "0" in the RPM box to "100" (as seen from Figure 4.54), and completely automated calculations lead to a reduced flow rate of 562.2 gpm, as shown in Figure 4.55 (page 170).

Example 4.8   **167**

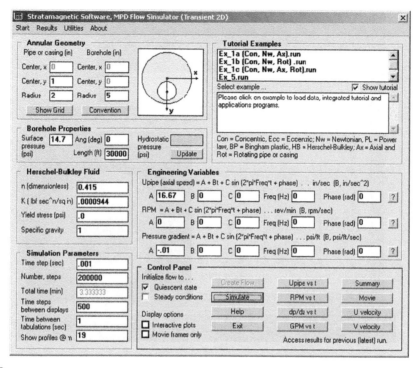

FIGURE 4.52

Transient 2D flow menu (no rotation).

As is well known, the location of maximum axial velocity moves azimuthally, and our results are consistent with this observation, a fact that may be useful in cuttings transport and hole-cleaning applications. Computed results also indicate that the time to reach equilibrium decreases with rotation. The results presented here, for pipe moving both axially and azimuthally, show that pressure gradient calculations are doable and straightforwardly performed for general Power law fluids in highly eccentric annuli.

Run C

In the next calculation, we repeat that in Figure 4.54, which included axial pipe movement and nonzero rotational speed in addition to borehole eccentricity and non-Newtonian Power law flow, but we now consider the additive effects of Herschel-Bulkley yield stress. In Figure 4.56, we modify the previous "0" to "0.002 psi" and leave all other parameters unchanged. As before, the calculations require about 30 seconds and are performed stably.

Figure 4.57 (page 172) shows that the volumetric flow rate is reduced from 562.2 gpm to 516.9 gpm, for a 9 percent reduction. One might ask what the required pressure gradient would be for our yield stress fluid if we needed to maintain a 562-gpm flow rate. For our steady flow solver, direct "pressure gradient specified" and inverse "flow rate specified" calculation modes were

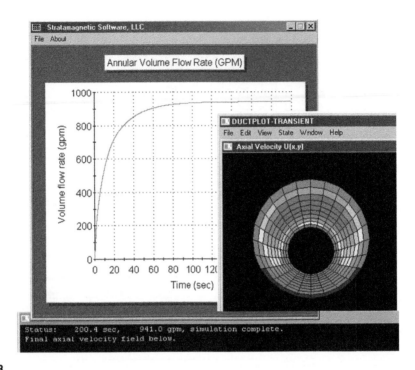

FIGURE 4.53

Eccentric Power law results without pipe rotation.

available. For mathematical reasons, this is not practical for transient simulations. A simple procedure requires us to manually attempt reasonable pressure gradient guesses. This procedure can be very efficient. For this example, the author determined that -0.011 psi/ft would yield 562 gpm after three tries, or about two minutes of desk time. In other words, the presence of yield stress steepened the pressure gradient by a substantial 10 percent.

Run D

Next, we reconsider the yield stress problem in Figure 4.56 and determine the consequences of increasing rotation rate from 100 to 200 rpm. The input screen is shown in Figure 4.58. The effect of doubling rotational speed is a decreased flow rate for the same -0.01 psi/ft, in this case a much smaller 443.3 gpm, as shown in Figure 4.59 (page 174). And what if we had insisted on 562 gpm? Then some simple manual "cut and try" calculations with different pressure gradient guesses lead to a substantially steepened -0.0131 psi/ft, a value that was obtained within two minutes with four different guesses.

Favorable effect of rotation on hole cleaning

The detailed effects of rotation and yield stress have been discussed in the context of eccentric borehole annuli with coupled axial drillstring movement. These calculations represent completely new industry capabilities. It is interesting to note that, from Figure 4.53 for nonrotating flow, the

Example 4.8   **169**

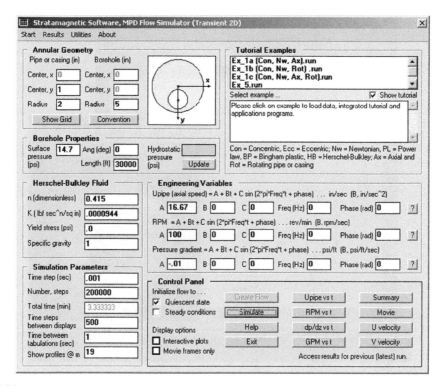

FIGURE 4.54

Modified flow with 100-rpm drillstring rotation.

location of maximum axial flow speed lies symmetrically at the top at the wide side of the eccentric annulus. When rotation exists, as shown in Figures 4.55, 4.57, and 4.59, the location of the maximum moves azimuthally as shown, consistently with other known investigations. (Note that in these three diagrams the "red," which can be seen in the online figures, denotes different speeds.) That increased relative speeds are achieved at the bottom of the annulus is consistent with the improved hole-cleaning ability of drillstrings under rotation observed under many field conditions. Of course, this improvement comes at the expense of steepened pressure gradients, a crucial trade-off whose value must be assessed by the drilling engineer. The end decision made at the rig site will depend on "the numbers," which can only be obtained computationally.

Run E

Here we study the effect of slowdown in drillstring rotational rate. Acceleration and deceleration are always encountered in start-up and shutdown. We repeat the calculation of Figure 4.58, starting with 200 rpm for our nonzero yield stress fluid. But as shown in Figure 4.60 (page 175), we allow our 200 rpm to slow down to 0, as seen from the "−0.5" deceleration rate selected under the RPM menu. Clicking on "?" to the right produces a plot of the assumed RPM versus time curve in

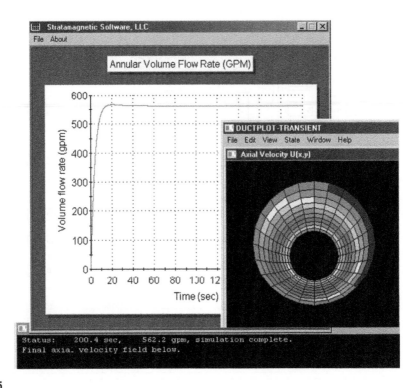

FIGURE 4.55

Reduced flow rate achieved in shorter time.

Figure 4.61. (Note that numerous time functions for axial pipe speed, rotational rate, and pressure gradient are permissible with the simulator.)

The calculated flow rate versus time response is shown in Figure 4.62. This flow rate increases as expected, with drillstring rotation rate decreasing. In this transient simulation, the location of maximum axial velocity is not stationary, but instead propagates azimuthally about the eccentric annulus. A "snapshot" at one instant in time is shown in Figure 4.63. Although this example is purely transient, we have included it in our steady eccentric annular flow chapter to highlight the importance (or, perhaps, unpredictability) of transient effects. The shape of the transient rate curve in Figure 4.62, we emphasize, is obtained for a simple Herschel-Bulkley fluid and not one with "memory" effects.

Run F

In this final example, we consider a complete steady swab-surge application with high annular eccentricity and a nonlinear yield stress fluid, and allow the drillpipe to move axially while simultaneously rotating. This demonstrates the capabilities in our math models and provides a complete summary of the software menu sequences needed to perform similar calculations. In order to proceed, the Swab-Surge Worksheet must be invoked from the main "MPD Flow Simulator (Steady 2D)" in Figure 4.64 (page 177).

Example 4.8   171

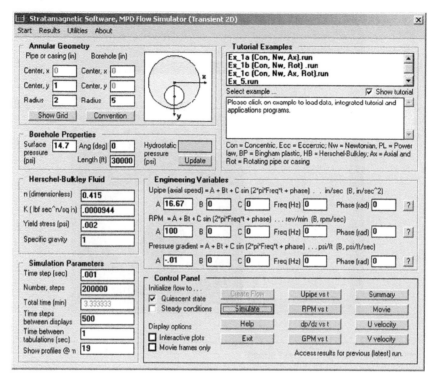

FIGURE 4.56

Flow at 100 rpm, now with 0.002-psi yield stress.

In the Worksheet, we consider a five-inch radius hole and a pipe trip-out speed of 5,000 ft/hr. During this operation, we wish to pump continuously, with the surface mud pump rate set at 856.9 gpm. Now, as the drillpipe is withdrawn from the hole, fluid must rush in to fill the bottom-hole void. The Worksheet indicates that the effective annular flow rate is 516.9 gpm and that the pipe speed is 16.67 in./sec.

Now, we wish to focus our study on the non-Newtonian flow of a Herschel-Bulkley fluid with $n = 0.415$, $K = 0.0000944$ lbf sec^n/in.^{2}, and $\tau_{yield} = 0.002$ psi, in an annulus formed by a 4-in.-diameter pipe in a 10-in.-diameter hole, with an eccentricity of 0.3333. We will demonstrate the solution process for flows without and with rotation. If we wish to consider axial movement only but *without* rotation, we can run the steady flow calculation shown in Figure 4.65 in "volumetric flow rate specified" mode. Clicking on "QuikSim" produces the screen output iteration history that follows.

```
SIMULATION STARTS ...
Herschel-Bulkley model, with exponent "n" equal
to 0.4150E +00 and consistency factor of 0.9440E-04
lbf sec^n/sq in.
A yield stress of 0.2000E-02 psi is taken.
```

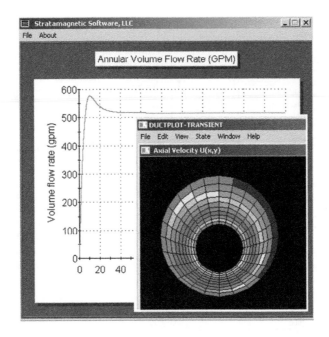

FIGURE 4.57

Flow at 100 rpm, now with 0.002-psi yield stress.

```
Borehole axis radius of curvature is 0.1000E +04 ft.
Axial speed of inner pipe is 0.1667E +02 in/sec.
Target flow rate of 0.5169E +03 gal/min specified.

Iterating on pressure gradient to match flow rate ...

Iteration    100, Error = .00672962
Iteration    200, Error = .00248959
Iteration    300, Error = .00119476
Iteration    400, Error = .00052236
Iteration    500, Error = .00019270
Iteration    600, Error = .00005923
Iteration    700, Error = .00001814
Iteration    800, Error = .00000521
Iteration    900, Error = .00000171
Iteration   1000, Error = .00000047

o  Axial pressure gradient of -.1000E +00 psi/ft
      yields volume flow rate of 0.4076E +06 gal/min.

Flow rate target error is 0.7876E +05 %

Iteration    100, Error = .00371665
Iteration    200, Error = .00067117
```

Example 4.8   **173**

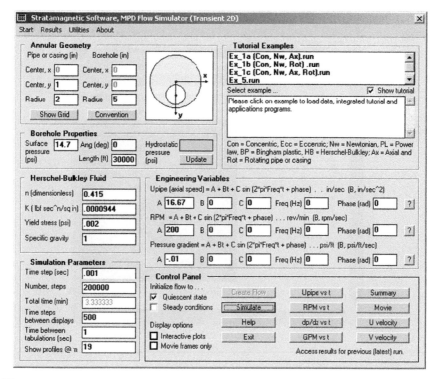

FIGURE 4.58

Flow at 200 rpm with 0.002-psi yield stress.

```
Iteration  300, Error = .00014123
Iteration  400, Error = .00002945
Iteration  500, Error = .00000702
Iteration  600, Error = .00000192
Iteration  700, Error = .00000038
Iteration  800, Error = .00000010
Iteration  900, Error = .00000010
Iteration 1000, Error = .00000010
```

0 Axial pressure gradient of -.5000E-01 psi/ft
yields volume flow rate of 0.4141E+05 gal/min.

Flow rate target error is 0.7911E+04 %
.
.
.
.

o Axial pressure gradient of -.6250E-02 psi/ft
yields volume flow rate of 0.6708E+03 gal/min.

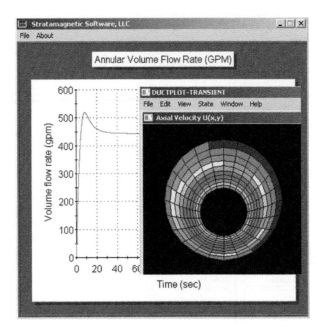

FIGURE 4.59

Flow at 200 rpm with 0.002-psi yield stress.

```
Flow rate target error is 0.2977E+02 %

Iteration    100, Error = .00000000
Iteration    200, Error = .00000011
Iteration    300, Error = .00000000
Iteration    400, Error = .00000011
Iteration    500, Error = .00000011
Iteration    600, Error = .00000011
Iteration    700, Error = .00000000
Iteration    800, Error = .00000021
Iteration    900, Error = .00000011
Iteration   1000, Error = .00000000

o  Axial pressure gradient of -.4688E-02 psi/ft
yields volume flow rate of 0.5217E+03 gal/min.

Pressure gradient found iteratively, -.4688E-02 psi/ft,
to yield 0.5217E+03 gal/min versus target 0.5169E+03 gal/min.

Note: Iterations terminate within 1% of target rate.
Refine result by manually changing pressure gradient.
```

In other words, the pressure gradient associated with the *nonrotating* flow is -0.004688 psi/ft. The corresponding axial velocity field is shown in Figure 4.66 (page 178) in a variety of available

Example 4.8   **175**

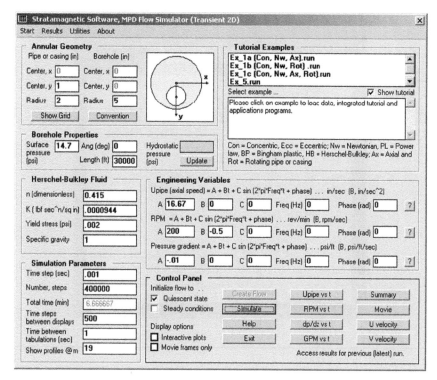

FIGURE 4.60

Decreasing rotational rate, from 200 to 0 rpm.

plots. Note that for nonrotating flows, our "Steady 2D" solver automatically computes the required pressure gradient using an internal inverse procedure. It has not been possible to develop a steady solver that allows rotation that is also unconditionally numerically stable. Fortunately, this does not mean that steady rotating flows cannot be computed.

We demonstrate how by considering the effect of a 100-rpm rotational rate. We use the "Transient 2D" solver in Figure 4.67, with input boxes completed for the same simulation parameters. Our strategy is to solve a fully transient problem until steady-state behavior is obtained. Because a "flow rate specified" mode is not available for transient calculations, one must resort to repeated guesses for pressure gradient, but we have found that three or four will usually lead to a flow rate within 1 to 2 percent of the target value. Since each trial calculation equilibrates quite rapidly, as shown in Figure 4.68, the total "desk time" required is often two minutes or less.

For this rotating flow run, a pressure gradient of -0.01 psi/ft is required, as compared to the $-.004688$ psi/ft obtained in the nonrotating case. In other words, pressure gradients are twice as severe because of rotation. The "Results" menu in Figure 4.67 provides numerous postprocessed results in addition to those of Figure 4.68 (page 180). For example, axial and azimuthal velocity distributions are available, as given in Figure 4.69, as are detailed color plots of different physical properties like apparent viscosity, shear rate, and viscous stress.

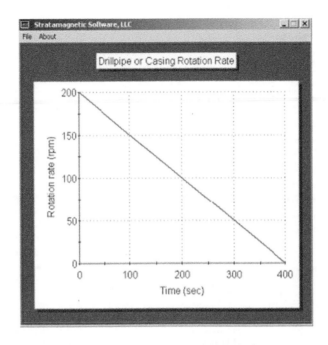

FIGURE 4.61

Linearly decreasing rpm, from 200 to 0.

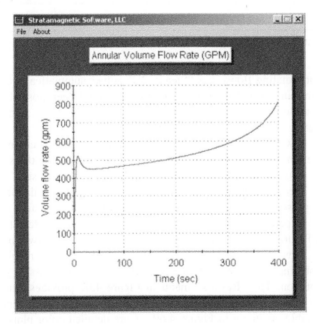

FIGURE 4.62

Transient increasing flow rate with decreasing rpm.

Example 4.9   **177**

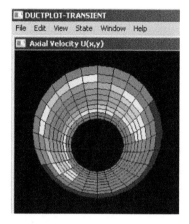

FIGURE 4.63

Transient movement of maximum point as rpm decreases.

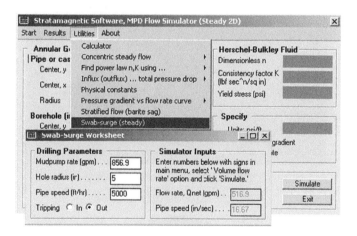

FIGURE 4.64

Running the "Swab-Surge Worksheet." (Areas that do not affect the Worksheet calculator are shaded.)

EXAMPLE 4.9

Transient Swab-Surge on a Steady-State Basis

Let us recall that the axial momentum equation takes the general functional form $\rho\,(\partial u/\partial t + v\,\partial u/\partial y + w\,\partial u/\partial x) = -\partial p/\partial z + \partial S_{zy}/\partial y + \partial S_{zx}/\partial x$ when body forces and variations in "z" are ignored. The resulting two-dimensional equation applies to transient flows with rotation and axial movement as well as to all rheological models. In later chapters, techniques are developed to integrate this in time and applications

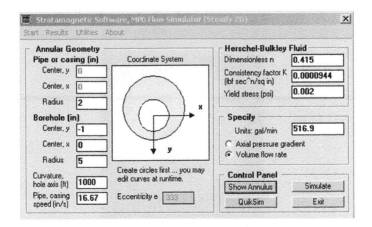

FIGURE 4.65

Steady 2D solver.

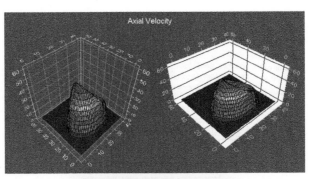

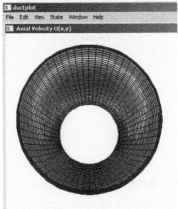

FIGURE 4.66

Computed axial velocity (nonrotating).

Example 4.10   **179**

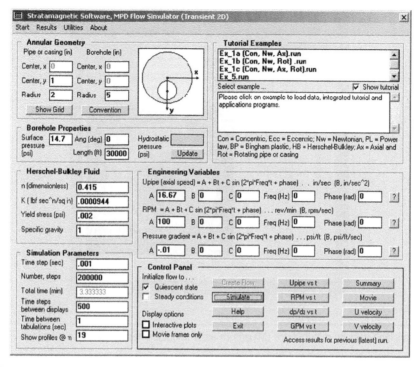

FIGURE 4.67

Transient 2D solver.

are given. If true transient effects (i.e., those modeled by the "$\partial u/\partial t$" term) can be ignored, the resulting $\rho \, (v \, \partial u/\partial y + w \, \partial u/\partial x) \approx -\partial p/\partial z + \partial S_{zy}/\partial y + \partial S_{zx}/\partial x$ underlies the work in this chapter.

If continuous but transient flow rate pumping is allowed during tripping, but under quasi-steady conditions, one might ask how the downhole "pressure response versus time" response is constructed. The answer is available in the illustrative procedures developed earlier. We recapitulate the basic ideas, which may or may not be obvious. First, the "flow rate versus pressure gradient curve" is constructed using, possibly, a combination of the steady-state models described—for example, one that might take the forms shown earlier in Figure 4.37 or 4.38. At any time $t = t_n$, we have an assumed volumetric flow rate Q_n for which a pressure gradient $(\partial P/\partial z)_n$ is now known. Then, the downhole pressure at the drillbit is simply $P_n = (\partial P/\partial z)_n \, L + P_{surf}(t_n)$, where L is the borehole length and $P_{surf}(t_n)$ is the surface choke pressure. This $P_n(t_n)$ can be plotted against t_n for display.

EXAMPLE 4.10

Equivalent Circulating Density Calculations

A formula is available for equivalent circulating density (ECD) calculation whose derivation is very straightforward. Again, we start from first principles with $\rho \, (\partial u/\partial t + v \, \partial u/\partial y + w \, \partial u/\partial x) = -\rho$

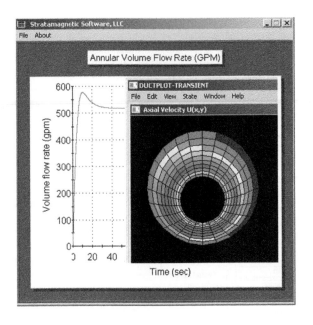

FIGURE 4.68

Flow rate history and velocity distribution. (Maximum axial velocities appear at annular bottom.)

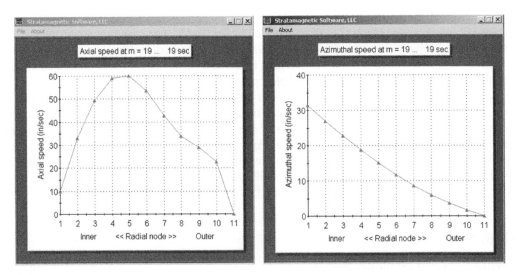

FIGURE 4.69

Axial and azimuthal velocities at cross section "m = 19."

Example 4.10   **181**

$g - \partial p/\partial z + \partial S_{zy}/\partial y + \partial S_{zx}/\partial x$, where we have now included the body force $-\rho g$ (where g is the acceleration due to gravity) and assumed "z" to be vertical. The left side is "ma," while the right is "F" in "F = ma." The first two terms on the right can be factored as $-g(\rho + 1/g\ \partial p/\partial z)$, from which it is clear that the combination $1/g\ \partial p/\partial z$ has the dimensions of the density, ρ. This is known as the "equivalent circulating density" because it provides an additive correction to ρ for hydrostatic applications.

When the pressure gradient $\partial p/\partial z$ is available from flow calculations, the ECD $= 1/g\ \partial p/\partial z$ formula applies. If the pressure gradient is expressed as N psi/ft, where N is dimensionless, then ECD $= 19.25$ N lbm/gal. For example, if a viscous non-Newtonian pipe or annular flow is associated with a pressure gradient of -0.01 psi/ft, then we have ECD $= 19.25\ (0.01)$ lbm/gal or 0.1925 lbm/gal. (This might be compared to the density of water, with a value of approximately 8.33 lbm/gal.). ECDs provide a useful way of appreciating the magnitude of any pressure gradient, but are, in themselves, not fundamentally important in fluid dynamics. They are, of course, useful in MPD job planning.

More Steady Flow Applications

5

In Chapters 2, 3, and 4, we developed the theory and mathematical methods needed to model steady, nonrotating, non-Newtonian flows in general borehole annular cross sections having arbitrary geometries. These included new topological concepts useful in creating boundary-conforming curvilinear grid systems, derivation of momentum equations transformed to these coordinates, plus introduction of iterative methods required for fast, robust, and numerically stable solutions. This work extends the early models reported in the books *Borehole Flow Modeling* and *Computational Rheology* by improving accuracy while reducing calculation times and computer memory resources.

In this chapter, we continue our focus on steady flows by presenting additional math models that are useful in dealing with specific aspects of annular flow simulation and engineering application. These simpler models, while not trivial in any mathematical sense, were also used to validate the more sophisticated models in this book for fluids with general rheologies in complicated flow domains. Examples include the first exact, closed-form solution for Herschel-Bulkley fluids in concentric annuli; Newtonian flow in concentric annuli with moving walls; solutions modeling flows in the presence of barite sag; and Newtonian flows in general rectangular ducts.

Importantly, we also address field and laboratory validations for the steady annular flows calculated using the new methods, in particular dealing with cuttings transport in deviated wells, evaluation of spotting fluid effectiveness in stuck pipe removal, the effect of non-Newtonian flow pressure drops in boreholes with bends, the effects of steady rotation in concentric systems, and so on. While this chapter addresses steady flow validations, it is important to emphasize that results of transient, three-dimensional, multiphase extensions of these methods are also consistent with experiments, as will be discussed in Chapters 8 and 9. These represent a major thrust of our research.

MODEL 5.1

Newtonian Flow in Concentric Annulus with Axially Moving (but Nonrotating) Pipe or Casing

We consider Newtonian annular flow between concentric cylinders, as shown in Figure 5.1, in which the inner cylinder moves with a constant speed V_∞ in either direction and an external pressure gradient dP/dz is imposed in the axial z direction.

Since both radial and azimuthal velocities are assumed to vanish, $\partial/\partial t = 0$ holds for steady flow, and $\partial/\partial z = 0$ holds for two-dimensional problems (applicable if cylinder lengths are sufficiently long), the Navier-Stokes equations reduce to a single one for the axial velocity v(r)—namely,

$$d^2v/dr^2 + r^{-1}\, dv/dr = \mu^{-1}P_z \tag{5.1a}$$

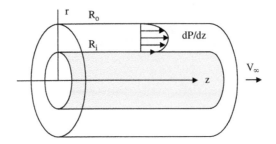

FIGURE 5.1

Steady, concentric, Newtonian flow with moving pipe.

The boundary conditions are

$$v(R_L) = V_\infty \text{ along inner radius} \tag{5.1b}$$

$$v(R_o) = 0 \text{ along outer radius (i.e., no-slip)} \tag{5.1c}$$

The exact solution for velocity at any position "r" is

$$v(r) = [(R_o^2 - R_i^2)P_z/(4\mu) + V_\infty] \log_e (R_o/r)/\log_e (R_o/R_i) + (r^2 - R_o^2)P_z/(4\mu) \tag{5.1d}$$

The volumetric flow rate is given by

$$Q = \int_{R_i}^{R_o} v(r) \, 2\pi r dr \tag{5.1e}$$

$$Q = \pi V_\infty [(R_o^2 - R_i^2)/(2\log_e (R_o/R_i)) - R_i^2] \\ + (\pi P_z/(8\mu))[(R_o^2 - R_i^2)^2 + (R_i^4 - R_o^4) \log_e (R_o/R_i)]/\log_e(R_o/R_i) \tag{5.1f}$$

Note that when the inner cylinder is stationary with $V_\infty = 0$, the relationship between Q and pressure gradient is linear. In fact, Q is directly proportional to P_z and varies inversely with the viscosity μ. The additional factor seen in Equation 5.1f depends only on geometrical details. The viscous shear stress at any position "r" is simply the product between viscosity and shear rate and is known from

$$\tau(r) = \mu \, dv(r)/dr = \mu\{ - [(R_o^2 - R_i^2)P_z/(4\mu) + V_\infty]/(r \, \log_e(R_o/R_i)) + r \, P_z/(2\mu)\} \tag{5.1g}$$

In particular, the shear stress at the moving cylinder $r = R_i$ is

$$\tau(R_L) = \mu \, dv(R_i)/dr = \mu\{ - [(R_o^2 - R_i^2)P_z/(4\mu) + V_\infty]/(R_i \log_e (R_o/R_i)) + R_iP_z/(2\mu)\} \tag{5.1h}$$

If the axial length of the cylinder is L, the total shear force acting on the inner cylinder is given by the product of $\tau(R_i)$ and the surface area $2\pi\tau R_i$ or

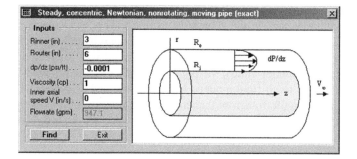

FIGURE 5.2

Software user interface.

$$F_{shear} = 2\pi R_i \, L \, \tau(R_t)$$
$$= 2\pi \, \mu \, R_i \, L\{ -[(R_o^2 - R_i^2)P_z/(4\mu) + V_\infty]/(R_i \, \log_e \, (R_o/R_i)) + R_i \, P_z/(2\mu)\}$$ (5.1i)

The shear force per unit length is

$$F_{shear}/L = 2\pi \, \mu \, R_i\{ -[(R_o^2 - R_i^2)P_z/(4\mu) + V_\infty]/(R_i \, \log_e \, (R_o/R_i)) + R_i \, P_z/(2\mu)\}$$ (5.1j)

A simple implementation for volumetric flow rate only is shown in Figure 5.2 and is accessible from the "Utilities" menu. The results obtained with this model are used to validate steady and transient implementations of our curvilinear grid, finite difference-based algorithms. A more detailed version of the software that evaluates the stress and force formulas derived here is available upon request.

MODEL 5.2

Density Stratification (Barite Sag) and Recirculating Annular Vortexes That Impede Fluid Flow

Problems with cuttings accumulation, flow blockage, and resultant stuck pipe in deviated wells are becoming increasingly important operational issues as interest in horizontal drilling continues. For small angles (ß) from the vertical, annular flows and hole cleaning are well understood; for example, cleaning efficiency is always improved by increasing velocity, viscosity, or both. But beyond 30 degrees, these issues are rife with challenging questions.

Many unexplained, confusing, and conflicting observations are reported by different investigators; however, it turns out that bottom viscous stress (which tends to erode cuttings beds having well-defined mechanical yield stresses) is the correlation parameter that explains many of these discrepancies. The author's books *Borehole Flow Modeling in Horizontal, Deviated and Vertical Wells* (Chin, 1992) and *Computational Rheology for Pipeline and Annular Flow* (Chin, 2001) address hole-cleaning applications in which cuttings and other drilling debris block mud transport. These issues are addressed later.

Here we will learn that flows can be blocked even when no externally introduced debris appears in the system. In other words, dangerous flow blockage can arise from fluid-dynamical effects alone. This possibility is very real whenever there exist density gradients in a direction perpendicular to the flow (e.g., "barite sag" in the context of drilling). This blockage is also possible in pipe flows of slurries—for instance, slurries carrying ground wax and hydrate particles or other debris.

Although we have focused on rheological effects associated with Newtonian, Power law, Bingham plastic, and Herschel-Bulkley fluids, the physical mechanisms considered in this section relate to inertial effects and apply to all fluid models. In work addressed elsewhere in this book, the pressure field is assumed to be uniform across the annulus; the velocity field is therefore unidirectional, with the fluid flowing axially from high-pressure regions to low-pressure regions. These assumptions are reasonable, since numerous flows do behave in this manner. Below we will relax these assumptions, but turn to more general flows with density stratification. A special class of annular and pipe flow lends itself to strange occurrences we call "recirculating vortex flows," to which we now turn our attention.

What are recirculating vortex flows?

In deviated holes where circulation has been temporarily interrupted, weighting material such as barite, drilled cuttings, or cement additives may fall out of suspension. Similarly, pipelines containing slurries with wax or hydrate particles can develop vertical density gradients when flow is temporarily slowed or halted. This gravity stratification has mass density increasing downward. And this stable stratification, which we refer to collectively as "barite sag," is thought to be responsible for the trapped, self-contained "recirculation zones" or "bubbles" observed by many experimenters.

Recirculation zones contain rotating, swirling, "ferris-wheel-like" motions within their interiors; the external fluid that flows around them "sees" these zones as stationary obstacles that impede their axial movement up the annulus or pipe. Excellent color videotapes showing these vortex-like motions in detail have been produced by M-I Drilling Fluids; they were viewed by the author in its Houston facilities prior to the initial printing of *Borehole Flow Modeling*.

These strange occurrences are just that; their appearances seem to be sporadic and unpredictable, as much myth as reality. However, once they are formed, they remain as stable fluid-dynamical structures that are extremely difficult to remove. They are dangerous and undesirable because of their tendency to entrain cuttings, block axial flow, and increase the possibility of stuck pipe. One might ask, "Why do these bubbles form? What are the controlling parameters? How can their occurrences be prevented?"

Detailed study of M-I's tapes suggests that the recirculating flows form independently of viscosity and rheology to leading order; that is, they do not depend primarily on "n" and "K." They appear to be inertia dominated, depending on density effects themselves, while nonconservative viscous terms play only a minor role in sustaining or damping the motion. This leaves the component of density stratification normal to the hole axis as the primary culprit; it alone is responsible for the highly three-dimensional pressure field that drives local pockets of secondary flow. It is possible, of course, to have multiple bubbles coexisting along a long deviated hole.

Again, these recirculating bubbles, observed near pipe bends, stabilizers, and possibly marine risers and other obstructions, are important for various practical reasons. First, they block the streamwise axial flow, resulting in the need for increased pressure to pass a given volumetric flow rate. Second, because they entrain the mud and further trap drilled cuttings, they are a likely cause of stuck pipe. Third, the external flow modified by these bubbles can also affect the very process of cuttings bed formation and removal itself.

Fortunately, these bubbles can be studied, modeled, and characterized in a rather simple manner; very instructive "snapshots" of streamline patterns covering a range of vortex effects are given later. This section identifies the *nondimensional* channel parameter **Ch** responsible for vortex bubble formation and describes the physics of these recirculating flows. The equations of motion are given and solved using finite difference methods for several practical flows. The detailed bubble development process is described and illustrated in a sequence of computer-generated pictures.

Motivating ideas and controlling variables

The general governing momentum equations are Euler's equations, which describe large-amplitude, inviscid shear flow in both stratified and unstratified media (Schlichting, 1968; Turner, 1973). The problem at hand can be modeled as two-dimensional stratified flow into a sink, following Yih (1960, 1969, 1980), who initially solved the problem for meteorological and geophysical applications.

The general Euler equations for two-dimensional flows in inclined boreholes simplify to the following three, for a variable density ρ, and the velocity components u and v. These nonetheless remain intractable.

$$\rho(uu_x + vu_y) = -p_x \tag{5.2}$$

$$\rho(uv_x + vv_y) = -p_y - g\rho\cos\alpha \tag{5.3}$$

$$(\rho u)_x + (\rho v)_y = 0 \tag{5.4}$$

In fluid mechanics, it is common to reduce the number of unknowns by introducing a "streamfunction" by virtue of mass conservation. Lines of constant streamfunction can be shown to describe streamlines of the flow, across which fluid motion is not possible. Yih shows that, in his problems, the formulation and solution that follow apply.

$$\Psi_{\xi\xi} + \Psi_{\eta\eta} = F^{-2}(\eta - \Psi) \tag{5.5a}$$

$$\Psi = 0 \text{ at } \eta = 0 \tag{5.5b}$$

$$\Psi = 1 \text{ at } \eta = 1 \text{ and at } \xi = 0, \eta \neq 0 \tag{5.5c}$$

$$\Psi = \eta \text{ at } \xi = -\infty \tag{5.5d}$$

$$\Psi = \eta + (2/\pi)\sum_{n=1}^{\infty} n^{-1}\exp[(n^2\pi^2 - F^{-2})^{1/2}\xi]\sin n\pi\eta \tag{5.5e}$$

$$-\infty < \xi \leq 0, 0 \leq \eta \leq 1 \tag{5.5f}$$

Here the sink is located at $\xi = 0$ and $\eta = 0$. A linear density variation is assumed far upstream, the oncoming flow is almost constant, $F > 1/\pi$ is a Froude number comparing inertia to gravity effects, and Ψ is a normalized streamfunction.

In other words, we have assumed that the steady vortex flow is contained in a two-dimensional rectangular box in the plane of the hole axis (x) and the direction of density stratification (y). This is based on experimental observation: The vortical flows do not wrap around the drillpipe. In the equations just given, u and v would be velocities in the x and y directions, respectively. Subscripts indicate partial derivatives. In our convention, ρ is a fluid density that varies linearly with y far upstream, p(x, y) is the unknown pressure field, g is the acceleration due to gravity, and α is the angle the borehole axis makes with the horizontal. Equations 5.2 and 5.3 are momentum equations in the x and y directions, while Equation 5.4 describes mass conservation.

Nondimensional parameters are important in understanding physical events. The Reynolds number, which measures the relative effects of inertia on viscous forces, is one example. It dictates the onset of turbulence; also, like Reynolds numbers imply dynamically similar flow patterns. Analogous dimensionless variables are used in different areas of physics—for example, the mobility ratio in reservoir engineering or the Mach number in high-speed aerodynamics.

Close examination of Equations 5.2 through 5.4 using affine transformations shows that the physics of bubble formation depends on a *single* nondimensional variable **Ch** characterizing the channel flow. It is constructed from the combination of two simpler ones. The first is a Froude number $U^2/gL \cos\alpha$, where U is the average oncoming speed and L is the channel height between the pipe and borehole walls. The second is a relative measure of stratification—say, $d\rho/\rho_{ref}$ ($d\rho$ might represent the density difference between the bottom and top of the annulus or pipe, and ρ_{ref} may be taken as their arithmetic average). The combined parameter **Ch** of practical significance is

$$\mathbf{Ch} = U^2 \rho_{ref}/gL \; d\rho \; \cos \alpha \tag{5.6}$$

We now summarize our findings. For large values of **Ch**, recirculation bubbles will *not* form; the streamlines of motion are essentially straight, and the rheology-dominated models apply. For small **Ch**'s of order unity, small recirculation zones *do* form and elongate in the streamwise dimension as **Ch** decreases. For still smaller values—that is, values below a critical value of 0.3183—solutions with wavy upstream flows are found, which may or may not be physically realistic. Equations 5.2 through 5.4 can be solved using "brute force" computational methods, but they are more cleverly treated by introducing the streamfunction used by aerodynamicists and reservoir engineers.

When the problem is reformulated in this manner, the result is a linear Poisson equation that can be integrated in closed analytical form as indicated above. Streamlines are obtained by connecting computed streamfunction elevations having like values. The arithmetic difference in Ψ between any two points is proportional to the volumetric flow rate passing through the two points. Velocity and pressure fields can be obtained by postprocessing computed Ψ solutions.

Detailed calculations

Solutions obtained for a 20×40 mesh require less than one second. In Figures 5.3 through 5.8, we allow the flow to disappear into a "mathematical sink" (in practice, the distance to this obstacle over the height L will appear as a second ratio). This sink simulates the presence of obstacles or pipe elbows located further upstream. With decreasing values of **Ch**, the appearance of an elongating recirculation bubble is seen. Streamfunction data as well as processed contour plots are given. The stand-alone vortexes so obtained are stable, since they represent patches of angular momentum that physical laws insist must be conserved.

In this sense, they are not unlike trailing aircraft tip vortices that persist indefinitely until dissipation renders them harmless. However, annular bubbles are worse: The channel flow itself is what drives them, perpetuates them, and increases their ability to do harm by further entraining solid

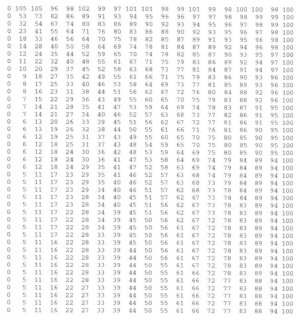

```
0 105 105  96  98 102  99  97 101 101  98  99 101  99  98 100 100  98 100
0  53  73  82  86  89  91  93  94  95  96  96  97  97  98  98  99  99 100
0  32  54  67  74  80  83  86  89  90  92  93  94  95  96  97  98     100
0  23  41  55  64  71  76  80  83  86  88  90  92  93  95  96  97  98 100
0  18  33  46  56  64  70  75  78  82  85  87  89  91  93  95  96  98 100
0  14  28  40  50  58  64  69  74  78  81  84  87  89  92  94  96  98 100
0  12  24  35  44  52  59  65  70  74  78  82  85  87  90  93  95  97 100
0  11  22  32  40  48  55  61  67  71  75  79  83  86  89  92  94  97 100
0  10  20  29  37  45  52  58  63  68  73  77  81  84  87  91  94  97 100
0   9  18  27  35  42  49  55  61  66  71  75  79  83  86  90  93  96 100
0   8  17  25  33  40  46  53  58  64  69  73  77  81  85  89  93  96 100
0   8  16  23  31  38  44  51  56  62  67  72  76  80  84  88  92  96 100
0   7  15  22  29  36  43  49  55  60  65  70  75  79  83  88  92  96 100
0   7  14  21  28  35  41  47  53  59  64  69  74  78  83  87  91  95 100
0   7  14  21  27  34  40  46  52  57  63  68  73  77  82  86  91  95 100
0   6  13  20  26  33  39  45  51  56  62  67  72  77  81  86  91  95 100
0   6  13  19  26  32  38  44  50  55  61  66  71  76  81  86  90  95 100
0   6  12  19  25  31  37  43  49  55  60  65  70  75  80  85  90  95 100
0   6  12  18  25  31  37  43  48  54  59  65  70  75  80  85  90  95 100
0   6  12  18  24  30  36  42  48  53  59  64  69  75  80  85  90  95 100
0   6  12  18  24  30  36  41  47  53  58  64  69  74  79  84  89  94 100
0   6  12  18  24  29  35  41  47  52  58  63  69  74  79  84  89  94 100
0   5  11  17  23  29  35  41  46  52  57  63  68  74  79  84  89  94 100
0   5  11  17  23  29  35  40  46  52  57  63  68  73  79  84  89  94 100
0   5  11  17  23  29  34  40  46  51  57  62  68  73  78  84  89  94 100
0   5  11  17  23  28  34  40  45  51  57  62  67  73  78  84  89  94 100
0   5  11  17  23  28  34  40  45  51  56  62  67  73  78  83  89  94 100
0   5  11  17  22  28  34  39  45  51  56  62  67  73  78  83  89  94 100
0   5  11  17  22  28  34  39  45  50  56  62  67  72  78  83  89  94 100
0   5  11  17  22  28  34  39  45  50  56  61  67  72  78  83  89  94 100
0   5  11  17  22  28  33  39  45  50  56  61  67  72  78  83  89  94 100
0   5  11  16  22  28  33  39  44  50  56  61  67  72  78  83  89  94 100
0   5  11  16  22  28  33  39  44  50  56  61  67  72  78  83  89  94 100
0   5  11  16  22  28  33  39  44  50  55  61  67  72  78  83  89  94 100
0   5  11  16  22  28  33  39  44  50  55  61  66  72  78  83  89  94 100
0   5  11  16  22  27  33  39  44  50  55  61  66  72  77  83  88  94 100
0   5  11  16  22  27  33  39  44  50  55  61  66  72  77  83  88  94 100
0   5  11  16  22  27  33  39  44  50  55  61  66  72  77  83  88  94 100
```

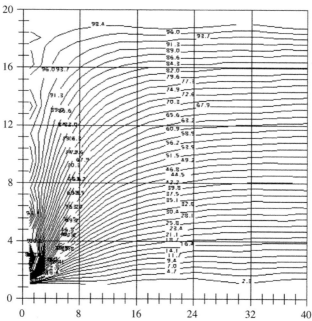

FIGURE 5.3

Ch = 1.0, straight streamlines without recirculation.

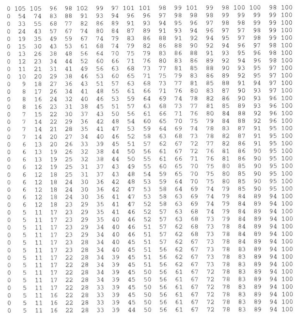

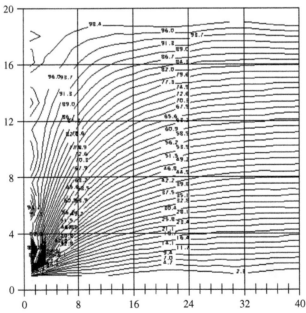

FIGURE 5.4

Ch = 0.5, straight streamlines without recirculation.

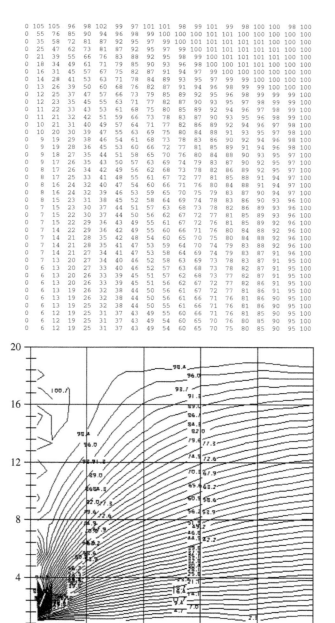

FIGURE 5.5

Ch = 0.35, minor recirculating vortex.

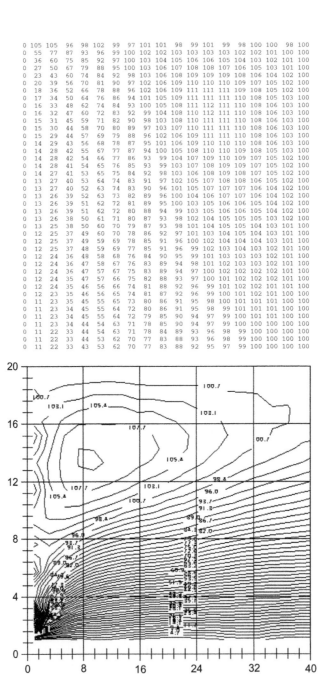

```
0 105 105  96  98 102  99  97 101 101  98  99 101  99  98 100 100  98 100
0  55  77  87  93  96  99 100 102 102 103 103 103 103 102 102 101 100 100
0  36  60  75  85  92  97 100 103 104 105 106 106 105 104 103 102 101 100
0  27  50  67  79  88  95 100 103 106 107 108 108 107 106 105 103 101 100
0  23  43  60  74  84  92  98 103 106 108 109 109 109 108 106 104 102 100
0  20  39  56  70  81  90  97 102 106 109 110 110 110 109 107 105 102 100
0  18  36  52  66  78  88  96 102 106 109 111 111 111 110 108 105 102 100
0  17  34  50  64  76  86  94 101 105 109 111 111 111 110 108 105 103 100
0  16  33  48  62  74  84  93 100 105 108 111 112 111 110 108 106 103 100
0  16  32  47  60  72  83  92  99 104 108 110 112 111 110 108 106 103 100
0  15  31  45  59  71  82  90  98 103 108 110 111 111 110 108 106 103 100
0  15  30  44  58  70  80  89  97 103 107 110 111 111 110 108 106 103 100
0  15  29  44  57  69  79  88  96 102 106 109 111 111 110 108 106 103 100
0  14  29  43  56  68  78  87  95 101 106 109 110 110 110 108 106 103 100
0  14  28  42  55  67  77  87  94 100 105 108 110 110 109 108 105 103 100
0  14  28  42  54  66  77  86  93  99 104 107 109 110 109 107 105 102 100
0  14  28  41  54  65  76  85  93  99 103 107 108 109 109 107 105 102 100
0  14  27  41  53  65  75  84  92  98 103 106 108 109 108 107 105 102 100
0  13  27  40  53  64  74  83  91  97 102 105 107 108 108 106 105 102 100
0  13  27  40  52  63  74  83  90  96 101 105 107 107 107 106 104 102 100
0  13  26  39  52  63  73  82  89  96 100 104 106 107 107 106 104 102 100
0  13  26  39  51  62  72  81  89  95 100 103 105 106 106 105 104 102 100
0  13  26  39  51  62  72  80  88  94  99 103 105 106 106 105 104 102 100
0  13  26  38  50  61  71  80  87  93  98 102 104 105 105 105 103 102 100
0  13  25  38  50  60  70  79  87  93  98 101 104 105 105 104 103 101 100
0  12  25  37  49  60  70  78  86  92  97 101 103 104 105 104 103 101 100
0  12  25  37  49  59  69  78  85  91  96 100 102 104 104 104 103 101 100
0  12  25  37  48  59  69  77  85  91  96  99 102 103 104 103 102 101 100
0  12  24  36  48  58  68  76  84  90  95  99 101 103 103 103 102 101 100
0  12  24  36  47  58  67  76  83  89  94  98 101 102 103 103 102 101 100
0  12  24  36  47  57  67  75  83  89  94  97 100 102 102 102 102 101 100
0  12  24  35  47  57  66  75  82  88  93  97 100 101 102 102 102 101 100
0  12  24  35  46  56  66  74  81  88  92  96  99 101 102 102 101 101 100
0  12  23  35  46  56  65  74  81  87  92  96  99 100 101 102 101 100 100
0  11  23  35  45  55  65  73  80  86  91  95  98 100 101 101 101 100 100
0  11  23  34  45  55  64  72  80  86  91  95  98  99 101 101 101 100 100
0  11  23  34  45  55  64  72  79  85  90  94  97  99 100 101 101 100 100
0  11  23  34  44  54  63  71  78  85  90  94  97  99 100 100 100 100 100
0  11  22  33  44  54  63  71  78  84  89  93  96  98  99 100 100 100 100
0  11  22  33  44  53  62  70  77  83  88  93  96  98  99 100 100 100 100
0  11  22  33  43  53  62  70  77  83  88  92  95  97  99 100 100 100 100
```

FIGURE 5.6

Ch = 0.320, large-scale recirculation.

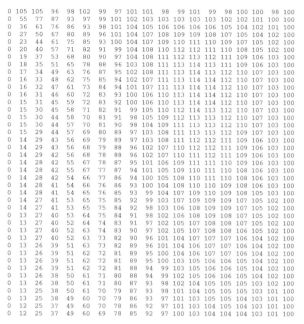

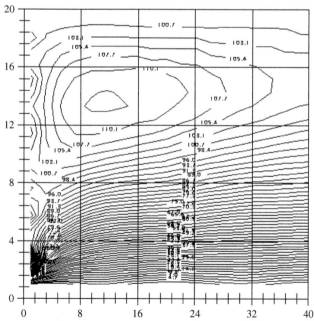

FIGURE 5.7

Ch = 0.319, major flow blockage.

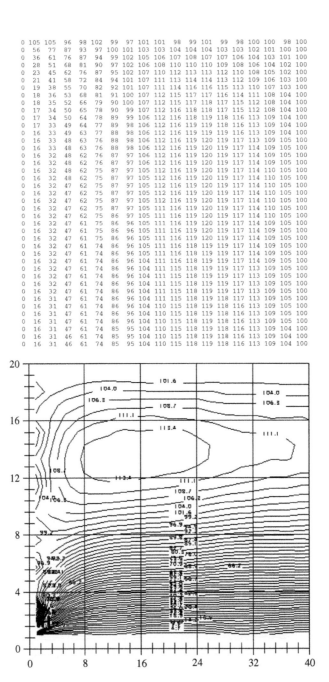

```
0 105 105  96  98 102  99  97 101 101  98  99 101  99  98 100 100  98 100
0  56  77  87  93  97 100 101 103 103 104 104 104 103 103 102 101 100 100
0  36  61  76  87  94  99 102 105 106 107 108 107 107 106 104 103 101 100
0  28  51  68  81  90  97 102 106 108 110 110 110 109 108 106 104 102 100
0  23  45  62  76  87  95 102 107 110 112 113 113 112 110 108 105 102 100
0  21  41  58  72  84  94 101 107 111 113 114 114 113 112 109 106 103 100
0  19  38  55  70  82  92 101 107 111 114 116 116 115 113 110 107 103 100
0  18  36  53  68  81  91 100 107 112 115 117 117 116 114 111 108 104 100
0  18  35  52  66  79  90 100 107 112 115 117 118 117 115 112 108 104 100
0  17  34  50  65  78  90  99 107 112 116 118 118 117 115 112 108 104 100
0  17  34  50  64  78  89  99 106 112 116 118 119 118 116 113 109 104 100
0  17  33  49  64  77  89  98 106 112 116 119 119 118 116 113 109 104 100
0  16  33  49  63  77  88  98 106 112 116 119 119 119 116 113 109 104 100
0  16  33  48  63  76  88  98 106 112 116 119 120 119 117 114 109 105 100
0  16  33  48  63  76  88  98 106 112 116 119 120 119 117 114 109 105 100
0  16  32  48  62  76  87  97 106 112 116 119 120 119 117 114 109 105 100
0  16  32  48  62  76  87  97 106 112 116 119 120 119 117 114 109 105 100
0  16  32  48  62  75  87  97 105 112 116 119 120 119 117 114 110 105 100
0  16  32  48  62  75  87  97 105 112 116 119 120 119 117 114 110 105 100
0  16  32  47  62  75  87  97 105 112 116 119 120 119 117 114 110 105 100
0  16  32  47  62  75  87  97 105 112 116 119 120 119 117 114 110 105 100
0  16  32  47  62  75  87  97 105 112 116 119 120 119 117 114 110 105 100
0  16  32  47  62  75  86  97 105 111 116 119 120 119 117 114 110 105 100
0  16  32  47  61  75  86  96 105 111 116 119 120 119 117 114 109 105 100
0  16  32  47  61  75  86  96 105 111 116 119 120 119 117 114 109 105 100
0  16  32  47  61  74  86  96 105 111 116 118 119 119 117 114 109 105 100
0  16  32  47  61  74  86  96 105 111 116 118 119 119 117 114 109 105 100
0  16  32  47  61  74  86  96 104 111 116 118 119 119 117 114 109 105 100
0  16  32  47  61  74  86  96 104 111 115 118 119 119 117 113 109 105 100
0  16  32  47  61  74  86  96 104 111 115 118 119 119 117 113 109 105 100
0  16  32  47  61  74  86  96 104 111 115 118 119 119 117 113 109 105 100
0  16  31  47  61  74  86  96 104 111 115 118 119 118 117 113 109 105 100
0  16  31  47  61  74  86  96 104 110 115 118 119 118 116 113 109 105 100
0  16  31  47  61  74  86  96 104 110 115 118 119 118 116 113 109 105 100
0  16  31  47  61  74  85  95 104 110 115 118 119 118 116 113 109 104 100
0  16  31  46  61  74  85  95 104 110 115 118 119 118 116 113 109 104 100
0  16  31  46  61  74  85  95 104 110 115 118 119 118 116 113 109 104 100
```

FIGURE 5.8

Ch = 0.3185, major flow blockage by elongated vortex structure.

debris in the annulus. In the pipeline context, slurry density gradients likewise promote flow blockage and vortex recirculation; the resulting "sandpapering," allowing continuous rubbing against pipe walls, can lead to metal erosion, decreased strength, and unexpected rupture.

How to avoid stagnant bubbles

We have shown that recirculating zones can develop from interactions between inertia and gravity forces. These bubbles form when density stratification, hole deviation, and pump rate fulfill certain special conditions. These are elegantly captured in a single channel variable, the nondimensional parameter $\mathbf{Ch} = U^2 \rho_{ref}/gLd\rho \cos\alpha$. Moreover, the resulting flow fields can be efficiently computed and displayed, thus allowing us to understand better their dynamical consequences.

Suppressing recirculating flows is simply accomplished: *Avoid small values of the nondimensional* **Ch** *parameter*. Small values, as is evident from Equation 5.6, can result from different isolated effects. For example, they decrease as the hole becomes more horizontal, as density differences become more pronounced, or as pumping rates decrease. But none of these factors alone control the physics; it is the combination *taken together* that controls bubble formation and perhaps the fate of a drilling program.

We have modeled the problem as the single-phase flow of a stratified fluid rather than as the combined motion of dual-phase fluid and solid continuum. This simplifies the mathematical issues without sacrificing the essential physical details. For practical purposes, the parameter **Ch** can be viewed as a "danger indicator" signaling impending cuttings transport or stuck pipe problems. It is the single most important parameter whenever interrupted circulation or poor suspension properties lead to gravity segregation and settling of weighting materials in drilling mud.

These considerations also apply to cementing, where density segregation due to gravity *and* slow velocities is likely. When recirculation zones form in either the mud or the cement above or beneath the casing, the displacement effectiveness of the cement is severely impeded. The result is mud left in place, an undesirable situation necessitating squeeze jobs. Similar remarks apply to pipeline applications. Recirculation zones are likely to be encountered at low flow rates that promote density stratification, and immediately prior to flow start-up, when slurry particles have been allowed to settle out.

We emphasize that the vortical bubbles considered here are *not* the "Taylor vortices" studied in the classical fluid mechanics of homogeneous flows. Taylor vortices are "doughnuts" that would normally "wrap around," in our case, the drillpipe; to the author's knowledge, these have not been observed in drilling applications. They can be created in the absence of density stratification—that is, they can be found in purely homogeneous fluids. Importantly, Taylor vortices would owe their existence to finite drillstring length effects and would represent completely different physical mechanisms.

A practical example

We have discussed the dynamical significance of the nondimensional parameter **Ch** that appears in the normalized equations of motion. For use in practical estimates, the channel variable may be written more clearly as a multiplicative sequence of dimensionless entities:

$$\mathbf{Ch} = U^2\rho_{ref}/gLd\rho \ \cos \ \alpha = (U^2/gL) \times (\rho_{ref}/d\rho) \times (1/\cos \ \alpha) \tag{5.7}$$

Let us consider an annular flow studied in the cuttings transport examples of *Borehole Flow Modeling*. For the 2-in. and 5-in. pipe and borehole radii, the cross-sectional area is $\pi (5^2 - 2^2)$ or 66 in.^{2}. The experimental data used in Discussions 1 and 2 of Chapter 5 in *Borehole Flow Modeling* assume oncoming linear velocities of 1.91, 2.86, and 3.82 ft/sec. Since 1 ft/sec corresponds to a volumetric flow rate of 1 ft/sec $\times$ 66 in.^{2} or 205.7 gpm, the flow rates are 393, 588, and 786 gpm, respectively. So, at the lowest flow rate of 393 gpm (a reasonable field number), the average linear speed over the entire annulus is approximately 2 ft/sec. But the low-side average will be much smaller—say, 0.5 ft/sec. And if the pipe is displaced halfway down, the length scale L will be roughly (5.2)/2 in. or 0.13 ft.

Thus, the first factor in Equation 5.7 takes the value $U^2/gL = (0.5)^2/(32.2 \times 0.13) = 0.06$. If we assume a 20 percent density stratification, then $\rho_{ref}/d\rho = 5.0$; the product of the two factors is 0.30. For a highly deviated well inclined 70° from the vertical axis, $\alpha = 90° - 70° = 20°$ and $\cos 20° = 0.94$. Thus, we obtain **Ch** = 0.30/0.94 = 0.32. This value, as Figures 5.3 through 5.8 show, lies just at the threshold of danger. Velocities lower than the assumed value are even more likely to sustain recirculatory flows; higher ones, in contrast, are safer.

Of course, the numbers used above are only estimates; a three-dimensional, viscous solution is required to establish true length and velocity scales. But these approximate results show that bottomhole conditions typical of those used in drilling and cementing *are* associated with low values of **Ch** near unity.

We emphasize that **Ch** is the only nondimensional parameter appearing in Equations 5.2 through 5.4. Another parameter describing the geometry of the annular domain would normally appear through boundary conditions. For convenience, though (and for the sake of argument only), we have replaced this requirement with an idealized "sink." In any real calculation, exact geometrical effects must be included to complete the formulation. Also note that our recirculating flows get worse as the borehole becomes more horizontal; that is, **Ch** decreases as α becomes smaller. This is in stark contrast to the unidirectional, homogeneous flows usually studied, which typically perform worst near 45°, at least with respect to cuttings transport efficiency. The structure of Equation 5.7 correctly shows that in near-vertical wells with α approaching 90°, **Ch** tends to infinity; thus, the effects of flow blockage due to the vortical bubbles considered here are relegated to highly deviated wells.

Again, flow properties such as local velocity, shear rate, and pressure can be obtained from the computed streamfunction straightforwardly. They may be useful correlation parameters for cuttings transport efficiency and local bed buildup. Continuing research is underway, exploring similarities between this problem and the density-dependent flows studied in dynamic meteorology and oceanography. Obvious extensions of our observations for annular flow apply to the pipeline transport of wax and hydrate slurries.

Software implementation

The stratified flow solution in this section can be accessed from the "Utilities" menu in Figure 5.9, which calls the program in Figure 5.10. Clicking "Results" yields numerical streamfunction results (e.g., those in Figure 5.3) and the three-dimensional color plots shown in Figures 5.11 through 5.13.

MODEL 5.3

Herschel-Bulkley Flow in Concentric Annulus with Axially Stationary and Nonrotating Drillpipe or Casing

Non-Newtonian fluids with yield stress are responsible for plug zones that move as solid bodies within sheared flows. Unlike circular pipe flow, the plug in a concentric annulus is defined by two

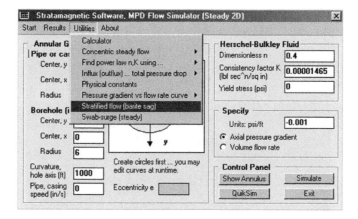

FIGURE 5.9

Stratified flow, user interface.

FIGURE 5.10

Stratified flow, user interface.

radial points where shear rates are discontinuous. Sudden changes in shear rate imply that the velocity derivative is not continuous everywhere, therefore precluding simple numerical solutions such as the finite difference methods used for Power law fluids. To date, yield stress solutions for concentric annuli are not available, except under simple slot flow approximations. In our exact solution that follows, we do not invoke slot or thin annulus assumptions but require that the pipe or casing remain immobile.

Here we consider Herschel-Bulkley fluids, which encompass Newtonian, Power law, and Bingham plastic yield flows. An exact analytical approach is developed that produces integral solutions in terms of a parameter "C." This "C" satisfies special kinematic and dynamic constraints, and is iteratively determined using a numerical scheme. Once solved, the velocity field is available and is used to calculate total volumetric flow rate. In addition, the size and location of the plug zone is accurately quantified. Typically, less than one second of computing time is required for convergence and solution display.

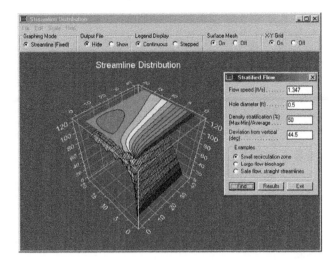

FIGURE 5.11

Small recirculation zone.

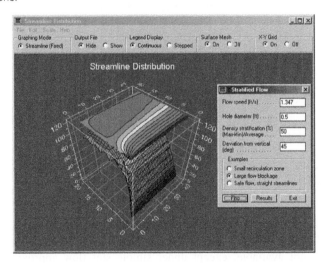

FIGURE 5.12

Large flow blockage.

Mathematical formulation

We consider the concentric annular flow of non-Newtonian Herschel-Bulkley fluids with yield stress, assuming the nomenclature in Figure 5.14. The axial pressure gradient satisfies $dp/dz < 0$ when the velocity $U(r) > 0$ flows to the right. The radial coordinate is "r."

It is not possible to formulate finite difference models because shear rate discontinuities at plug boundaries imply nonexistent derivatives. Thus, we will use less restrictive integral representations

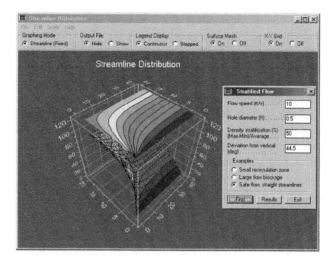

FIGURE 5.13

Safe flow, straight streamlines.

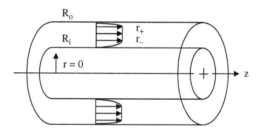

FIGURE 5.14

Concentric annular flow.

of the solution. Fortunately, we are able to develop an exact solution. The detailed equations and constraints are developed in the following discussion.

Axial momentum balance

Let r and z denote radial and axial coordinates in the annulus of Figure 5.14. If $\tau(r)$ and $p(z)$ represent viscous stress and pressure acting on a fluid element, a simple momentum balance requires

$$d(r\ \tau)/dr = -r\ dp/dz \qquad (5.8a)$$

Equation 5.8a can be integrated to give $r\ \tau(r) = -1/2\ dp/dz\ r^2 + C$, where C is an integration constant, so that

$$\tau(r) = -1/2\ dp/dz\ r + C/r \qquad (5.8b)$$

In non-Newtonian flows with yield stress τ_{yield}, a plug flow moving with constant speed is found in regions where $\tau < \tau_{yield}$. For circular *pipe* flow, we set $C = 0$ because shear stresses cannot be infinite along the axis $r = 0$. The circular plug is defined by $0 \leq r \leq r_{plug}$, where the plug radius follows on setting $\tau = \tau_{yield}$ with $r = r_{plug}$ in $\tau(r) = -1/2 \ dp/dz \ r$. It is clear that $r_{plug} = -2\tau_{yield}/(dp/dz) > 0$ always exists, a radius that separates the plug from the shearing flow. For pipe flows, this $C = 0$ requirement renders analysis straightforward.

Now, from Figure 5.14, it is clear that *two* plug radii characterize annular flows, falling between the inner and outer values R_i and R_o. The argument for vanishing C no longer applies, because $r = 0$ does not fall in the radial domain: stress is never infinite. A nonzero C now plays an important role in the analysis and satisfies several physical constraints.

Formulas for plug radii

Note that Equation 5.8b, which states that "$\tau(r) = -1/2 \ dp/dz \ r + C_1/r$," applies to the *outer annulus*, where $dU(r)/dr < 0$ and $\tau > 0$. Now, if we rewrite "$\tau_{yield} = -1/2 \ dp/dz \ r + C_1/r$" as the quadratic equation "$1/2 \ dp/dz \ r^2 + \tau_{yield} \ r - C_1 = 0$," we can determine possible plug radii by solving

$$r_a = \{-\tau_{yield} - \sqrt{(\tau_{yield}^2 + 2C_1 \ dp/dz)}\}/(dp/dz) \tag{5.9a}$$

$$r_b = \{-\tau_{yield} + \sqrt{(\tau_{yield}^2 + 2C_1 \ dp/dz)}\}/(dp/dz) \tag{5.9b}$$

Next, consider the *inner annulus*, where $dU(r)/dr > 0$ and $\tau < 0$, and we now have the formula "$-\tau_{yield} = -1/2 \ dp/dz \ r + C_2/r$." This leads to the quadratic equation "$1/2 \ dp/dz \ r^2 - \tau_{yield} \ r - C_2 = 0$," for which possible plug radii are

$$r_c = \{+\tau_{yield} - \sqrt{(\tau_{yield}^2 + 2C_2 \ dp/dz)}\}/(dp/dz) \tag{5.9c}$$

$$r_d = \{+\tau_{yield} + \sqrt{(\tau_{yield}^2 + 2C_2 \ dp/dz)}\}/(dp/dz) \tag{5.9d}$$

In the zero yield limit $\tau_{yield} \rightarrow 0$, we find that $r_a = -\sqrt{(2C_1 \ dp/dz)}/(dp/dz)$ and $r_b = +\sqrt{(2C_1 \ dp/dz)}/(dp/dz)$, and also that $r_c = -\sqrt{(2C_2 \ dp/dz)}/(dp/dz)$ and $r_d = +\sqrt{(2C_2 \ dp/dz)}/(dp/dz)$. If all the $\sqrt{\ }$'s are positive, then on noting that dp/dz is negative, it follows that $r_a > 0$ and $r_b < 0$, and $r_c > 0$ and $r_d < 0$. Thus, only the positive radii in Equations 5.9a and 5.9c are meaningful. But these formulas must give the same value. This is possible by taking $C_1 = C_2 = C$, where C is now a single unknown. Hence, we write Equations 5.9a and 5.9c as

$$r_+ = \{-\tau_{yield} - \sqrt{(\tau_{yield}^2 + 2C \ dp/dz)}\}/(dp/dz) \tag{5.9e}$$

$$r_- = \{+\tau_{yield} - \sqrt{(\tau_{yield}^2 + 2Cdp/dz)}\}/(dp/dz) \tag{5.9f}$$

for the required plug radii, satisfying $r_+ > r_-$ as required in Figure 5.14.

Kinematic constraints

For physical solutions to exist, the discriminant must be non-negative; that is, we seek $\tau_{yield}^2 + 2C \ dp/dz \geq 0$. Since $dp/dz < 0$, negative C values will be valid, but positive values are allowed if

$$C \leq \tau_{yield}^2/(2 \ |dp/dz|) \tag{5.10a}$$

which decreases as yield stress decreases. This provides the upper bound. A lower bound is obtained by considering the limit in which $\tau_{yield} = 0$ vanishes altogether, in which case $0 = +1/2$ $dp/dz\ r_{plug} - C/r_{plug}$ implies that the constant satisfies $C = 1/2\ dp/dz\ r_{plug}^2 < 0$. Thus, C can be negative, as noted above, but it will be at the very least, equal to the negative value $1/2\ dp/dz\ R_o^2$, where R_o is the outer annular radius. Hence, we may write

$$1/2\ dp/dz\ R_o^2 \le C \le \tau_{yield}^2/(2\ |dp/dz|) \tag{5.10b}$$

Furthermore, C must be chosen such that

$$r_-(\tau_{yield}, dp/dz, C) > R_i \tag{5.10c}$$

$$r_+(\tau_{yield}, dp/dz, C) < R_o \tag{5.10d}$$

$$u(r, \tau_{yield}, dp/dz, C) > 0 \tag{5.10e}$$

Dynamic constraints

The physical properties of the annular velocity field are introduced through a suitable constitutive stress-strain relationship. We assume the classical Herschel-Bulkley model with n, K, and τ_{yield} values as shown in Equation 5.11a. For the *outer annulus*, where $dU/dr < 0$, we write

$$\tau = \tau_{yield} + (-K\ dU/dr)^n \tag{5.11a}$$

where "$-dU/dr$" and τ are both positive, so that all of the ( ) brackets are positive. We substitute this into Equation 5.8b—that is, $\tau(r) = -1/2\ dp/dz\ r + C/r$—to obtain

$$\tau_{yield} + (-K\ dU/dr)^n = -1/2\ dp/dz\ r + C/r \tag{5.11b}$$

from which

$$dU(r)/dr = -(1/K)(-1/2\ dp/dz\ r + C/r - \tau_{yield})^{1/n} \tag{5.12a}$$

In order for solutions to exist, the quantity within the brackets must be positive, so that

$$-1/2\ dp/dz\ r + C/r - \tau_{yield} \ge 0 \tag{5.12b}$$

When this constraint is satisfied, Equation 5.12a can be integrated over (r, R_o) to give

$$U(r) = +(1/K) \int_r^{R_o} (-1/2\ dp/dz\ r + C/r - \tau_{yield})^{1/n}\ dr \tag{5.12c}$$

for $r_+ < r < R_o$, where we have used the outer no-slip axial velocity condition $U(R_o) = 0$.

For the *inner annulus*, we require Equation 5.11a in a form suitable for $dU/dr > 0$ and $\tau < 0$. This is achieved by taking "$-\tau = \tau_{yield} + (K\ dU/dr)^n$" so that all of the ( ) brackets are positive. If we substitute "$\tau(r) = -1/2\ dp/dz\ r + C/r$," we obtain

$$dU(r)/dr = +(1/K)(1/2\ dp/dz\ r - C/r - \tau_{yield})^{1/n} \tag{5.13a}$$

for which we require

$$1/2\ dp/dz\ r - C/r - \tau_{yield} \ge 0 \tag{5.13b}$$

If this is satisfied, we integrate over (R_i, r) and apply the no-slip axial velocity condition $U(R_i) = 0$, to obtain

$$U(r) = +(1/K) \int_{R_i}^{r} (1/2 \; dp/dz \; r - C/r - \tau_{yield})^{1/n} dr \tag{5.13c}$$

for $R_i < r < r_-$. Now the plug moves with a constant speed U_{plug} in $r_- < r < r_+$. Its value from Equation 5.12c at $r = r_+$ must equal that using Equation 5.13c at the location $r = r_-$ if there is no slippage kinematically. In other words,

$$\frac{\displaystyle\int_{R_i}^{r_-} (1/2 \; dp/dz \; r - C/r - \tau_{yield})^{1/n} \; dr}{\displaystyle\int_{r_+}^{R_o} (-1/2 \; dp/dz \; r + C/r - \tau_{yield})^{1/n} \; dr} = 1 \tag{5.14}$$

Numerical evaluation of constraints

For a flow (corresponding to given R_i, R_o, n, K, τ_{yield}, and dp/dz) to exist, C must satisfy all of the conditions derived. Solutions for C may not exist, for instance, when pressure gradients cannot overcome fluid yield resistance. For such flows, $U(r)$ vanishes, although the viscous stress may not. Equation 5.14 is very useful. The left side defines a dimensionless function $T(R_i, R_o, n, K, \tau_{yield}, dp/dz; C)$ that increases monotonically as C decreases when all other parameters are fixed. In our iterative solution, we start with the largest C in Equation 5.10b and incrementally decrease C values by one-thousandth of the total C interval.

Our earlier constraints are tested first to reduce computing times, since the simple logic tests required only involve "$<$" and "$>$." Finally, if a value of C^* exists such that $T(C^*) < 1$ and $T(C^* - \Delta C) > 1$, the solution is C^*. This value is used to evaluate Equation 5.12c for the outer annular velocity and Equation 5.13c for the inner annular velocity. The plug velocity is obtained by evaluating Equation 5.12c at r_+ or Equation 5.13c at r_-. Shear rates and viscous stresses are obtained by using the equations for $dU(r)/dr$ and $\tau(r)$. With $U(r)$ available, the total volumetric flow rate Q can be determined from the integral

$$Q = \int_{R_i}^{R_o} U(r) \; 2\pi r \; dr \tag{5.15}$$

For the integrals in Equations 5.12c, 5.13c, and 5.15, the trapezoidal rule was taken. Several simple checks were used. First, numerical solutions for Q were validated against exact solutions available for Newtonian flow. Second, for a "narrow annulus," the maximum speed is always found at the center of the channel, with or without yield stress. Third, with all parameters, particularly dp/dz fixed, the flow rate Q decreases as τ_{yield} increases. We emphasize that although we have used numerical evaluations for our integrals, our solutions are exact from a theoretical perspective, since the integrations can be made as accurate as desired by decreasing integration step size.

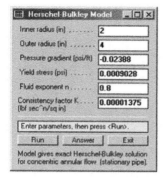

FIGURE 5.15

Exact Herschel-Bulkley concentric model.

Software interface and typical results

The model derived here is called and executed from the interface in Figure 5.15 accessed under the "Utility" menu in our steady flow solver. The complete output file under the assumptions shown is duplicated in Figure 5.16.

Limitations in the derivation

While the Herschel-Bulkley model derived here provides exact and accurate results for all yield stress values, extending the classic Power law and Bingham plastic solutions of Fredrickson and Bird (1958), the concentric flow derivation is not generalizable to eccentric annuli because simple formulas using circular symmetries are not available. This section uses exact relationships to define plug and sheared zones unambiguously and to apply the respective velocity formulas accordingly.

In eccentric problems, the size, shape, and location of the plug zone cannot be determined a priori, and a solution to the flow problem has remained elusive. Authors have typically resorted to slot flow or narrow annulus assumptions, but the limitations of these models are obvious, in particular when real-world effects like high eccentricities, cuttings beds, and washouts are important.

Next, we address the limitations inherent in the standard Herschel-Bulkley model itself and provide a practical, comprehensive, and mathematically rigorous solution for yield stress flows in general eccentric annuli, with or without axial pipe or casing movement and with or without inner body rotation. The solutions developed for yield stress fluids in this book apply to eccentric annuli not only under steady conditions but also when the pipe or casing undergoes general combined transient reciprocation and rotation.

MODEL 5.4

Extended Herschel-Bulkley Flow in Eccentric Annulus with Axially Moving But Nonrotating Drillpipe or Casing

In fluid flows where yield stresses exist, "plug zones" are to be found. These plugs move as solid bodies within the flowing system. For pipes with circular cross sections and for concentric annuli,

```
Herschel-Bulkley (Concentric) Annulus Model:
Exact solution to differential equations ...

INPUT SUMMARY
Inner annular radius  (in):  0.2000E+01
Outer annular radius  (in):  0.4000E+01
Pressure gradient (psi/ft):  -.2388E-01
Fluid exponent n (dimless):  0.8000E+00
Fluid yield stress   (psi):  0.9028E-03
K factor (lbf sec^n/sq in):  0.1375E-04

Plug is between R =   2.5 and   3.4 in.

R =  2.0 in, U = 0.0000E+00 ft/s
R =  2.1 in, U = 0.2076E+01 ft/s
R =  2.2 in, U = 0.3513E+01 ft/s
R =  2.3 in, U = 0.4399E+01 ft/s


R =  3.6 in, U = 0.4396E+01 ft/s
R =  3.7 in, U = 0.3762E+01 ft/s
R =  3.8 in, U = 0.2827E+01 ft/s
R =  3.9 in, U = 0.1577E+01 ft/s
R =  4.0 in, U = 0.0000E+00 ft/s

Volume flow rate BPM: 0.1124E+02
               cuft/s: 0.1052E+01
                  GPM: 0.4719E+03

     Rad (in)  Speed  (ft/s)    0
       4.00    0.0000E+00    *
       3.90    0.1577E+01    |        *
       3.80    0.2827E+01    |          *
       3.70    0.3762E+01    |            *
       3.60    0.4396E+01    |              *
       3.50    0.4750E+01    |               *
       3.40    0.4863E+01    |                *
       3.30    0.4882E+01    |                *
       3.20    0.4882E+01    |                *
       3.10    0.4882E+01    |                *
       3.00    0.4882E+01    |                *
       2.90    0.4882E+01    |                *
       2.80    0.4882E+01    |                *
       2.70    0.4882E+01    |                *
       2.60    0.4882E+01    |                *
       2.50    0.4882E+01    |                *
       2.40    0.4822E+01    |                *
       2.30    0.4399E+01    |              *
       2.20    0.3513E+01    |          *
       2.10    0.2076E+01    |      *
       2.00    0.0000E+00    *
```

FIGURE 5.16

Exact velocity profile result.

we have derived exact analytical solutions for plug zone size and shape assuming Herschel-Bulkley fluids in the previous section. For circular pipes, the cross-sectional plug is simply a circle; for concentric annuli, the plug is a ring.

The appearance of solid plugs within moving streams results from the rheological model used by mathematicians to idealize the physics. Since the shear rate is $\Gamma = [(\partial u/\partial y)^2 + (\partial u/\partial x)^2]^{1/2}$, the usual idealization takes the form

$$N = K \ \Gamma^{n-1} + S_{yield}/\Gamma \ \text{ if } \ \{1/2 \ \text{trace} \ (\underline{\underline{S}} \bullet \underline{\underline{S}})\}^{1/2} > \tau_0$$
$$\underline{\underline{D}} = 0 \ \text{ if } \ \{1/2 \ \text{trace} \ (\underline{\underline{S}} \bullet \underline{\underline{S}})\}^{1/2} < \tau_0 \tag{5.16a}$$

where the general extra stress tensor is denoted $\underline{\underline{S}}$ and the deformation tensor is given by $\underline{\underline{D}}$. Here, τ_0 is the so-called yield stress. The discontinuous "if-then" character behind Equation 5.16a, somewhat artificial, is responsible for the sudden transition from shear flow to plug flow commonly quoted.

As noted, for flows with azimuthal symmetry—that is, circular pipes and concentric annuli—exact, rigorous mathematical solutions are in fact possible. For noncircular ducts and eccentric annuli, which describe a large number of practical engineering problems, it has not been possible to characterize plug zone size and shape, even approximately. Thus, one of the most significant petroleum engineering problems important to both drilling and cementing cannot be modeled at all, let alone accurately. In order to remedy this situation, we observe that the discontinuity offered in Equation 5.16a is really an artificial one, introduced by theorists for, of all reasons, "simplicity."

This unfortunately leads to the solution difficulties noted. In reality, practical engineering flows do not suddenly turn from shear to plug flow; the transition may be rapid, but it will occur continuously over finite measurable distances. Moreover to the author's knowledge, ideal plugs have never been observed experimentally. We therefore turn to more realistic rheological models that apply continuously throughout the *entire* problem domain and that, if the underlying flow parameters permit, lead to plug zones naturally during the computational solution process.

The conventional Herschel-Bulkley viscoplastic model, which includes Bingham plastics as a special limit when the exponent "n" is unity, requires that $\tau = \tau_0 + K(d\gamma/dt)^n$ if $\tau > \tau_0$ and $d\gamma/dt = 0$ otherwise. Here τ is the shear stress, τ_0 is the yield stress, K is the consistency factor, n is the exponent, and $d\gamma/dt$ is the shear rate. As explained, this model is far from perfect. For example, both Herschel-Bulkley and Bingham plastic models predict fictitious infinite viscosities in the limit of vanishing shear rate, a fact that often leads to numerical instabilities. This same infinity also precludes numerical methods, which typically assume that derivatives exist. In addition, the behavior is not compatible with conservation laws that govern many complex flows.

An alternative to the standard Herschel-Bulkley model is the use of continuous functions that apply to sheared regimes and, in addition, through and into the plug zone. One such model is suggested by Souza, Mendez and Dutra (2004): $\tau = \{1 - \exp(-\eta_0\, d\gamma/dt/\tau_0)\}\{\tau_0 + K\,(d\gamma/dt)^n\}$, which applies *everywhere* in the problem domain. Its apparent viscosity function is

$$\eta = \tau/(d\gamma/dt) = \{1-\exp(-\eta_0 d\gamma/dt/\tau_0)\}\{\tau_0/(d\gamma/dt) + K(d\gamma/dt)^{n-1}\} \tag{5.16b}$$

The "apparent viscosity versus shear stress" and "shear stress versus shear rate" diagrams, from Souza et al. (2004), are duplicated in Figure 5.17. What are the physical consequences of this model? Equation 5.16b, in fact, represents an "extended Herschel-Bulkley" model in this sense. For

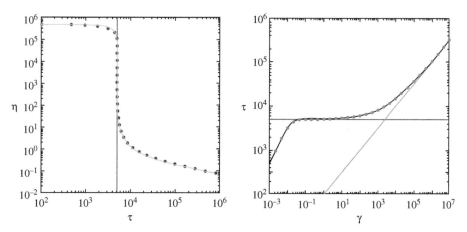

FIGURE 5.17

Extended Herschel-Bulkley law.

infinite shear rates, one would recover $\tau = \tau_0 + K \, (d\gamma/dt)^n$. But for low shear rates, a simple Taylor expansion leads to $\eta \approx \{\eta_0(d\gamma/dt)/\tau_0\}\{\tau_0/(d\gamma/dt) + K \, (d\gamma/dt)^{n-1}\} \approx \eta_0$, where it is clear now that η_0 represents a very high (assumed) viscosity for the plug zone.

The use of Equation 5.16b in numerical algorithms simplifies both formulation and coding, since internal boundaries and plug domains do not need to be determined as part of the solution. A single constitutive law (as opposed to the use of two relationships in Equation 5.16a) applies everywhere, thus simplifying computational logic; moreover, the continuous function assumed also possesses continuous derivatives everywhere and allows the use of standard difference formulas. Cumbersome numerical matching across internal boundaries is completely avoided. In a practical computer program, the plug zone viscosity might be assumed, for example, as 1,000 cp. In fact, we choose high values of η_0 that will additionally stabilize the numerical integration schemes used. This strategy is applied throughout our work, both to our iterative relaxation schemes for steady-state problems and to our transient integration schemes for more complicated formulations.

It is important to recognize that the standard and extended Herschel-Bulkley models here are *not* identical and will not give identical results even in the concentric case, although they will be close. Thus, it is of interest to consider typical numbers. Yield stresses in drilling and cementing applications are often quoted as multiples of "lbf/100 ft^2" or 0.00006944 psi. An order-of-magnitude correct yield might be 0.0001 psi. In Figure 5.18, we compare concentric results obtained from our exact Herschel-Bulkley solver with that produced by the steady, curvilinear grid simulator with the generalized constitutive relation. The exact flow rate is 1,387 gpm, while the finite difference solution gives 1,364 gpm, incurring less than a 2 percent error. If the yield stress is increased to 0.0005 psi, then the exact flow rate is 884.7 gpm, while the approximate value is 933.0 gpm, for a 5.5 percent difference. The difference increases as yield stress increases.

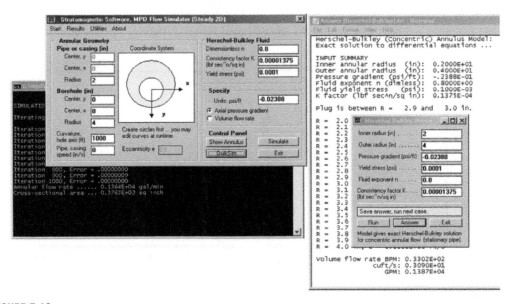

FIGURE 5.18

Comparison for 0.0001-psi yield stress.

In Figure 5.19, we apply our steady, curvilinear grid, finite difference simulator to a water-like Bingham plastic fluid and show that the volumetric flow rate for the parameters that are shown is 1,632 gpm. For this test case, the flow rate is 8,461 gpm without yield stress. Calculations require about one second of computing, including screen display time. The "flat top" profiles associated with yield stress flows appear naturally even for highly eccentric annuli. Again, the steady flow simulator supports constant speed drillpipe or casing movement.

In Figure 5.20, for instance, eccentric annular plug flow solutions are obtained for stationary pipe (left), pipe moving opposite to the flow (middle) and pipe moving with the flow (right). Our use of a generalized Herschel-Bulkley fluid applies to both steady and unsteady formulations. Yield stress applications in transient flow are pursued in Chapter 7.

Finally, we emphasize that the term "exact" refers to exact solutions of the discontinuous model in Equation 5.16a. The model in Equation 5.16b, on the other hand, describes a continuous velocity field with continuous derivatives. The "exact" discontinuous formulation, of course, is less applicable in an engineering sense than our continuous model, since physical properties may not change suddenly within a flow field. In this sense, the value of a yield stress model should not be judged by how consistent it is with an unnatural formulation but by how consistent it is with reality.

MODEL 5.5

Steady Non-Newtonian Flow in Boreholes with Bends

Bends in pipelines and annuli are interesting because they are associated with additional losses; that is, to maintain a prescribed volumetric flow rate, a greater pressure drop is required in ducts with bends than in those without. (This book does *not* deal with "secondary flows" such as rotating vortical eddies attached to solid walls or corners, separated viscous flows, and so on.) This is true because the net fluid stresses that act along duct walls are higher. We will first discuss the problem analytically, in the context of Newtonian flow; in this limit, exact solutions are derived for Poiseuille flow between curved concentric plates, but we will also focus on the form of the new differential equation used.

The closed-form expressions derived for Newtonian flow, which contain the required centrifugal force modifications, are new. Their derivation motivates our methodology for non-Newtonian flows in steady, three-dimensional, curved, closed, simple, and annular ducts, which can only be analyzed computationally. This simulation feature is built into the "Steady 2D" solver. (It does not appear in our "Transient 2D" simulator because of time and budget constraints.) As shown in the user interface of Figure 1.8 at the bottom left, only the radius of curvature needs to be entered in the "Curvature, hole axis (ft)" text input box. For straight ducts, a large value, say 1,000 ft, can be used.

Straight, closed ducts

This book provides analytical solutions for Newtonian flows in *straight* circular ducts (i.e., Eqs. 1.1b and 1.1c) and in rectangular conduits (see Model 5.10), as shown in Figures 5.21 and 5.22. We also developed a general non-Newtonian viscous flow solver applicable to arbitrary cross section but straight eccentric annuli utilizing curvilinear meshes (Figure 5.23). These solutions have been validated in detail in a number of application examples. We now ask, "How are these methodologies extended to handle finite-radius bends along borehole axes?" These extensions, important to modeling borehole curvature in directional wells, are motivated by the parallel plate solutions derived next.

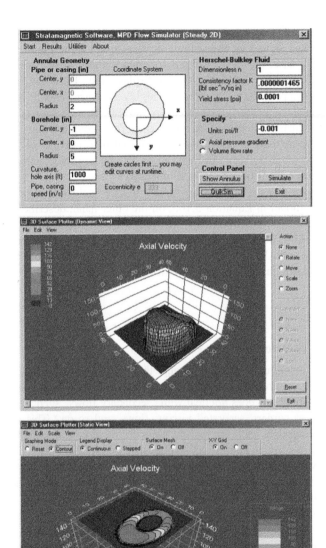

FIGURE 5.19

Bingham plastic run (note large plug zone).

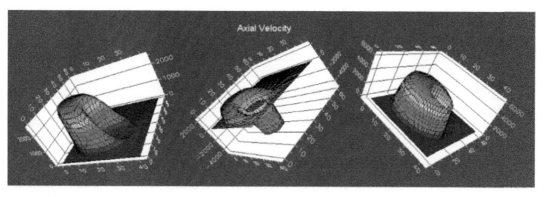

FIGURE 5.20

Non-Newtonian plug velocity profiles with pipe movement.

FIGURE 5.21

Viscous flow in a circular pipe.

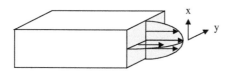

FIGURE 5.22

Viscous flow in a rectangular duct.

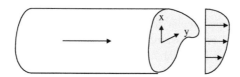

FIGURE 5.23

Viscous flow in a general duct.

FIGURE 5.24

Flow between parallel plates.

Hagen-Poiseuille flow between planes

Let us consider here the plane Poiseuille flow between parallel plates shown in Figure 5.24, but for simplicity, we will restrict ourselves first to Newtonian fluids.

Let y be the coordinate perpendicular to the flow, with $y = 0$ and H representing the walls of a duct of height H. If u(y) represents the velocity, the Navier-Stokes equations reduce to Equation 5.17—that is,

$$d^2u(y)/dy^2 = 1/\mu \ dP/dz \tag{5.17}$$

$$u(0) = u(H) = 0 \tag{5.18}$$

which is solved with the no-slip conditions in Equation 5.18. Again, μ is the Newtonian viscosity and dP/dz is the constant axial pressure gradient. The velocity solution is the well-known parabolic profile

$$u(y) = 1/2(1/\mu \ dP/dz) \ y \ (y - H) \tag{5.19}$$

which yields the volumetric flow rate "Q/L" (per unit length "L" out of the page):

$$Q/L = \int_0^H u(y) \ dy = -(1/\mu \ dP/dz)H^3/12 \tag{5.20}$$

Flow between concentric plates

Now suppose that the upper and lower walls are bent so that they conform to the circumferences of concentric circles with radii "R" and "R + H," where R is the radius of curvature of the smaller circle. We ask, "How are corrections to Equations 5.19 and 5.20 obtained?"

It is instructive to turn to the exact momentum law in the azimuthal "θ" direction used in Model 5.6 for our analysis of rotating concentric flow—that is, Equation 5.49. There, v_θ represents the velocity in the circumferential direction. We now draw upon that azimuthal equation but apply it to the flow between the concentric curved plates shown in Figure 5.25.

Since there is no flow perpendicular to the page, $v_z = 0$; also, $v_r = 0$ because the velocity is directed only tangentially, and F_θ is assumed to be zero. In these coordinates, the flow is steady, and θ and "z" variations vanish identically. Thus, the azimuthal equation reduces to the ordinary differential equation

$$d^2v_\theta/dr^2 + 1/r \ dv_\theta/dr - v_\theta/r^2 = 1/\mu \ \{1/r \ dP/d\theta\} \tag{5.21}$$

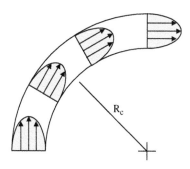

FIGURE 5.25

Opened duct flow between concentric curved plates.

where the right side, containing the axial pressure gradient "$1/r\ dP/d\theta$," is approximately constant. It must be solved together with the no-slip conditions

$$v_\theta(R) = v_\theta(R+H) = 0 \tag{5.22}$$

A closed-form solution can be obtained as

$$v_\theta(r)/[1/\mu\ \{1/r\ dP/d\theta\}] = \{(R+H)^3 - R^3\}/\{R^2 - (R+H)^2\} \times (r/3)$$
$$- R^2(R+H)^2/\{R^2 - (R+H)^2\} \times \{H/(3r)\} + 1/3\ r^2 \tag{5.23}$$

Then the volumetric flow rate "Q/L" per unit length (out of the page) is

$$Q/L = \int_{R}^{R+H} v_\theta(r)\ dr \tag{5.24}$$

$$Q/L = (1/18)(1/\mu\ dP/dz)\{-6R^3H - 9R^2H^2 - 5RH^3 - H^4 + 6R^4\ \ln(R+H)$$
$$+ 12HR^3\ \ln(R+H) + 6H^2R^2\ \ln(R+H) - 6R^4\ \ln(R)$$
$$- 12R^3H\ \ln(R) - 6R^2H^2\ \ln(R)\}/(2R+H) \tag{5.25}$$

where we have replaced "$1/r\ dP/d\theta$" by "dP/dz." Now, in the limit $R \gg H$, Equation 5.25 simplifies to

$$Q/L \approx (1/\mu\ dP/dz)\{-H^3/12 + H^5/(180R^2) + O(1/R^3)\} \tag{5.26}$$

The first term in Equation 5.26 is the result in Equation 5.20—that is, the asymptotic contribution of the straight parallel plate solution. Subsequent terms represent corrections for finite R. In general, Equation 5.25 applies to all R and H combinations without restriction.

Typical calculations
It is interesting to ask, "How does total volumetric flow rate in such a curved "pipe" compare with classical parallel plate theory?" For this purpose, consider the ratio obtained by dividing Equation 5.25 by Equation 5.20. It is plotted in Figure 5.26, where we have set H = 1 and varied R.

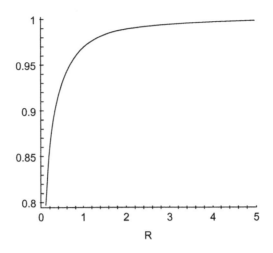

FIGURE 5.26

Volumetric flow rate ratio, with H = 1.

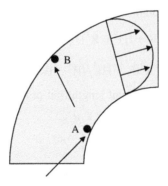

FIGURE 5.27

Particles impinging at duct walls.

   This ratio tends to "1" quickly, when $R > 5$. We also ask, "What is the worst flow rate penalty possible?" If we take $R \to 0$, it can be shown that the ratio approaches 2/3. Thus, for Newtonian flow between concentric plates, the volumetric flow rate is at worst equal to 2/3 of the value obtained between parallel plates for the same H. This assumes that the flow is steady and laminar, with no secondary viscous flow in the cross-sectional plane.

   We also use the velocity solution in Equation 5.23 to study the viscous stresses at the walls of our concentric channel. Consider Figure 5.27, which shows two impinging particles lodged at A and B, which may represent wax, hydrate, cuttings, or other debris. The likelihood that they will dislodge depends on the local viscous stress, among other factors. In this problem, $v_z = v_r = 0$ and $\partial/\partial\theta = \partial/\partial z = 0$, leaving the single stress component $\tau_{r\theta}(r) = \mu \, r \, \partial(v_\theta/r)/\partial r$. In particular, we plot "Stress Ratio" $= - \tau_{r\theta}(R + H)/\tau_{r\theta}(R)$ in Figure 5.28, with $H = 1$ and varying R. The "minus" is

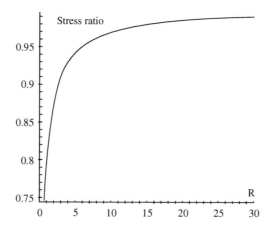

FIGURE 5.28

"Stress Ratio," H = 1.

used to keep the ratio positive, since the signs of the opposing stresses are opposite. The result is shown in Figure 5.28.

This graph shows that stresses at the outer wall are less whenever axis curvatures are finite. Thus, with all parameters equal, there is less likelihood that B will dislodge more quickly than A. The velocity and stress solutions obtained here are also useful in determining how and where debris settles within the duct. Numerous factors enter, of course, among them particle size, shape, and distribution; buoyancy effects; local velocities and gradients; and so on. Such studies follow lines established in the sedimentary transport literature.

Flows in closed curved ducts

Our analysis shows that corrections for bends along the axis can be obtained by solving $v_\theta(r)$ in cylindrical coordinates. It is apparent that the extension of Equations 5.25 and 5.26 to cover *closed* rectangular ducts (versus "opened" concentric plates) with finite radius of curvature (e.g., Figure 5.22) only requires the solution of Equation 5.21 *with* the "$\partial^2 v_\theta/\partial z^2$" term in the earlier azimuthal equation, leading to

$$\partial^2 v_\theta/\partial r^2 + 1/r \,\partial v_\theta/\partial r - v_\theta/r^2 + \partial^2 v_\theta/\partial z^2 = 1/\mu \,\partial P/\partial z_{axial} \tag{5.27}$$

where the notation "$\partial P/\partial z_{axial}$" for axial pressure gradient replaces the "$\partial P/\partial z$" used previously, noting that the "z" in the $v_\theta(r,z)$ of Equation 5.27 is now perpendicular to the page. For the duct in Figure 5.22, the no-slip conditions are $v_\theta(R,z) = v_\theta(R + H,z) = 0$ and $v_\theta(r,z_1) = v_\theta(r,z_2) = 0$, where $z = z_1$ and z_2 are end planes parallel to the page.

With our extension to rectangular geometries clear, the passage to *bent ducts with arbitrary closed cross sections* (e.g., Figure 5.29) is obtained by taking Equation 5.27 again, but with no-slip conditions applied along the perimeter of the shaded duct area or annular domain. Ducts with multiple bends are studied by combining multiple ducts with piecewise constant radii of curvature. Since the total flow rate is fixed, each section will be characterized by different axial pressure gradients.

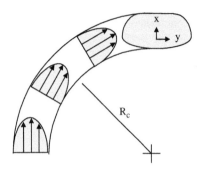

FIGURE 5.29

Arbitrary closed duct with curved axis.

Of course, Equation 5.27 is quite different from the original equation for straight flows—that is, from "$\partial^2 u/\partial y^2 + \partial^2 u/\partial x^2 \approx N(\Gamma)^{-1} \partial P/\partial z + \ldots$" To use the previous algorithm, we rewrite Equation 5.27 in the form

$$\partial^2 v_\theta/\partial r^2 + \partial^2 v_\theta/\partial z^2 = 1/\mu \ \partial P/\partial z_{axial} - 1/R \ \partial v_\theta/\partial r + v_\theta/R^2 \tag{5.28}$$

where we have transferred the new terms to the right side and replaced the variable "r" coefficients with constants, assuming $R >> H$ so that $r \approx R$.

The "r, z" in Equation 5.28 are just the "y, x" cross-sectional variables used earlier. In our iterative solution for this Newtonian flow, the right-side velocity terms of Equation 5.28 are evaluated using latest values, with the relaxation method continuing until convergence.

For non-Newtonian flows, a similar procedure applies; that is, μ is replaced by the apparent viscosity $N(\Gamma)$. To demonstrate the basic ideas, consider the general θ momentum equation

$$\rho(\partial v_\theta/\partial t + v_r \ \partial v_\theta/\partial r + v_\theta/r \ \partial v_\theta/\partial\theta + v_\theta v_r/r + v_z \ \partial v_\theta/\partial z)$$
$$= -1/r \ \partial P/\partial\theta + 1/r^2 \ \partial(r^2 S_{\theta r})/\partial r + 1/r \ \partial S_{\theta\theta}/\partial\theta + \partial S_{\theta z}/\partial z + \text{body forces} \tag{5.29}$$

If $\partial/\partial t = \partial/\partial\theta = v_r = v_z = 0$ and body forces vanish, then

$$\partial P/\partial z_{axial} \approx 1/r \ \partial P/\partial\theta = 1/r^2 \ \partial(r^2 S_{\theta r})/\partial r + 1/r \ \partial S_{\theta\theta}/\partial\theta + \partial S_{\theta z}/\partial z \tag{5.30}$$

If we now substitute $\underline{S} = 2 \ N(\Gamma) \ \underline{D}$ from Equation 2.39, we obtain

$$\partial^2 v_\theta/\partial r^2 + \partial^2 v_\theta/\partial z^2 + 1/r \ \partial v_\theta/\partial r - v_\theta/r^2$$
$$= 1/N(\Gamma) \ \partial P/\partial z_{axial} + \ldots \tag{5.31}$$

where "$+\ldots$" represents terms containing derivatives of $N(\Gamma)$. These are completely retained in our steady flow solution process but, for brevity, are not written out. The revised partial differential equation, of course, applies whether or not the inner pipe moves. The constant positive, zero, or negative speed is entered at the bottom left of the "Steady 2D" user interface in Figure 1.8.

These changes are easily implemented in software. For example, our straight-duct "line relaxation" Fortran source code previously included the lines that incorporate "$N(\Gamma)^{-1} \partial P/\partial z$," where the other terms shown are related to the Thompson mapping. To introduce borehole curvature, the term shown in bold is simply replaced as indicated by comparing Figures 5.30 and 5.31.

```
WW(J)  =  -ALPHA(I,J)*(U(I-1,J)+U(I+1,J))/DPSI2
1            +GAKOB(I,J)*GAKOB(I,J)* PGRAD/APPVIS(I,J)
2            +2.0*BETA(I,J)*
3            (U(I+1,J+1)-U(I-1,J+1)-U(I+1,J-1)+U(I-1,J-1))/
4            (4.*DPSI*DETA)
```

FIGURE 5.30

Original straight-duct Fortran source code.

```
       CHANGE = PGRAD/APPVIS(I,J)
1            -(YETA(I,J)*(U(I+1,J)-U(I-1,J))/(2.*DPSI)
2            - YPSI(I,J)*(U(I,J+1)-U(I,J-1))/(2.*DETA))/
3            (GAKOB(I,J)*RCURV)
4            + U(I,J)/(RCURV**2.)
C
       WW(J)  = -ALPHA(I,J)*(U(I-1,J)+U(I+1,J))/DPSI2
1            +GAKOB(I,J)*GAKOB(I,J)*CHANGE
2            +2.0*BETA(I,J)*
3            (U(I+1,J+1)-U(I-1,J+1)-U(I+1,J-1)+U(I-1,J-1))/
4            (4.*DPSI*DETA)
```

FIGURE 5.31

Modified Fortran source code.

The second and third lines of our Fortran source code for "CHANGE" represent the "r" velocity derivative in transformed coordinates. Newtonian calculations similar to those performed for "concentric plate Poiseuille flow" show that, when pressure gradient is prescribed, volumetric flow rate again decreases as the radius of curvature R_c tends to zero. For a circular cross section of radius R, the decrease is roughly 20 percent relative to Hagen-Poiseuille flow when R_c and R are comparable. We have focused on Newtonian flows because exact solutions were available and, importantly, our results applied to all viscosities and pressure gradients. However, results will vary for pipelines with noncircular cross sections, non-Newtonian flow, or both; general conclusions, of course, cannot be offered, but computations can now be easily performed with the numerically stable implementation just derived.

MODEL 5.6

Newtonian and Power Law Flow in Concentric Annulus with Rotating (But Axially Stationary) Pipe or Casing

Analytical solutions for nonlinearly coupled axial and circumferential velocities, their deformation, stress, and pressure fields, are obtained for concentric annular flow in an inclined borehole with a centered, rotating drillstring or casing. The closed-form solutions are used to derive formulas for volumetric flow rate, maximum borehole wall stress, apparent viscosity, and other quantities as functions of "r." The analysis is restricted to Newtonian and Power law fluids. Our Newtonian results are exact solutions to the viscous Navier-Stokes equations without geometric approximation.

For Power law fluids, the analytical results reduce to the Newtonian solutions in the "n = 1" limit. All solutions satisfy no-slip viscous boundary conditions at both the rotating drillstring and the borehole wall. Our pipe is assumed to be stationary axially. The formulas are explicit; they require no iteration and are easily programmed on calculators and computers. Extensive analytical and calculated results are given, which elucidate the physical differences between the two fluid types.

General governing equations

The equations governing general fluid motion are available from many excellent textbooks on continuum mechanics (Schlichting, 1968; Slattery, 1981). We will cite these equations without proof. Let v_r, v_θ, and v_z denote Eulerian fluid velocities, and F_r, F_θ, and F_z the body forces, in the r, θ, and z directions, respectively. Here (r, θ, z) are standard circular cylindrical coordinates.

Also, let ρ be the *constant* fluid density and p be the pressure, and denote by S_{rr}, $S_{r\theta}$, $S_{\theta\theta}$, S_{rz}, $S_{\theta r}$, $S_{\theta z}$, S_{zr}, $S_{z\theta}$, and S_{zz} the nine elements of the general extra stress tensor $\underline{\underline{S}}$. If t is time and ∂'s represent partial derivatives, the complete equations obtained from Newton's law and mass conservation are

Momentum equation in r:

$$\rho(\partial v_r/\partial t + v_r \partial v_r/\partial r + v_\theta/r\, \partial v_r/\partial\theta - v_\theta{}^2/r + v_z\, \partial v_r/\partial z)$$
$$= F_r - \partial p/\partial r + 1/r\, \partial(rS_{rr})/\partial r + 1/r\, \partial(S_{r\theta})/\partial\theta + \partial(S_{rz})/\partial z - S_{\theta\theta}/r \tag{5.32}$$

Momentum equation in θ:

$$\rho(\partial v_\theta/\partial t + v_r\, \partial v_\theta/\partial r + v_\theta/r\, \partial v_\theta/\partial\theta + v_r v_\theta/r + v_z\, \partial v_\theta/\partial z)$$
$$= F_\theta - 1/r\, \partial p/\partial\theta + 1/r^2\, \partial(r^2 S_{\theta r})/\partial r + 1/r\, \partial(S_{\theta\theta})/\partial\theta + \partial(S_{\theta z})/\partial z \tag{5.33}$$

Momentum equation in z:

$$\rho(\partial v_z/\partial t + v_r\, \partial v_z/\partial r + v_\theta/r\, \partial v_z/\partial\theta + v_z \partial v_z/\partial z)$$
$$= F_z - \partial p/\partial z + 1/r\, \partial(rS_{zr})/\partial r + 1/r\, \partial(S_{z\theta})/\partial\theta + \partial(S_{zz})/\partial z \tag{5.34}$$

Mass continuity equation:

$$1/r\, \partial(rv_r)/\partial r + 1/r\, \partial v_\theta/\partial\theta + \partial v_z/\partial z = 0 \tag{5.35}$$

These equations apply to all Newtonian and non-Newtonian fluids. In continuum mechanics, the most common class of empirical models for isotropic, incompressible fluids assumes that $\underline{\underline{S}}$ can be related to the rate of deformation tensor $\underline{\underline{D}}$ by a relationship of the form

$$\underline{\underline{S}} = 2N(\Gamma)\underline{\underline{D}} \tag{5.36}$$

where the elements of $\underline{\underline{D}}$ are

$$D_{rr} = \partial v_r/\partial r \tag{5.37}$$

$$D_{\theta\theta} = 1/r\, \partial v_\theta/\partial\theta + v_r/r \tag{5.38}$$

$$D_{zz} = \partial v_z/\partial z \tag{5.39}$$

$$D_{r\theta} = D_{\theta r} = [r\, \partial(v_\theta/r)/\partial r + 1/r\, \partial v_r/\partial\theta]/2 \tag{5.40}$$

$$D_{rz} = D_{zr} = [\partial v_r/\partial z + \partial v_z/\partial r]/2 \tag{5.41}$$

$$D_{\theta z} = D_{z\theta} = [\partial v_\theta/\partial z + 1/r\, \partial v_z/\partial\theta]/2 \tag{5.42}$$

In Equation 5.36, $N(\Gamma)$ is the "apparent viscosity function" satisfying

$$N(\Gamma) > 0 \tag{5.43}$$

$\Gamma(r, \theta, z)$ being the scalar functional of v_r, v_θ, and v_z defined by the tensor operation

$$\Gamma = \{2 \text{ trace } (\underline{\mathbf{D}} \bullet \underline{\mathbf{D}})\}^{1/2} \tag{5.44}$$

These considerations are still very general. Let us examine an important and practical simplification. The Ostwald-de Waele model for two-parameter "Power law fluids" assumes that the apparent viscosity satisfies

$$N(\Gamma) = K \, \Gamma^{n-1} \tag{5.45}$$

where the exponent "n" and consistency factor "K" are constants. Power law fluids are "pseudoplastic" when $0 < n < 1$; Newtonian when $n = 1$, and "dilatant" when $n > 1$. Most drilling fluids are pseudoplastic. In the limit, taking "$n = 1$, $K = \mu$," Equation 5.45 reduces to a Newtonian model with $N(\Gamma) = \mu$, where μ is the laminar viscosity; here stress is linearly proportional to shear rate.

On the other hand, when n and K take on general values, the apparent viscosity function becomes somewhat complicated. For isotropic, rotating flows without velocity dependence on the azimuthal coordinate θ, the function Γ in Equation 5.45 takes the form

$$\Gamma = [(\partial v_z/\partial r)^2 + r^2(\partial\{v_\theta/r\}/\partial r)^2]^{1/2} \tag{5.46}$$

as we will show, so that Equation 5.45 becomes

$$N(\Gamma) = K[(\partial v_z/\partial r)^2 + r^2(\partial\{v_\theta/r\}/\partial r)^2]^{(n-1)/2} \tag{5.47}$$

This apparent viscosity reduces to the conventional $N(\Gamma) = K \, (\partial v_z/\partial r)^{(n-1)}$ for "axial only" flows without rotation; and to $N(\Gamma) = K \, (r\partial\{v_\theta/r\}/\partial r)^{(n-1)}$ for "rotation only" viscometer flows without axial velocity. When both axial and circumferential velocities are present, as in annular flows with drillstring rotation, neither of these simplifications applies. This leads to mathematical difficulty. Even though "v_θ (max)" is known from the rotational rate, the magnitude of the nondimensional "v_θ (max)/v_z(max)" ratio cannot be accurately estimated because v_z is highly sensitive to n, K, rotational rate, and pressure drop. Thus, it is impossible to determine beforehand whether or not rotational effects will be weak; simple "axial flow only" formulas cannot be used a priori.

Our result for Newtonian flow, an *exact* solution to the Navier-Stokes equations, is considered first, *without* geometric approximation. Then a closed form analytical solution for pseudoplastic and dilatant Power law fluids is developed for more general n's; we will derive results for rotating flows using Equation 5.47 in its entirety, which lead to useful formulas that can be evaluated explicitly without iteration. Because the mathematical manipulations are complicated, the Newtonian limit is examined first to gain insight into the general case. This is instructive because it allows us to highlight the physical differences between Newtonian and Power law flows.

The annular geometry is shown in Figure 5.32. A drillpipe (or casing) and borehole combination is inclined at an angle α relative to the ground, with $\alpha = 0°$ for horizontal wells and $\alpha = 90°$ for vertical wells. "Z" denotes any point within the drillpipe or annular fluid; section AA is cut normal to the local z axis. What is shown in Figure 5.33 resolves the vertical body force at Z, due to

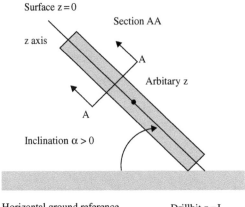

Surface z = 0

Section AA

z axis

A

Arbitary z

A

Inclination α > 0

Horizontal ground reference   Drillbit z = L

FIGURE 5.32

Borehole configuration.

gravity, into components parallel and perpendicular to the axis. Figure 5.34 further breaks the latter into vectors in the radial and azimuthal directions of the cylindrical coordinate system at section AA. Physical assumptions about the drillstring and borehole flow in these coordinates are developed next. Their engineering and mathematical consistency will be evaluated, and application formulas and detailed calculations will be given.

Exact Newtonian flow solution

For Newtonian flows, the stress is linearly proportional to shear rate; the proportionality constant is the laminar viscosity μ. We assume for simplicity that μ is constant (temperature or pressure dependencies would complicate the solution by coupling additional energy balance and material equations). Thus, Equations 5.32 through 5.34 become

Momentum equation in r:

$$\rho\{\partial v_r/\partial t + v_r\, \partial v_r/\partial r + v_\theta/r\, \partial v_r/\partial\theta - v_\theta^2/r + v_z \partial v_r/\partial z\} = F_r - \partial p/\partial r$$
$$+ \mu\{\partial^2 v_r/\partial r^2 + 1/r\, \partial v_r/\partial r - v_r/r^2 + 1/r^2\, \partial^2 v_r/\partial\theta^2 - 2/r^2\, \partial v_\theta/\partial\theta + \partial^2 v_r/\partial z^2\} \qquad (5.48)$$

Momentum equation in θ:

$$\rho\{\partial v_\theta/\partial t + v_r\, \partial v_\theta/\partial r + v_\theta/r\, \partial v_\theta/\partial\theta + v_r v_\theta/r + v_z\, \partial v_\theta/\partial z\} = F_\theta - 1/r\, \partial p/\partial\theta$$
$$+ \mu\{\partial^2 v_\theta/\partial r^2 + 1/r\, \partial v_\theta/\partial r - v_\theta/r^2 + 1/r^2\, \partial^2 v_\theta/\partial\theta^2 + 2/r^2\, \partial v_r/\partial\theta + \partial^2 v_\theta/\partial z^2\} \qquad (5.49)$$

Momentum equation in z:

$$\rho\{\partial v_z/\partial t + v_r\, \partial v_z/\partial r + v_\theta/r\, \partial v_z/\partial\theta + v_z\, \partial v_z/\partial z\} = F_z - \partial p/\partial z$$
$$+ \mu\{\partial^2 v_z/\partial r^2 + 1/r\, \partial v_z/\partial r + 1/r^2\, \partial^2 v_z/\partial\theta^2 + \partial^2 v_z/\partial z^2\} \qquad (5.50)$$

In this section, it is convenient to rewrite Equation 5.35 in the expanded form.

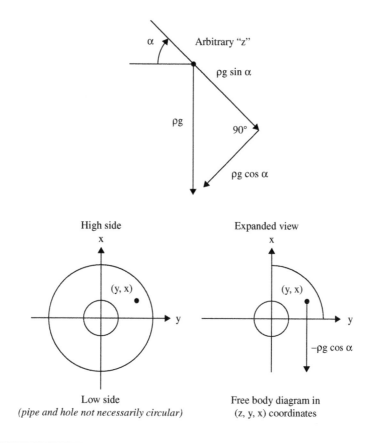

FIGURE 5.33

Gravity vector components.

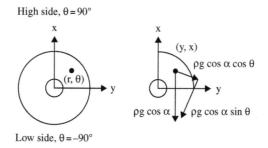

FIGURE 5.34

Free body diagram, gravity in (r, θ, z) coordinates.

Mass continuity equation:

$$\partial v_r/\partial r + v_r/r + 1/r\ \partial v_\theta/\partial\theta + \partial v_z/\partial z = 0 \tag{5.51}$$

Now consider the free body diagrams in Figures 5.32 through 5.34. Figure 5.32 shows a straight borehole with a centered, rotating drillstring inclined at an angle α relative to the ground. Figure 5.33, referring to this geometry, resolves the gravity vector **g** into components parallel and perpendicular to the hole axis. Figure 5.34 applies to the circular cross section AA in Figure 5.32 and introduces local cylindrical coordinates (r, θ). The "low side, $\theta = -90°$" marks the position where cuttings beds would normally form. The force $\rho g \cos\alpha$ of Figure 5.33 is resolved into orthogonal components $\rho g \cos\alpha \sin\theta$ and $\rho g \cos\alpha \cos\theta$.

Physical assumptions about the flow are now given. First, it is expected that at any section AA along the borehole axis z, the velocity fields will appear to be the same; they are invariant, so z derivatives of v_r, v_θ, and v_z vanish. Also, since the drillpipe and borehole walls are assumed to be impermeable, $v_r = 0$ throughout. (In formation invasion modeling, this would not apply.) While we do have pipe rotation, the use of circular cylindrical coordinates (with constant v_θ at the drillstring) renders the mathematical formulation steady. Thus, all time derivatives vanish. These assumptions imply that

$$-\rho v_\theta^2/r = F_r - \partial p/\partial r - 2\mu/r^2\ \partial v_\theta/\partial\theta \tag{5.52}$$

$$\rho v_\theta/r\ \partial v_\theta/\partial\theta = F_\theta - 1/r\ \partial p/\partial\theta$$
$$+ \mu\{\partial^2 v_\theta/\partial r^2 + 1/r\ \partial v_\theta/\partial r - v_\theta/r^2 + 1/r^2\ \partial^2 v_\theta/\partial\theta^2\} \tag{5.53}$$

$$\rho v_\theta/r\ \partial v_z/\partial\theta = F_z - \partial p/\partial z$$
$$+ \mu\{\partial^2 v_z/\partial r^2 + 1/r\ \partial v_z/\partial r + 1/r^2\ \partial^2 v_z/\partial\theta^2\} \tag{5.54}$$

$$\partial v_\theta/\partial\theta = 0 \tag{5.55}$$

Equation 5.55 is useful in further simplifying Equations 5.52 through 5.54. We straightforwardly obtain

$$\rho v_\theta^2/r = \rho g \cos\alpha \sin\theta + \partial p/\partial r \tag{5.56}$$

$$0 = -\rho g \cos\alpha \cos\theta - 1/r\ \partial p/\partial\theta + \mu\{\partial^2 v_\theta/\partial r^2 + 1/r\ \partial v_\theta/\partial r - v_\theta/r^2\} \tag{5.57}$$

$$\rho v_\theta/r\ \partial v_z/\partial\theta = \rho g \sin\alpha - \partial p/\partial z + \mu\{\partial^2 v_z/\partial r^2$$
$$+ 1/r\ \partial v_z/\partial r + 1/r^2\ \partial^2 v_z/\partial\theta^2\} \tag{5.58}$$

where we have substituted the body force components of Figures 5.33 and 5.34. Now, since Equation 5.58 does not explicitly contain θ, it follows that v_z is independent of θ. Since we already showed that there is no z dependence, we find that $v_z = v_z (r)$ is a function of r only. Equation 5.58 therefore becomes

$$0 = \rho g \sin\alpha - \partial p/\partial z + \mu\{\partial^2 v_z/\partial r^2 + 1/r\ \partial v_z/\partial r\} \tag{5.59}$$

To achieve further simplicity, we resolve (without loss of generality) the pressure p(r, θ, z) into its component dynamic pressures P(z) and P*(r), and its hydrostatic contribution, through the separation of variables:

$$p(r, \theta, z) = P(z) + P^*(r) + z\rho g \sin\alpha - r\,\rho g\,\cos\alpha\,\sin\theta \tag{5.60}$$

This reduces the governing Navier-Stokes equations to the simpler but mathematically equivalent system

$$\partial^2 v_z/\partial r^2 + 1/r\,\partial v_z/\partial r = 1/\mu\,dP(z)/dz = \text{constant} \tag{5.61}$$

$$\partial^2 v_\theta/\partial r^2 + 1/r\,\partial v_\theta/\partial r - v_\theta/r^2 = 0 \tag{5.62}$$

$$\rho\,v_\theta^2/r = dP^*(r)/dr \tag{5.63}$$

The separation of variables introduced in Equation 5.60 and the explicit elimination of "g" in Equations 5.61 through 5.63 do not mean that gravity is unimportant; the effects of gravity are simply tracked in the dP(z)/dz term of Equation 5.61. The function P*(r) will depend on the velocity solution to be obtained. Equations 5.61 through 5.63 are also significant in another respect. The velocity fields $v_z(r)$ and $v_\theta(r)$ can be obtained *independently* of each other, despite the nonlinearity of the Newtonian Navier-Stokes equations, because Equations 5.61 and 5.62 physically uncouple. This decoupling occurs because the nonlinear convective terms in the original momentum equations identically vanish. Equation 5.63 is only applied (after the fact) to calculate the radial pressure field P*(r) for use in Equation 5.60. This decoupling applies only to Newtonian flows. For non-Newtonian flows, $v_z(r)$ and $v_\theta(r)$ are strongly coupled mathematically, and different solution strategies are needed.

This degeneracy with Newtonian flows means that their physical properties will be completely different from those of Power law fluids. For Newtonian flows, changes in the rotational rate will not affect properties in the axial direction, in contrast to non-Newtonian flows. Cuttings transport recommendations deduced, for example, using water as the working medium cannot be extrapolated to general drilling fluids having fractional values of n, using any form of dimensional analysis.

Similarly, observations for Power law fluids need not apply to water. This uncoupling was apparently first observed by Savins and Wallick (1966), and the author is indebted to J. Savins for bringing this earlier result to his attention. Savins and Wallick noted that in Newtonian flows, no coupling among the discharge rate, axial pressure gradient, relative motion, and torque through viscosity exists. But we emphasize that the coupling between v_z and v_θ reappears in eccentric geometries even for Newtonian flows.

Because Equations 5.61 and 5.62 are linear, it is possible to solve for the complete flow field using exact classical methods. We will give all required solutions without proof, since they can be verified by direct substitution. For the *inside* of the drillpipe, the axial flow solution to Equation 5.61 satisfying no-slip conditions at the pipe radius r = R_P and zero shear stress at the centerline defined by r = 0 is

$$v_z(r) = (r^2 - R_P^2)/4\mu\,dP(z)/dz \tag{5.64}$$

The rotating flow solution to Equation 5.62 satisfying bounded flow at $r = 0$ and $v_\theta/r = \omega$ at $r = R_P$ is

$$v_\theta = \omega r \qquad (5.65)$$

This is just the expected equation for solid body rotation. Here, "ω" is a constant drillstring rotational rate. These velocity results, again, can be linearly superposed despite the nonlinearity of the underlying equations.

Now let L denote the length of the pipe, P_{mp} denote the constant pressure at the "mud pump" $z = 0$, and P^- denote the drillpipe pressure at $z = L$ just upstream of the bit nozzles. Direct integration of Equation 5.63 and substitution in Equation 5.60 yield the complementary solution for pressure:

$$\begin{aligned} p(r, \theta, z) = P_{mp} + (P^- - P_{mp})\, z/L + \rho\omega^2 r^2/2 \\ + \rho g(z \sin \alpha - r \cos \alpha \sin \theta) + \text{constant} \end{aligned} \qquad (5.66)$$

For the annular region between the rotating drillstring and the stationary borehole wall, the solution of Equation 5.61 satisfying no-slip conditions at the pipe radius $r = R_P$ and at the borehole radius $r = R_B$ is

$$v_z(r) = \{r^2 - R_P{}^2 + (R_B{}^2 - R_P{}^2)(\log\ r/R_P)/\log\ R_P/R_B\}1/4\mu\ dP(z)/dz \qquad (5.67)$$

where "log" denotes the natural logarithm. The solution to Equation 5.62 satisfying $v_\theta = 0$ at $r = R_B$ and $v_\theta = \omega r$ at $r = R_P$ is

$$v_\theta(r) = \omega R_P(R_B/r - r/R_B)/(R_B/R_P - R_P/R_B) \qquad (5.68)$$

Now let P^+ be the pressure at $z = L$ just outside the bit nozzles, and P_{ex} be the surface exit pressure at $z = 0$. The solution for pressure from Equation 5.63 is

$$\begin{aligned} p(r, \theta, z) = P^+ + (P_{ex} - P^+)(L - z)/L + \rho g(z \sin \alpha - r \cos \alpha \sin \theta) \\ + \rho\omega^2 R_P{}^2\{-1/2(R_B/r)^2 + 1/2(r/R_B)^2 - 2\ \log\ (r/R_B) \\ + \text{constant}\}/(R_B/R_P - R_P/R_B)^2 \end{aligned} \qquad (5.69)$$

Observe that the pressure $p(r, \theta, z)$ depends on all three coordinates, even though $v_z(r)$ depends only on r. The pressure gradient $\partial p/\partial r$, for example, throws cuttings through centrifugal force; it likewise depends on r, θ, and z and on ρ, g, and α. It may be an important correlation parameter in cuttings transport and bed formation studies. The additive constants in Equations 5.66 and 5.69 have no dynamical significance. Equations 5.64 through 5.69 describe completely and *exactly* the internal drillpipe flow and the external annular borehole flow. No geometrical simplifications have been made. The solution applies to an inclined, centered drillstring rotating at a constant angular rate ω, but it is restricted to a Newtonian fluid.

Again, these concentric solutions show that in the Newtonian limit, the velocities $v_z(r)$ and $v_\theta(r)$ uncouple; this is not the case for eccentric flows. And this is never so with non-Newtonian drilling flows, whether concentric or eccentric. Thus, the analysis methods developed here must be extended to account for the physical coupling.

Non-Newtonian Power law solution

For general non-Newtonian flows, the Navier-Stokes equations (see Equations 5.48 through 5.50) do not apply; direct recourse to Equations 5.32 through 5.34 must be made. However, many of the physical assumptions used and justified above still hold. If we again assume a constant density flow and assume that velocities do not vary with z, θ, and t and that $v_r = 0$, we again obtain our Equation 5.55. This implies mass conservation. It leads to further simplifications in Equations 5.32 through 5.34, and in the tensor definitions given by Equations 5.36 through 5.45. The result is the reduced system of equations:

$$0 = \rho g \sin \alpha - \partial p/\partial z + 1/r \; \partial(Nr \; \partial v_z/\partial r)/\partial r \tag{5.70}$$

$$0 = -\rho g \cos \alpha \cos \theta - 1/r \; \partial p/\partial \theta + 1/r^2 \; \partial(Nr^3 \; \partial(v_\theta/r) \; \partial r)/\partial r \tag{5.71}$$

$$-\rho v_\theta^2/r = -\rho g \cos \alpha \sin \theta - \partial p/\partial r \\ + 1/r \; \partial(Nr \; \partial(v_\theta/r)/\partial r)/\partial \theta + \partial(N \; \partial v_z/\partial r)/\partial z \tag{5.72}$$

At this point, we introduce the same separation of variables for pressure used for Newtonian flows, Equation 5.60, so that Equations 5.70 through 5.72 become

$$0 = -\partial P/\partial z + 1/r \; \partial(Nr \; \partial v_z/\partial r)/\partial r \tag{5.73}$$

$$0 = \partial(Nr^3 \; \partial(v_\theta/r)/\partial r)/\partial r \tag{5.74}$$

$$-\rho v_\theta^2/r = -\partial P^*/\partial r + 1/r \; \partial(Nr \; \partial(v_\theta/r)/\partial r)/\partial \theta + \partial(N \; \partial v_z/\partial r)/\partial z \tag{5.75}$$

Of course, the P*(r) applicable to non-Newtonian flows will follow from the solution to Equation 5.75; Equations 5.66 and 5.69 for Newtonian flows do not apply. Since θ does not explicitly appear in Equation 5.75, v_z and v_θ do not depend on θ or on z either, as previously assumed. Thus, all partial derivatives with respect to θ and z vanish. Without approximation, the final set of *ordinary* differential equations takes the form

$$1/r \; d(Nr \; dv_z/dr)/dr = dP/dz = \text{constant} \tag{5.76}$$

$$d(Nr^3 d(v_\theta/r)dr)/dr = 0 \tag{5.77}$$

$$dP^*/dr = \rho v_\theta^2/r \tag{5.78}$$

where $N(\Gamma)$ is the complete velocity functional given in Equation 5.46. The application of Equation 5.47 couples our axial and azimuthal velocities, and is the source of mathematical complication.

The solutions to Equations 5.76 through 5.78 may appear to be simple. For example, the unknowns v_θ and v_z are governed by two second-order ordinary differential equations—namely, Equations 5.76 and 5.77; the four constants of integration are completely determined by four no-slip conditions at the rotating drillstring surface and the stationary borehole wall. Moreover, the radial pressure (governed by Equation 5.78) is obtained after the fact only, once v_θ is available.

In reality, the difficulty lies with the fact that Equations 5.76 and 5.77 are nonlinearly coupled through Equation 5.47. It is not possible to solve for either v_z or v_θ sequentially, as we did for the "simpler" Navier-Stokes equations. Because the actual physical coupling is strong at the leading order, it is incorrect to solve for non-Newtonian effects using perturbation series methods—say, expanded about decoupled Newtonian solutions. The method described here required tedious trial and error; 24 ways to implement no-slip conditions were possible, and not all yielded equations that can be integrated.

We successfully derived closed-form, explicit, analytical solutions for the coupled velocity fields. However, the desire for closed-form solutions required an additional "narrow annulus" assumption. Still, the resulting solutions are useful, since they yield explicit answers for rotating flows, thus providing key physical insight into the role of different flow parameters.

The method devised for arbitrary n that follows does *not* apply to the Newtonian limit where $n = 1$, for which solutions are already available. But in the $n \to 1 \pm$ limit of our Power law results, we will show that we recover the Navier-Stokes solution. Thus, the physical dependence on n is continuous, and the results obtained in this chapter cover all values of n. With these preliminary remarks said and done, we proceed with the analysis.

Let us multiply Equation 5.76 by r throughout. Next we integrate the result and also integrate Equation 5.77 once with respect to r, to yield

$$Nr \ dv_z/dr = r^2/2 \ dP/dz + E_1 \tag{5.79}$$

$$Nr^3 \ d(v_\theta/r)dr = E_2 \tag{5.80}$$

where E_1 and E_2 are integration constants. Division of Equation 5.79 by Equation 5.80 gives a result (independent of the apparent viscosity $N(\Gamma)$) relating v_z to v_θ/r:

$$dv_z/dr = (r^4/2 \ dP/dz + E_1 r^2)/E_2 \ d(v_\theta/r)dr \tag{5.81}$$

At this point, it is convenient to introduce the angular velocity

$$\Omega(r) = v_\theta/r \tag{5.82}$$

Substitution of the tensor elements $\underline{\underline{D}}$ in Equation 5.44 leads to

$$\Gamma = \{2 \ \text{trace} \ (\underline{\underline{D}} \bullet \underline{\underline{D}})\}^{1/2} = [(\partial v_z/\partial r)^2 + r^2(\partial\{v_\theta/r\}/\partial r)^2]^{1/2} \tag{5.83}$$

so that the Power law apparent viscosity given by Equation 5.45 becomes

$$N(\Gamma) = K \left[(\partial v_z/\partial r)^2 + r^2(\partial\{v_\theta/r\}/\partial r)^2 \right]^{(n-1)/2} \tag{5.84}$$

These results were stated without proof in Equations 5.46 and 5.47. Now we combine Equations 5.80 and 5.84 so that

$$K \left[(\partial v_z/\partial r)^2 + r^2(\partial\Omega/\partial r)^2 \right]^{(n-1)/2} \ d\Omega/dr = E_2/r^3 \tag{5.85}$$

If dv_z/dr is eliminated using Equation 5.81, we obtain, after very lengthy manipulations,

$$\begin{aligned} d\Omega/dr = (E_2/K)^{1/n} [r^{(2n+4)/(n-1)} \\ + r^{(4n+2)/(n-1)} \{(E_1 + r^2/2dP/dz)/E_2\}^2]^{(1-n)/2n} \end{aligned} \tag{5.86}$$

Next, we integrate Equation 5.86 over the interval (r, R_B), where R_B is the borehole radius. If we apply the first no-slip boundary condition

$$\Omega(R_B) = 0 \tag{5.87}$$

(there are four no-slip conditions altogether) and invoke the Mean Value Theorem of differential calculus, using as the appropriate mean the arithmetic average, we obtain

$$\Omega(r) = (E_2/K)^{1/n}(r - R_B)[((r + R_B)/2)^{(2n+4)/(n-1)}$$
$$+ ((r + R_B)/2)^{(4n+2)/(n-1)}\{(E_1 + (r + R_B)^2/8 \ dP/dz)/E_2\}^2]^{(1-n)/2n} \tag{5.88}$$

At this point, though, we do not yet apply any of the remaining three no-slip velocity boundary conditions. We turn our attention to v_z instead. We can derive a differential equation independent of Ω by combining Equations 5.81, 5.82, and 5.86 as follows:

$$dv_z/dr = (r^4/2 \ dP/dz + E_1 r^2)/E_2 \ d\Omega/dr$$
$$= r^2(E_1 + r^2/2 \ dP/dz)/E_2 \times (E_2/K)^{1/n}[r^{(2n+4)/(n-1)} \tag{5.89}$$
$$+ r^{(4n+2)/(n-1)}\{(E_1 + r^2/2 \ dP/dz)/E_2\}^2]^{(1-n)/2n}$$

We next integrate Equation 5.89 over (R_P, r), where R_P is the drillpipe radius, subject to the second no-slip condition

$$v_z(R_P) = 0 \tag{5.90}$$

An integration similar to that used for Equation 5.86, again invoking the Mean Value Theorem, leads to a result analogous to Equation 5.88:

$$v_z(r) = ((r + R_P)/2)^2(E_1 + ((r + R_P)/2)^2/2 \ dP/dz)/E_2$$
$$\times (E_2/K)^{1/n}[((r + R_P)/2)^{(2n+4)/(n-1)} + ((r + R_P)/2)^{(4n+2)/(n-1)} \tag{5.91}$$
$$\{(E_1 + ((r + R_P)/2)^2/2 \ dP/dz)/E_2\}^2]^{(1-n)/2n}(r - R_P)$$

Very useful results are obtained if we now apply the third no-slip condition:

$$v_z(R_B) = 0 \tag{5.92}$$

With this constraint, Equation 5.91 leads to a somewhat unwieldy combination of terms:

$$0 = ((R_B + R_P)/2)^2(E_1 + ((R_B + R_P)/2)^2/2) \ dP/dz)/E_2$$
$$\times (E_2/K)^{1/n}[((R_B + R_P)/2)^{(2n+4)/(n-1)} + ((R_B + R_P)/2)^{(4n+2)/(n-1)} \tag{5.93}$$
$$\{(E_1 + ((R_B + R_P)/2)^2/2 \ dP/dz)/E_2\}^2]^{(1-n)/(2n)}(R_B - R_P)$$

But if we observe that the quantity contained within the square brackets "[ ]" is positive definite, and that $(R_B - R_P)$ is nonzero, it follows that the left-hand side "0" can be obtained only if

$$E_1 = - (R_B + R_P)^2/8 \ dP/dz \tag{5.94}$$

holds identically. The remaining integration constant E_2 is determined from the last of the four no-slip conditions

$$\Omega(R_P) = \omega \tag{5.95}$$

Equation 5.95 requires fluid at the pipe surface to move with the rotating surface. Here, without loss of generality, $\omega < 0$ is the constant drillstring angular rotational speed. Combination of Equations 5.88, 5.94, and 5.95, after lengthy manipulations, leads to the surprisingly simple result that

$$E_2 = K \ (\omega/(R_P - R_B))^n((R_P + R_B)/2)^{n+2} \tag{5.96}$$

With all four no-slip conditions applied, the four integration constants, and hence the analytical solution for our Power law model, are completely determined. We next perform validation checks before deriving applications formulas.

Analytical validation

Different analytical procedures were required for Newtonian flows and Power law flows with general n's. This is related to the decoupling between axial and circumferential velocities in the singular $n = 1$ limit. On physical grounds, we expect that the Power law solution, if correct, would behave "continuously" through $n = 1$ as the fluid passes from the dilatant to the pseudoplastic state. That is, the solution should change smoothly when n varies from $1 - \delta$ to $1 + \delta$, where $|\delta| << 1$ is a small number. This continuous dependence and physical consistency will be demonstrated next. The validation also guards against error, given the quantity of algebraic manipulations involved.

The formulas derived above for general Power law fluids will be checked against exact Newtonian results where $K = \mu$ and $n = 1$. For consistency, we will take the narrow annulus limit of those formulas, a geometric approximation used in the Power law derivation. We will demonstrate that the closed form results obtained for non-Newtonian fluids are indeed "continuous in n" through the singular point $n = 1$.

We first check our results for the stresses $S_{r\theta}$ and $S_{\theta r}$. From Equations 5.36, 5.40, and 5.88, we find that

$$S_{r\theta} = S_{\theta r} = K(\omega/(R_P - R_B))^n((R_P + R_B)/2)^{n+2} \ r^{-2} \qquad (5.97)$$

In the limit $K = \mu$ and $n = 1$, Equation 5.97 for Power law fluids reduces to

$$S_{r\theta} = S_{\theta r} = \mu \ \omega/\{(R_P - R_B)r^2\} \times ((R_P + R_B)/2)^3 \qquad (5.98)$$

On the other hand, the definition $S_{r\theta} = S_{\theta r} = \mu \ d\Omega/dr$ inferred from Equations 5.36 and 5.40 becomes, using Equations 5.68 and 5.82 for Newtonian flow,

$$S_{r\theta} = S_{\theta r} = \mu \ \omega/\{(R_P - R_B)r^2\} \times 2(R_P R_B)^2/(R_P + R_B) \qquad (5.99)$$

Are the two second factors "$((R_P + R_B)/2)^3$" and "$2(R_P R_B)^2/(R_P + R_B)$" in Equations 5.98 and 5.99 consistent? If we evaluate these expressions in the narrow annulus limit, setting $R_P = R_B = R$, we obtain R^3 in *both* cases, providing the required validation. This consistency holds for all values of dP/dz.

For our second check, consider the Power law stresses S_{rz} and S_{zr} obtained from Equations 5.36, 5.41, and 5.89:

$$S_{rz} = S_{zr} = E_1/r + 1/2 \ r \ dP/dz = \{1/2 \ r - (R_P + R_B)^2/(8r)\} \ dP/dz \qquad (5.100)$$

The corresponding formula in the Newtonian limit is

$$S_{rz} = S_{zr} = \mu \ dv_z/dr = \{1/2 \ r - (R_P^2 - R_B^2)/(4r \ \log \ R_P/R_B)\} \ dP/dz \qquad (5.101)$$

where we have used Equation 5.67 shown earlier. Now, is "$(R_P + R_B)^2/8$" consistent with "$(R_P^2 - R_B^2)/(4 \log R_P/R_B)$"? As before, consider the narrow annulus limit, setting $R_P = R_B = R$.

The first expression easily reduces to $R^2/2$. For the second, we expand $\log R_B/R_P = \log \{1 + (R_B - R_P)/R_P\} = (R_B - R_P)/R_P$ and retain only the first term of the Taylor expansion. Direct substitution yields $R^2/2$ again. Therefore, Equations 5.100 and 5.101 are consistent for all rotational rates ω. From our checks on both S_{rz} and $S_{r\theta}$, then, we find good physical consistency and consequently reliable algebraic computations.

Newtonian and Power law flow differences

Equations 5.60, 5.82, 5.88, 5.91, 5.94, and 5.96 specify the velocity fields v_z and $v_\theta = r\Omega(r)$ as functions of wellbore geometry, fluid rheology, pipe inclination, rotational rate, pressure gradient, and gravity, respectively. We emphasize that Equation 5.78, which is to be evaluated using the non-Newtonian solution for v_θ, provides only a partial solution for the complete radial pressure gradient. The remaining part is obtained by adding the "$-\rho g \cos \alpha \sin \theta$" contribution of Equation 5.60. As in Newtonian flows, the pressure and its spatial gradients depend on all the coordinates r, θ, and z and the parameters ρ, g, and α.

There are fundamental differences between these solutions and the Newtonian solutions. For example, in the latter the solutions for v_z and v_θ completely decouple despite the nonlinearity of the Navier-Stokes equations. The governing equations become linear. But for Power law flows, both v_z and v_θ remain highly coupled and nonlinear. In this sense, Newtonian results are singular; but the degeneracy disappears for eccentric geometries when the convective terms reappear. Cuttings transport experimenters working with *concentric Newtonian* flows will *not* be able to extrapolate their findings to practical eccentric geometries or non-Newtonian fluids.

Also, the expression for v_z in the Newtonian limit is directly proportional to dP/dz; however, as Equation 5.91 for Power law fluids shows, the dependence of v_z (and hence of total volumetric flow rate) on pressure gradient is nonlinear. Similarly, while Equation 5.68 shows that v_θ is directly proportional to the rotational rate ω, Equations 5.82, 5.88, and 5.96 illustrate a more complicated nonlinear dependence for Power law fluids. It is important to emphasize that, for a fixed annular flow geometry in Newtonian flow, v_z depends only on dP/dz and not on ω, and v_θ depends only on ω and not on dP/dz. But for Power law flows, v_z and v_θ each depend on both dP/dz and ω. Thus, "axial quantities" like net annular volumetric flow rate cannot be calculated without considering both dP/dz and ω.

Interestingly, though, the stresses $S_{r\theta}$ and S_{rz} in the non-Newtonian case preserve their "independence" as in Newtonian flows. That is, $S_{r\theta}$ depends only on ω and not on dP/dz, while S_{rz} depends only on dP/dz and not on ω (see Equations 5.102 through 5.105). The Power law stress values themselves, of course, are different from their Newtonian counterparts. Also, the "maximum stress" $(S_{r\theta}^2 + S_{rz}^2)^{1/2}$, important in borehole stability and cuttings bed erosion, depends on both ω and dP/dz, as it does in Newtonian flow.

An important question is the significance of rotation in practical calculations. Can "ω" be safely neglected in drilling and cementing applications? This depends on a nondimensional ratio of circumferential to axial momentum flux. While the "maximum v_θ" is easily obtained as "$\omega_{rpm} \times R_P$," the same estimate for v_z is difficult to obtain, since axial velocity is sensitive to both n and K, not to mention v_θ and dP/dz. In general, one needs to consider the full problem without approximation.

Of course, since the analytical solution is now available, the use of approximate "axial flow only" solutions is really a moot point. The Power law results and the formulas derived next are "explicit" in that they require no iteration. And although the software described later is written in Fortran, our equations are just as easily programmed on calculators. The important dependence of annular flows on "ω" will be demonstrated in calculated results.

More applications formulas

The cylindrical geometry of the present problem renders all stress tensor components except $S_{r\theta}$, $S_{\theta r}$, S_{zr}, and S_{rz} zero. From our Power law results, the required formulas for viscous stress can be shown to be

$$S_{r\theta} = S_{\theta r} = K(\omega/(R_P - R_B))^n((R_P + R_B)/2)^{n+2} \, r^{-2} \tag{5.102}$$

$$S_{rz} = S_{zr} = E_1/r + 1/2 \, r \, dP/dz = \{1/2 \, r - (R_P + R_B)^2/(8r)\} \, dP/dz \tag{5.103}$$

Their Newtonian counterparts take the form

$$S_{r\theta} = S_{\theta r} = \mu \, \omega/\{(R_P - R_B)r^2\} \times 2(R_P R_B)^2/(R_P + R_B) \tag{5.104}$$

$$S_{rz} = S_{zr} = \mu \, dv_z/dr = \{1/2 \, r - (R_P^2 - R_B^2)/(4r \log R_P/R_B)\} \, dP/dz \tag{5.105}$$

In studies on borehole erosion, annular velocity plays an important role, since drilling mud carries abrasive cuttings. The magnitude of fluid shear stress may also be important in unconsolidated sands, where tangential surface forces assist in wall erosion. Stress considerations also arise in cuttings bed transport analysis in highly deviated or horizontal holes (see Model 5.7). The individual components can be obtained by evaluating Equations 5.102 and 5.103 at $r = R_B$ for Power law fluids, and Equations 5.104 and 5.105 for Newtonian fluids. And since these stresses act in orthogonal directions, the "maximum stress" can be obtained by writing

$$S_{max}(R_B) = \{S_{r\theta}^2(R_B) + S_{rz}^2(R_B)\}^{1/2} \tag{5.106}$$

The shear force associated with this stress acts in a direction offset from the borehole axis by an angle

$$\Theta_{max \ shear} = \arctan \{S_{r\theta}(R_B)/S_{rz}(R_B)\} \tag{5.107}$$

Opposing the erosive effects of shear may be the stabilizing effects of hydrostatic and dynamic pressure. Explicit formulas for the pressures $P(z)$, $P^*(r)$, and the hydrostatic background level were given earlier.

To obtain the corresponding elements of the deformation tensor, we rewrite Equation 5.36 in the form

$$\underline{\underline{D}} = \underline{\underline{S}}/2N(\Gamma) \tag{5.108}$$

and substitute S_{rz} or $S_{r\theta}$ as required. In the Newtonian case, $N(\Gamma) = \mu$ is the laminar viscosity; for Power law fluids, Equation 5.47 applies. Stresses are important to transport problems; fluid deformations are useful for the kinematic studies often of interest to rheologists.

Annular volumetric flow rate, Q, which depends on pressure gradient, is important in determining mud pump power requirements and the cuttings transport capabilities of the drilling fluid. It is obtained by evaluating

$$Q = \int_{R_p}^{R_B} v_z(r) \, 2\pi r \, dr \tag{5.109}$$

In the preceding integrand, Equation 5.67 for $v_z(r)$ must be used for Newtonian flows, while Equation 5.91 will apply to Power law fluids.

Borehole temperature may play an important role in drilling. Problem areas include formation temperature interpretation and mud thermal stability (e.g., the "thinning" of oil-based muds with temperature limits cuttings transport efficiency). Many studies do not consider the effects of heat generation by internal friction, which may be non-negligible; in closed systems, temperature increases over time may be significant. Ideally, temperature effects due to fluid type and cumulative effects related to total circulation time should be identified.

The starting point is the equation describing energy balance within the fluid, that is, the PDE for the temperature field $T(r, \theta, z, t)$. Even if the velocity field is steady, temperature effects will typically not be, since irreversible thermodynamic effects cause continual increases of T with time. If temperature increases are large enough, the changes in viscosity, consistency factor, or fluid exponent as functions of T must be considered. Then the momentum and energy equations will be coupled. We will not consider this complicated situation yet, so that the velocity fields can be obtained independently of T. For Newtonian flows, we have $n = 1$ and $K = \mu$. The temperature field satisfies

$$
\begin{aligned}
\rho c(\partial T/\partial t &+ v_r\, \partial T/\partial r + v_\theta/r\, \partial T/\partial \theta + v_z\, \partial T/\partial z) \\
&= K_{th}\, [1/r\, \partial(r\, \partial T/\partial r)/\partial r + 1/r^2\, \partial^2 T/\partial \theta^2 + \partial^2 T/\partial z^2] \\
&+ 2\mu\, \{(\partial v_r/\partial r)^2 + [1/r(\partial v_\theta/\partial \theta + v_r)]^2 + (\partial v_z/\partial z)^2\} \\
&+ \mu\, \{(\partial v_\theta/\partial z + 1/r\, \partial v_z/\partial \theta)^2 + (\partial v_z/\partial r + \partial v_r/\partial z)^2 \\
&+ [1/r\, \partial v_r/\partial \theta + r\, \partial(v_\theta/r)/\partial r]^2\} + \rho Q^*
\end{aligned}
\tag{5.110}
$$

where c is the heat capacity, K_{th} is the thermal conductivity, and Q^* is an energy transmission function. The terms on the first line represent transient and convective effects; those on the second line model heat conduction; and those on the third through fifth are positive definite and represent heat generation due to internal fluid friction.

These irreversible thermodynamic effects are referred to collectively as the "dissipation function" or "heat generation function." The dissipation function Φ is in effect a distributed heat source within the moving fluid medium. If we employ the same assumptions as used in our solution of the Navier-Stokes equations for Newtonian flows, this expression reduces to

$$
\Phi = \mu\, \{(\partial v_z/\partial r)^2 + r^2(\partial \Omega/\partial r)^2\} > 0
\tag{5.111}
$$

which can be easily evaluated using Equations 5.67, 5.68, and 5.82. It is important to recognize that Φ depends on spatial velocity gradients only and not on velocity magnitudes. In a closed system, the fact that $\Phi > 0$ leads to increases in temperature in time if the borehole walls cannot conduct heat away quickly.

Equations 5.110 and 5.111 assume Newtonian flow. For general fluids, it is possible to show that the dissipation function now takes the form

$$
\begin{aligned}
\Phi = S_{rr}\, \partial v_r/\partial r &+ S_{\theta\theta}\, 1/r(\partial v_\theta/\partial \theta + v_r) \\
&+ S_{zz}\, \partial v_z/\partial z + S_{r\theta}[r\, \partial(v_\theta/r)/\partial r + 1/r\, \partial v_r/\partial \theta] \\
&+ S_{rz}(\partial v_z/\partial r + \partial v_r/\partial z) + S_{\theta z}(1/r\, \partial v_z/\partial \theta + \partial v_\theta/\partial z)
\end{aligned}
\tag{5.112}
$$

The geometrical simplifications used earlier reduce Equation 5.112 to

$$\Phi = K \{(\partial v_z/\partial r)^2 + r^2(\partial \Omega/\partial r)^2\}^{(n+1)/2} > 0 \qquad (5.113)$$

In the Newtonian limit with $K = \mu$ and $n = 1$, Equation 5.113 consistently reduces to Equation 5.111. Equations 5.86 and 5.89 are used to evaluate the expression for Φ. As before, Φ depends upon velocity gradients only and not on magnitudes; it largely arises from high shear at solid boundaries.

Detailed calculated results

The Power law results derived previously were coded in a Fortran algorithm designed to provide a suite of output "utility" solutions for any set of input data. These may be useful in determining operationally important quantities like volumetric flow rate and axial speed. However, they also provide research utilities needed, for example, to correlate experimental cuttings transport data or to interpret formation temperature data.

The core code resides in 30 lines of Fortran. It runs on a "stand-alone" basis or as an embedded subroutine for specialized applications. The formulas used are also programmable on calculators. Inputs include pipe or casing outer diameter, borehole diameter, axial pressure gradient, rotational rate, fluid exponent n, and consistency factor K. Outputs include tables, line plots, and ASCII character plots versus "r" for a number of useful functions, as follows:

- Axial velocity v_z (r)
- Angular velocity gradient $d\omega(r)/dr$
- Circumferential velocity v_θ (r)
- Radial pressure gradient
- Fluid rotational rate $\omega(r)$, "local rpm"
- Apparent viscosity versus "r"
- Total absolute speed
- Local frictional heat generation
- Angle between $v_z(r)$ and $v_\theta(r)$
- All stress tensor components
- Axial velocity gradient $dv_z(r)/dr$
- Maximum wellbore stress
- Azimuthal velocity gradient dv_θ/dr
- All deformation tensor components

We emphasize that the "radial pressure gradient" listed refers to the partial contribution in Equation 5.78, which depends on "r" only. For the complete gradient, Equation 5.60 shows that the term "$-\rho g \cos \alpha \sin \theta$" must be appended to the value calculated here. This contribution depends on ρ, g, α, and θ. In addition to the foregoing arrays, total annular volumetric flow rate and radial averages of all of the listed quantities are computed. Before proceeding to detailed computations, let us compare our concentric, rotating pipe, *narrow annulus* results in the limit of zero rotation with an exact solution.

MODEL 5.6, EXAMPLE 5.1

East Greenbriar No. 2

A mud hydraulics analysis was performed for "East Greenbriar No. 2" using a computer program offered by a service company. This program, which applies to nonrotating flows only, is based on the *exact* Fredrickson and Bird (1958) solution. In this example, the drillpipe outer radius is 2.5 in., the borehole radius is 5.0 in., the axial pressure gradient is 0.00389 psi/ft, the fluid exponent is 0.724, and the consistency factor is 0.268 lbf sec$^{0.724}$/(100 ft^2) (that is, 0.1861×10^{-4} lbf sec$^{0.724}$/in.^{2} in the units employed by our program).

The exact results computed using this data are an annular volumetric flow rate of 400 gal/min and an average axial speed of 130.7 ft/min. The same input data was used in our program, with an assumed drillstring "rpm" of 0.001. We computed 373.6 gal/min and 126.9 ft/min for this nonrotating flow, agreeing to within 7 percent for the not so narrow annulus.

Our model was designed, of course, to include the effects of drillstring rotation. We first considered an extremely large rpm of 300, with the same pressure gradient, to evaluate qualitative effects. The corresponding results were 526.1 gal/min and 175.6 ft/min. The ratio of the average circumferential speed to the average axial speed is 1.06, indicating that rotational effects are important. At 150 rpm, our volumetric flow rate of 458.7 gal/min exceeds 373.6 gal/min by 23 percent.

In this case, the ratio of average circumferential speed to axial speed is still a non-negligible 65 percent. These results suggest that static models tend to overestimate the pressure requirements needed by a rotating drillstring to produce a prescribed flow rate. Our hydraulics model indicates that including rotational effects, for a fixed pressure gradient, is likely to increase the volumetric flow rate over static predictions. These considerations may be important in planning long deviated wells where one needs to know, for a given rpm, what maximum borehole length is possible with the pump at hand.

MODEL 5.6, EXAMPLE 5.2

Detailed Spatial Properties versus "r"

Our algorithm does more than calculate annular volumetric flow rate and average axial speed. This section includes the entire output file from a typical run, in this case "East Greenbriar No. 2," with annotated comments. The input menu is nearly identical to the summary in Table 5.1. Because the numerical results are based on analytical, closed-form results, there are no computational inputs; the grid reference in Table 5.1 is a print control parameter.

At the present, the volumetric flow rate is the only quantity computed numerically; a second-order scheme is applied to our $v_z(r)$'s. All inputs are in plain English and are easily understandable. Outputs are similarly user friendly. All output quantities are defined, along with units, in a printout that precedes tabulated and plotted results. This printout is duplicated in Table 5.2.

The defined quantities are first tabulated, as shown in Table 5.3, as a function of the radial position "r." At this point, the total volumetric flow rate is computed and presented in textual form:

$$\text{Total volume flow rate (cubic in/sec)} = .2026E + 04$$
$$(\text{gal/min}) = .5261E + 03$$

A run-time menu prompts the user about quantities needed for display in ASCII file plots. The complete list of quantities was given previously. Plots corresponding to "East Greenbriar No. 2" are shown in Figures 5.35 through 5.50.

Finally, the computer algorithm calculates radially averaged quantities using the definition

$$F_{avg} = \int_{R_P}^{R_B} F(r)dr/(R_B - R_P) \tag{5.114}$$

Table 5.1 Summary of Input Parameters

- 0   Drillpipe outer radius (inches) = 2.5000
- 0   Borehole radius (inches) = 5.0000
- 0   Axial pressure gradient (psi/ft) = 0.0039
- 0   Drillstring rotation rate (rpm) = 300.0000
- 0   Drillstring rotation rate (rad/sec) = 31.4159
- 0   Fluid exponent "n" (nondimensional) = 0.7240
- 0   Consistency factor (lbf sec^n/sq in.) = 0.1861E-04
- 0   Mass density of fluid (lbf^{2}sec^4/ft) = 1.9000 (e.g., about 1.9 for water)
- 0   Number of radial "grid" positions = 18

Table 5.2 Analytical (Noniterative) Solutions Tabulated versus "r," Nomenclature, and Units

Symbols	Nomenclature	Units
r	Annular radial position	(in.)
V_z	Velocity in axial z direction	(in./sec)
V_θ	Circumferential velocity	(in./sec)
$d\theta/dt$ or W	θ velocity	(rad/sec) (1 rad/ sec = 9.5493 rpm)
dV_z/dr	Velocity gradient	(1/sec)
dV_θ/dr	Velocity gradient	(1/sec)
dW/dr	Angular speed gradient	(1/(sec × in.))
$S_{r\theta}$	rθ stress component	(psi)
S_{rz}	rz stress component	(psi)
S_{max}	Sqrt (S_{rz}**2 + $S_{r\theta}$**2)	(psi)
dP/dr	Radial pressure gradient	(psi/in.)
App-Vis	Apparent viscosity	(lbf sec /sq in.)
Dissip	Dissipation function (indicates frictional heat produced)	(lbf/(sec × sq in.))
Atan V_θ/V_z	Angle between V_θ and V_z vectors	(deg)
Net Spd	Sqrt (V_z**2 + V_θ**2)	(in./sec)
$D_{r\theta}$	rθ deformation tensor component	(1/sec)
D_{rz}	rz deformation tensor component	(1/sec)

and a second-order accurate integration scheme. Note that this is not a volume-weighted average. When properties vary rapidly over r, the linear average (or *any average*) may not be meaningful as a correlation or analysis parameter. Table 5.4 (page 240) displays computed average results.

Table 5.3 Calculated Quantities versus "r"

r	V_z	V_θ	W	$d(V_z)/dr$	$d(V_\theta)/dr$	dW/dr
5.00	.601E−04	.279E−04	.559E−05	−.610E+02	−.293E+02	−.586E+01
4.86	.848E+01	.407E+01	.837E+00	−.534E+02	−.293E+02	−.620E+01
4.72	.164E+02	.814E+01	.172E+01	−.460E+02	−.294E+02	−.659E+01
4.58	.237E+02	.122E+02	.266E+01	−.390E+02	−.295E+02	−.702E+01
4.44	.304E+02	.163E+02	.366E+01	−.321E+02	−.297E+02	−.751E+01
4.31	.365E+02	.203E+02	.472E+01	−.256E+02	−.302E+02	−.811E+01
4.17	.418E+02	.244E+02	.585E+01	−.193E+02	−.310E+02	−.885E+01
4.03	.462E+02	.284E+02	.705E+01	−.131E+02	−.324E+02	−.980E+01
3.89	.497E+02	.325E+02	.835E+01	−.672E+01	−.345E+02	−.110E+02
3.75	.521E+02	.366E+02	.975E+01	.273E−04	−.374E+02	−.126E+02
3.61	.533E+02	.407E+02	.113E+02	.738E+01	−.412E+02	−.145E+02
3.47	.532E+02	.449E+02	.129E+02	.157E+02	−.461E+02	−.170E+02
3.33	.516E+02	.492E+02	.148E+02	.251E+02	−.521E+02	−.201E+02
3.19	.483E+02	.536E+02	.168E+02	.358E+02	−.594E+02	−.239E+02
3.06	.432E+02	.582E+02	.190E+02	.480E+02	−.682E+02	−.286E+02
2.92	.361E+02	.630E+02	.216E+02	.619E+02	−.787E+02	−.344E+02
2.78	.266E+02	.680E+02	.245E+02	.778E+02	−.914E+02	−.417E+02
2.64	.147E+02	.732E+02	.277E+02	.959E+02	−.107E+03	−.510E+02
2.50	.000E+00	.785E+02	.314E+02	.117E+03	−.126E+03	−.628E+02

r	$S_{r\theta}$	S_{rz}	S_{max}	dP/dr	App-Vis	Dissip
5.00	.170E−03	−.355E−03	.393E−03	.143E−13	.582E−05	.266E−01
4.86	.180E−03	−.319E−03	.366E−03	.312E−03	.598E−05	.225E−01
4.72	.191E−03	−.283E−03	.341E−03	.128E−02	.614E−05	.190E−01
4.58	.203E−03	−.246E−03	.318E−03	.298E−02	.630E−05	.161E−01
4.44	.216E−03	−.208E−03	.299E−03	.545E−02	.646E−05	.139E−01
4.31	.230E−03	−.168E−03	.285E−03	.878E−02	.658E−05	.123E−01
4.17	.245E−03	−.128E−03	.277E−03	.131E−01	.665E−05	.115E−01
4.03	.262E−03	−.869E−04	.277E−03	.184E−01	.665E−05	.115E−01
3.89	.282E−03	−.442E−04	.285E−03	.248E−01	.658E−05	.124E−01
3.75	.303E−03	.175E−09	.303E−03	.326E−01	.643E−05	.143E−01
3.61	.327E−03	.459E−04	.330E−03	.420E−01	.622E−05	.175E−01
3.47	.353E−03	.936E−04	.365E−03	.532E−01	.598E−05	.223E−01
3.33	.383E−03	.144E−03	.409E−03	.665E−01	.573E−05	.292E−01
3.19	.417E−03	.196E−03	.461E−03	.824E−01	.547E−05	.388E−01
3.06	.456E−03	.251E−03	.520E−03	.102E+00	.523E−05	.518E−01
2.92	.501E−03	.309E−03	.588E−03	.125E+00	.499E−05	.693E−01
2.78	.552E−03	.370E−03	.665E−03	.152E+00	.476E−05	.928E−01
2.64	.611E−03	.436E−03	.751E−03	.186E+00	.455E−05	.124E+00
2.50	.681E−03	.507E−03	.849E−03	.226E+00	.434E−05	.166E+00

Table 5.3 Calculated Quantities versus "r" (Continued)

z	V_z	V_θ	Atan V_θ/V_z	Net Spd	$D_{r\theta}$	D_{rz}
5.00	.601E−04	.279E−04	.249E+02	.663E−04	.146E+02	−.305E+02
4.86	.848E+01	.407E+01	.256E+02	.940E+01	.151E+02	−.267E+02
4.72	.164E+02	.814E+01	.264E+02	.183E+02	.155E+02	−.230E+02
4.58	.237E+02	.122E+02	.272E+02	.267E+02	.161E+02	−.195E+02
4.44	.304E+02	.163E+02	.281E+02	.345E+02	.167E+02	−.161E+02
4.31	.365E+02	.203E+02	.291E+02	.417E+02	.175E+02	−.128E+02
4.17	.418E+02	.244E+02	.303E+02	.483E+02	.184E+02	−.965E+01
4.03	.462E+02	.284E+02	.316E+02	.542E+02	.197E+02	−.653E+01
3.89	.497E+02	.325E+02	.332E+02	.593E+02	.214E+02	−.336E+01
3.75	.521E+02	.366E+02	.351E+02	.636E+02	.236E+02	.137E−04
3.61	.533E+02	.407E+02	.374E+02	.670E+02	.262E+02	.369E+01
3.47	.532E+02	.449E+02	.402E+02	.696E+02	.295E+02	.783E+01
3.33	.516E+02	.492E+02	.436E+02	.713E+02	.334E+02	.125E+02
3.19	.483E+02	.536E+02	.480E+02	.722E+02	.381E+02	.179E+02
3.06	.432E+02	.582E+02	.534E+02	.725E+02	.436E+02	.240E+02
2.92	.361E+02	.630E+02	.602E+02	.726E+02	.502E+02	.309E+02
2.78	.266E+02	.680E+02	.686E+02	.730E+02	.579E+02	.389E+02
2.64	.147E+02	.732E+02	.786E+02	.746E+02	.673E+02	.480E+02
2.50	.000E+00	.785E+02	.900E+02	.785E+02	.785E+02	.584E+02

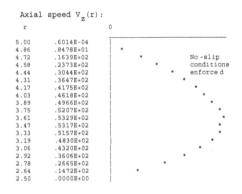

FIGURE 5.35

Axial speed.

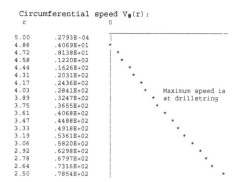

FIGURE 5.36

Circumferential speed.

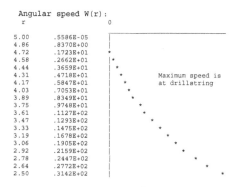

FIGURE 5.37

Angular speed.

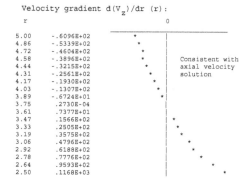

FIGURE 5.38

Velocity gradient.

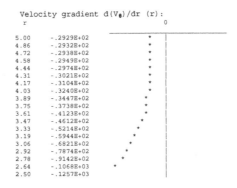

```
           Velocity gradient d(V₀)/dr (r):
        r                              0
     5.00    -.2929E+02                       *  |
     4.86    -.2932E+02                       *  |
     4.72    -.2938E+02                       *  |
     4.58    -.2949E+02                       *  |
     4.44    -.2974E+02                       *  |
     4.31    -.3021E+02                       *  |
     4.17    -.3104E+02                      *   |
     4.03    -.3240E+02                      *   |
     3.89    -.3447E+02                     *    |
     3.75    -.3738E+02                     *    |
     3.61    -.4123E+02                    *     |
     3.47    -.4612E+02                   *      |
     3.33    -.5214E+02                  *       |
     3.19    -.5944E+02                 *        |
     3.06    -.6821E+02               *          |
     2.92    -.7874E+02              *           |
     2.78    -.9142E+02            *             |
     2.64    -.1068E+03         *                |
     2.50    -.1257E+03                          |
```

FIGURE 5.39

Velocity gradient.

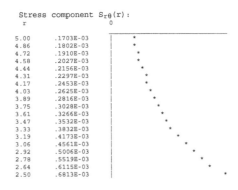

```
           Angular speed gradient dW/dr (r):
        r                              0
     5.00    -.5857E+01                     *  |
     4.86    -.6204E+01                     *  |
     4.72    -.6586E+01                     *  |
     4.58    -.7016E+01                     *  |
     4.44    -.7514E+01                     *  |
     4.31    -.8112E+01                     *  |
     4.17    -.8853E+01                    *   |
     4.03    -.9796E+01                    *   |
     3.89    -.1101E+02                    *   |
     3.75    -.1257E+02                   *    |
     3.61    -.1454E+02                   *    |
     3.47    -.1701E+02                  *     |
     3.33    -.2007E+02                 *      |
     3.19    -.2386E+02                *       |
     3.06    -.2856E+02              *         |
     2.92    -.3440E+02            *           |
     2.78    -.4172E+02          *             |
     2.64    -.5098E+02       *                |
     2.50    -.6283E+02                        |
```

FIGURE 5.40

Angular speed gradient.

```
           Stress component S_rθ(r):
        r                       0
     5.00    .1703E-03    |    *
     4.86    .1802E-03    |     *
     4.72    .1910E-03    |      *
     4.58    .2027E-03    |       *
     4.44    .2156E-03    |        *
     4.31    .2297E-03    |         *
     4.17    .2453E-03    |          *
     4.03    .2625E-03    |           *
     3.89    .2816E-03    |            *
     3.75    .3028E-03    |              *
     3.61    .3266E-03    |               *
     3.47    .3532E-03    |                 *
     3.33    .3832E-03    |                  *
     3.19    .4173E-03    |                    *
     3.06    .4561E-03    |                      *
     2.92    .5006E-03    |                       *
     2.78    .5519E-03    |                         *
     2.64    .6115E-03    |                           *
     2.50    .6813E-03    |                              *
```

FIGURE 5.41

Viscous stress.

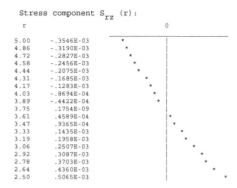

```
      Stress component S  (r):
                        rz
       r                             0
     5.00    -.3546E-03          *     |
     4.86    -.3190E-03         *      |
     4.72    -.2827E-03        *       |
     4.58    -.2456E-03         *      |
     4.44    -.2075E-03          *     |
     4.31    -.1685E-03           *    |
     4.17    -.1283E-03            *   |
     4.03    -.8694E-04             *  |
     3.89    -.4422E-04              * |
     3.75     .1754E-09                |
     3.61     .4589E-04                |*
     3.47     .9365E-04                | *
     3.33     .1435E-03                |  *
     3.19     .1958E-03                |   *
     3.06     .2507E-03                |    *
     2.92     .3087E-03                |     *
     2.78     .3703E-03                |      *
     2.64     .4360E-03                |       *
     2.50     .5065E-03                |        *
```

FIGURE 5.42

Viscous stress.

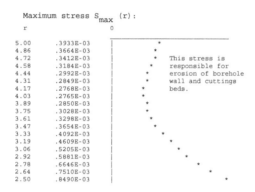

```
      Maximum stress S    (r):
                       max
       r                   0
     5.00     .3933E-03   |       *
     4.86     .3664E-03   |      *
     4.72     .3412E-03   |     *        This stress is
     4.58     .3184E-03   |    *         responsible for
     4.44     .2992E-03   |    *         erosion of borehole
     4.31     .2849E-03   |    *         wall and cuttings
     4.17     .2768E-03   |   *          beds.
     4.03     .2765E-03   |   *
     3.89     .2850E-03   |   *
     3.75     .3028E-03   |   *
     3.61     .3298E-03   |    *
     3.47     .3654E-03   |     *
     3.33     .4092E-03   |      *
     3.19     .4609E-03   |       *
     3.06     .5205E-03   |        *
     2.92     .5881E-03   |         *
     2.78     .6646E-03   |          *
     2.64     .7510E-03   |           *
     2.50     .8490E-03   |             *
```

FIGURE 5.43

Maximum viscous stress.

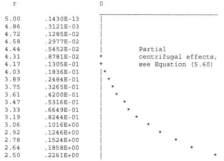

```
      Radial pressure gradient dP/dr (r):
       r                   0
     5.00     .1430E-13   |
     4.86     .3121E-03   |
     4.72     .1285E-02   |
     4.58     .2977E-02   |
     4.44     .5452E-02   |        Partial
     4.31     .8781E-02   *        centrifugal effects,
     4.17     .1305E-01   *        see Equation (5.60)
     4.03     .1836E-01   |*
     3.89     .2484E-01   | *
     3.75     .3265E-01   |  *
     3.61     .4200E-01   |   *
     3.47     .5316E-01   |    *
     3.33     .6649E-01   |     *
     3.19     .8244E-01   |       *
     3.06     .1016E+00   |        *
     2.92     .1246E+00   |          *
     2.78     .1524E+00   |           *
     2.64     .1858E+00   |             *
     2.50     .2261E+00   |               *
```

FIGURE 5.44

Radial pressure gradient.

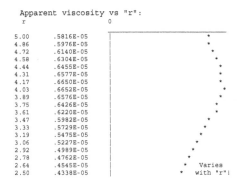

FIGURE 5.45

Apparent viscosity.

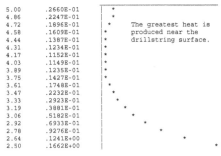

FIGURE 5.46

Dissipation function.

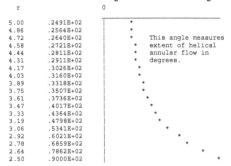

FIGURE 5.47

Velocity angle.

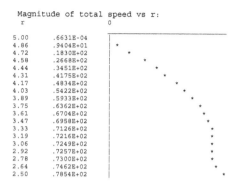

```
        Magnitude of total speed vs r:
     r                          0
  5.00     .6631E-04      |
  4.86     .9404E+01      |   *
  4.72     .1830E+02      |     *
  4.58     .2668E+02      |       *
  4.44     .3451E+02      |         *
  4.31     .4175E+02      |           *
  4.17     .4834E+02      |             *
  4.03     .5422E+02      |              *
  3.89     .5933E+02      |               *
  3.75     .6362E+02      |                *
  3.61     .6704E+02      |                 *
  3.47     .6958E+02      |                  *
  3.33     .7126E+02      |                   *
  3.19     .7216E+02      |                   *
  3.06     .7249E+02      |                   *
  2.92     .7257E+02      |                   *
  2.78     .7300E+02      |                   *
  2.64     .7462E+02      |                    *
  2.50     .7854E+02      |                      *
```

FIGURE 5.48

Total speed.

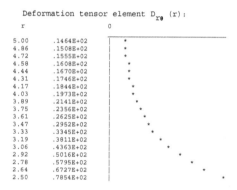

```
        Deformation tensor element D_rθ (r):
     r                          0
  5.00     .1464E+02      |  *
  4.86     .1508E+02      |  *
  4.72     .1555E+02      |  *
  4.58     .1608E+02      |   *
  4.44     .1670E+02      |   *
  4.31     .1746E+02      |   *
  4.17     .1844E+02      |    *
  4.03     .1973E+02      |    *
  3.89     .2141E+02      |     *
  3.75     .2356E+02      |      *
  3.61     .2625E+02      |       *
  3.47     .2952E+02      |        *
  3.33     .3345E+02      |         *
  3.19     .3811E+02      |           *
  3.06     .4363E+02      |            *
  2.92     .5016E+02      |              *
  2.78     .5795E+02      |               *
  2.64     .6727E+02      |                 *
  2.50     .7854E+02      |                    *
```

FIGURE 5.49

Deformation tensor element.

```
        Deformation tensor element D_rz (r):
     r                              0
  5.00    -.3048E+02        *       |
  4.86    -.2669E+02         *      |
  4.72    -.2302E+02          *     |
  4.58    -.1948E+02          *     |
  4.44    -.1607E+02           *    |
  4.31    -.1281E+02            *   |
  4.17    -.9648E+01             *  |
  4.03    -.6535E+01              * |
  3.89    -.3362E+01               *|
  3.75     .1365E-04              |
  3.61     .3689E+01              |
  3.47     .7828E+01              |  *
  3.33     .1253E+02              |   *
  3.19     .1788E+02              |    *
  3.06     .2398E+02              |     *
  2.92     .3094E+02              |      *
  2.78     .3888E+02              |       *
  2.64     .4796E+02              |         *
  2.50     .5839E+02              |          *
```

FIGURE 5.50

Deformation tensor element.

Table 5.4 Averaged Values of Annular Quantities

Apparent Viscosities	Averaged Parameter Values
Average V_z (in./sec)	.3512E+02
Average V_z (ft/min)	.1756E+03
Average V_θ (in./sec)	.3737E+02
Average W (rad/sec)	.1160E+02
Average total speed (in./sec)	.5379E+02
Average angle between V_z and V_θ (deg)	.4189E+02
Average $d(V_z)/dr$ (1/sec)	.0000E+00
Average $d(V_\theta)/dr$ (1/sec)	−.5028E+02
Average dW/dr (1/(sec X in.))	−.1906E+02
Average dP/dr (psi/in.)	.5718E−01
Average $S_{r\theta}$ (psi)	.3410E−03
Average S_{rz} (psi)	.2432E−04
Average S_{max} (psi)	.4146E−03
Average dissipation function (lbf/(sec sq in.))	.3753E−01
Average apparent viscosity (lbf sec/sq in.)	.5876E−05
Average $D_{r\theta}$ (1/sec)	.3094E+02
Average D_{rz} (1/sec)	.4445E+01

MODEL 5.6, EXAMPLE 5.3

More of East Greenbriar

We repeated the calculations for "East Greenbriar No. 2" with all parameters unchanged except for the fluid exponent, which we increased to a near-Newtonian level of 0.9 (again, 1.0 is the Newtonian value). In the first run, we considered a static, nonrotating drillstring with an "rpm" of 0.001, and obtained a volumetric flow rate of 196.2 gal/min. This is quite different from our earlier 373.6 gal/min, which assumed a fluid exponent of n = 0.724. That is, a 24 percent increase in the fluid exponent n resulted in a 47 percent decrease in flow rate; these numbers show how sensitive results are to changes in n.

The axial speeds, apparent viscosities, and averaged parameter values obtained are given in Figures 5.51 (page 241) and 5.52 (page 241), and in Table 5.5 (page 241). Note how the apparent viscosity is almost constant everywhere with respect to radial position; the well-known localized "pinch" is found near the center of the annulus, where the axial velocity gradient vanishes.

For our second run, we retained the foregoing parameters with the exception of drillstring rpm, which we increased for test purposes from 0.001 to 300 (the fluid exponent was still 0.9). The volumetric flow rate computed was 232.9 gpm, which was higher than the 196.2 gpm obtained above by a significant 18.7 percent. Thus, even for "almost Newtonian" Power law fluids, the effect of rotation allows a higher flow rate for the same pressure drop. Thus, to produce the lower flow rate, a pump having less pressure output than normal would suffice. Computed results are shown in Figures 5.53 and 5.54 (page 242) and Table 5.6 (page 243).

The effect of increasing drillstring rpm is an increase in the average borehole maximum stress by 3.42 times; this may be of interest to wellbore stability. The apparent viscosity in this example, unlike the previous example, is nearly constant everywhere and does not "pinch out." The analytical solutions derived in this chapter are of fundamental rheological interest. However, they are particularly useful in drilling and production applications, insofar as the effect of rotation on "volumetric flow rate versus pressure drop" is concerned. They allow us to study various operational "what-if" questions quickly and efficiently.

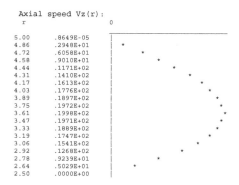

```
Axial speed Vz(r):
  r                        0
5.00    .8649E-05      |
4.86    .2948E+01      |     *
4.72    .6058E+01      |        *
4.58    .9010E+01      |          *
4.44    .1171E+02      |             *
4.31    .1410E+02      |               *
4.17    .1613E+02      |                 *
4.03    .1776E+02      |                  *
3.89    .1897E+02      |                   *
3.75    .1972E+02      |                    *
3.61    .1998E+02      |                    *
3.47    .1971E+02      |                    *
3.33    .1889E+02      |                   *
3.19    .1747E+02      |                 *
3.06    .1541E+02      |                *
2.92    .1268E+02      |             *
2.78    .9239E+01      |         *
2.64    .5029E+01      |      *
2.50    .0000E+00      |
```

FIGURE 5.51

Axial speed.

```
Apparent viscosity vs "r":
  r                        0
5.00    .1341E-04      |    *
4.86    .1357E-04      |    *
4.72    .1375E-04      |    *
4.58    .1397E-04      |    *
4.44    .1424E-04      |    *
4.31    .1457E-04      |    *
4.17    .1502E-04      |     *
4.03    .1568E-04      |      *
3.89    .1690E-04      |       *
3.75    .4468E-04      |                        *
3.61    .1683E-04      |       *
3.47    .1555E-04      |      *
3.33    .1483E-04      |    *
3.19    .1433E-04      |    *
3.06    .1394E-04      |    *
2.92    .1362E-04      |    *
2.78    .1335E-04      |   *
2.64    .1311E-04      |   *
2.50    .1289E-04      |   *
```

FIGURE 5.52

Apparent viscosity.

Table 5.5 Averaged Values of Annular Quantities

Apparent Viscosities	Averaged Parameter Values
Average V_z (in./sec)	.1305E+02
Average V_z (ft/min)	.6523E+02
Average V_θ (in./sec)	.3535E−03
Average W (rad/sec)	.1074E−03
Average total speed (in./sec)	.1305E+02
Average angle between V_z and $V_{\bar{O}}$ (deg)	.2441E+01
Average $d(V_z)/dr$ (1/sec)	.0000E+00
Average $d(V_\theta)/dr$ (1/sec)	−.4288E−03
Average dW/dr (1/(sec X in.))	−.1630E−03
Average dP/dr (psi/in.)	= .4719E−11

(Continued)

Table 5.5 (Continued)

Apparent Viscosities	Averaged Parameter Values
Average $S_{r\theta}$ (psi)	.7903E−08
Average S_{rz} (psi)	.2432E−04
Average S_{max} (psi)	.2088E−03
Average dissipation function (lbf/(sec sq in.))	.4442E−02
Average apparent viscosity (lbf sec/sq in.)	.1617E−04
Average $D_{r\theta}$ (1/sec) = .2681E−03	.2688E−03
Average D_{rz} (1/sec) = .9912E+00	.9912E+'00

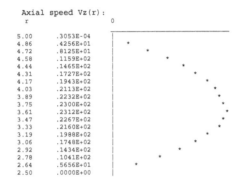

FIGURE 5.53

Axial speed.

FIGURE 5.54

Apparent viscosity.

Table 5.6 Averaged Values of Annular Quantities

Apparent Viscosities	Averaged Parameter Values
Average V_z (in./sec)	.1539E+02
Average V_z (ft/min)	.7693E+02
Average V_θ (in./sec)	.3730E+02
Average W (rad/sec)	.1162E+02
Average total speed (in./sec)	.4150E+02
Average angle between V_z and $V_{\hat{\theta}}$ (deg) = .5993E+02	
Average $d(V_z)/dr$ (1/sec)	.0000E+00
Average $d(V_\theta)/dr$ (1/sec)	−.4303E+02
Average dW/dr (1/(sec X in.))	−.1654E+02
Average dP/dr (psi/in.)	.5811E−01
Average $S_{r\theta}$ (psi)	.6717E−03
Average S_{rz} (psi)	.2432E−04
Average S_{max} (psi)	.7149E−03
Average dissipation function (lbf/(sec sq in.))	.4851E−01
Average apparent viscosity (lbf sec/sq in.)	.1251E−04
Average $D_{r\theta}$ (1/sec)	.2733E+02
Average D_{rz} (1/sec)	.1350E+01

These solutions also provide a means to correlate experimental data nondimensionally. Still, we emphasize that the rotation can induce opposite effects depending on how eccentric or concentric the annulus is for the non-Newtonian fluid assumed.

MODEL 5.7

Cuttings Transport Flow Correlations in Deviated Wells

Industry interest in horizontal and highly deviated wells has heightened the importance of annular flow modeling as it relates to hole cleaning. Cuttings transport to the surface is generally impeded by hole orientation; this is worsened by decreased "low-side" annular velocities due to pipe eccentricity. In addition, the blockage created by bed buildup decreases overall flow rate, further reducing cleaning efficiency. In what could possibly be a self-sustaining, destabilizing process, stuck pipe is a likely end result. This section discloses new cuttings transport correlations and suggests simple predictive measures to avoid bed buildup. Good hole cleaning and bed removal, of course, are important to cementing as well.

Few useful annular flow models are available despite their practical importance. The nonlinear equations governing Power law viscous fluids, for example, must be solved with difficult no-slip conditions for highly eccentric geometries. Recent slot flow models offer some improvement over parallel plate approaches. However, because they unrealistically require slow radial variations in the circumferential direction, large errors are possible. Even when they apply, these models can be cumbersome; they involve "elliptic integrals," which are too awkward for field use.

Recently developed bipolar coordinate models accurately simulate eccentric flows with circular pipes and boreholes; however, they cannot be extended to real-world applications containing washouts and cuttings beds. In this section, the eccentric flow model is used to interpret field and laboratory results. Because the model actually simulates reality, it has been possible to correlate problems associated with cuttings transport and stuck pipe to unique average mechanical properties of the computed flow field. These correlations are discussed next.

Water-based muds

Detailed computations using the eccentric model are described, assuming a Power law fluid, which correspond to the comprehensive suite of cuttings transport experiments conducted at the University of Tulsa (see Becker, Azar, and Okrajni (1989)). For a fixed inclination and oncoming flow rate, we demonstrate that "cuttings concentration" *correlates linearly with the mean viscous shear stress averaged over the lower half of the annulus*. Thus, impending cuttings problems can be eased by first determining the existing average stress level and then adjusting n, K, and gpm values to increase that stress. Physical arguments supporting our correlations will be given. We emphasize that the present approach is completely predictive and deterministic; it does *not* require empirical assumptions related to the "equivalent hydraulic radius" with questionable "pipe to annulus conversion factors."

Detailed experimental results for cuttings concentration, a useful indicator of transport efficiency and carrying capacity, were obtained at the University of Tulsa's large-scale flow loop. Fifteen bentonite-polymer water-based muds for three average flow rates (1.91, 2.86, and 3.82 ft/sec) at three borehole inclinations from vertical (30, 45, and 70 degrees) were tested. In Becker et al. (1989), Table 1 summarizes all measured mud properties, along with specific Power law exponents n and consistency factors K.

We emphasize that "water-based" does *not* imply Newtonian flow; in fact, the reported values of n differed substantially from unity. The annular geometry consisted of a 2-in.-radius pipe, displaced downward by 1.5 in. in a 5-in.-radius borehole; also, the pipe rotated at 50 rpm. Note that 50 rpm corresponds to a tangential surface speed of about 1 ft/sec, so that values in the annulus were much lower. Since the ratio $v_\theta/v_z << 1$, we neglected rotation in the this correlation study.

With flow rate and hole inclination fixed, the authors cross-plotted the nondimensional cuttings concentration, C, versus particular rheological properties for each mud type used. These included apparent viscosity, plastic viscosity (PV), yield point (YP), YP/PV, initial and 10-minute gel strength, "effective viscosity," K, and Fann dial readings at various rpms. Typically, the correlations obtained were poor, with one exception, to be discussed. That good correlations were not possible, of course, is not surprising; the "fluid properties" in Becker et al. (1989) were rotational viscometer readings describing the test instrument only. That is, they had no real bearing to the actual annular geometry and the corresponding downhole flow.

These cross-plots and tables, numbering more than 20, were nevertheless studied in detail; using them, the entire laboratory database was reconstructed. The steady eccentric annular model was then executed for each of the 135 experimental points; detailed results for *calculated* apparent viscosity, shear rate, viscous stress, and axial velocity, all of which varied spatially, were tabulated and statistically analyzed along with the experimental data.

Numerous cross-plots were produced, examined, and interpreted. The most meaningful correlation parameter found was the *mean viscous shear stress*, obtained by averaging computed values over the

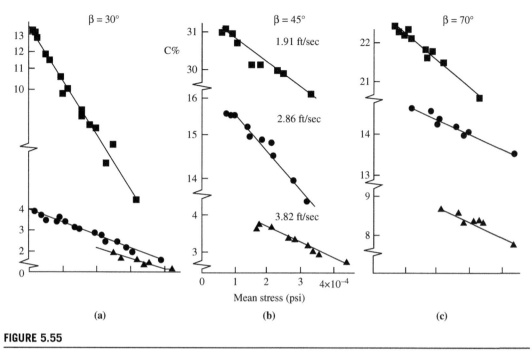

$\beta = 30°$ $\beta = 45°$ $\beta = 70°$

FIGURE 5.55

Cuttings transport correlation.

bottom half of the annulus, where cuttings in directional wells are known to form beds. Figure 5.55 (a−c) display cuttings concentration versus our mean shear stress for different average flow speeds and inclination angles ß from the vertical. Each plotted symbol represents a distinct test mud. Calculated correlation coefficients averaged a high 0.91 value. Our correlations apply to laminar flow only.

The program produces easily understood text-mode information. Figure 5.56 displays, for example, calculated areal results for viscous shear stress in the visual format described earlier. Tabulated results, in this case for "Mud No. 10" at 1.91 ft/sec, show that the "24" at the bottom refers to "0.00024 psi" (the numbers in the plot, when multiplied by 10^{-5}, thus give the actual psi level). A high value of "83" is seen on the upper pipe surface; lows are generally obtained away from solid surfaces and at the annular floor. The average of these calculated values, taken over the bottom half of the annulus, supply the mean stress points on the horizontal axes of Figures 5.55.

Becker et al. (1989) noted that the best data fit, obtained through trial and error, was obtained with low shear rate parameters, in particular Fann dial (stress) readings at low rotary speeds such as 6 rpm. This corresponds to a shear rate of 10/sec. Our exact, computed results gave averaged rates of 7 to 9/sec for *all* mud samples at 1.91 ft/sec; similarly, 11 to 14/sec at 2.86 ft/sec and 14 to 19/sec at 3.82 ft/sec. Since these are in the 10/sec range, they explain why a 6-rpm correlation worked, at least in their particular test setup. But, in general, the Becker "low rpm" recommendation will *not* apply a priori; each nonlinear annular flow presents a unique physical problem with its own characteristic shears. In general, pipe-to-hole diameter ratio, as well as eccentricity, enters the equation. But this poses no difficulty, since downhole properties *can* be obtained with minimal effort with the present computer model.

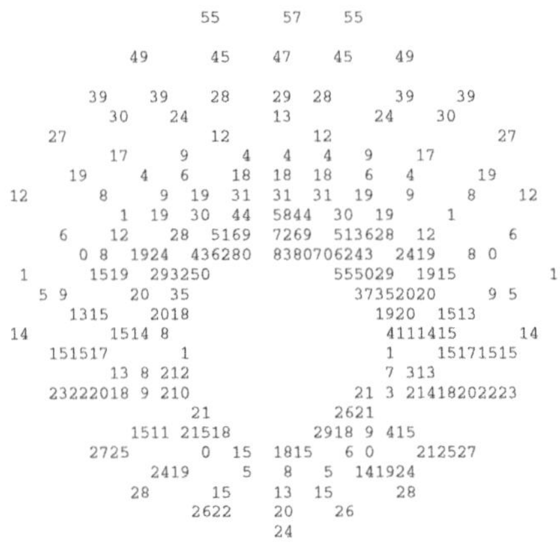

```
                        55      57    55
                49     45    47    45    49
        39    39    28    29  28      39    39
           30    24         13      24    30
    27              12         12                    27
        17      9      4    4   4    9    17
      19      4   6    18   18  18   6    4      19
  12        8      9   19   31  31  31   19   9      8      12
         1   19   30   44   5844  30   19      1
      6    12    28   5169  7269  513628  12         6
        0  8  1924  436280  8380706243  2419   8  0
  1        1519  293250           555029  1915          1
     5  9      20   35          37352020       9  5
       1315     2018            1920   1513
  14        1514  8          4111415        14
     151517       1          1     15171515
          13  8  212          7  313
     23222018  9  210         21  3  21418202223
              21              2621
          1511  21518      2918  9  415
       2725        0   15  1815   6  0     212527
          2419      5    8   5  141924
             28        15   13  15      28
                2622       20    26
                           24
```

FIGURE 5.56

Viscous stress.

Cuttings removal in near-vertical holes with $\beta < 10°$ is well understood; cleaning efficiency is proportional to annular velocity or, more precisely, the "Stokes product" between relative velocity and local viscosity. This product appears naturally in Stokes' original low Reynolds flow solutions for flows past spheres, forming part of the coefficient describing net viscous drag. For inclined wells, the usual notions regarding unimpeded settling velocities do not apply because different processes are at work. Cuttings travel almost immediately to the low side of the annulus, a consequence of gravity segregation; they remain there and form beds that may or may not slide downward. These truss- or lattice-like structures have well-defined mechanical yield stresses; the right amount of viscous friction will erode the cuttings bed, in the same way mud circulation limits dynamic filter cake growth. This explains our success in using bottom-averaged viscous stress as the correlation parameter. The straight line fit also indicates that bed properties are linear in an elastic sense.

These ideas are not entirely new. Slurry pipeline designers, for example, routinely consider "boundary shear" and "critical tractive force." They have successfully modeled sediment beds as "series of superposed layers" with distinct yield strengths (Streeter, 1961). However, these studies are usually restricted to Newtonian carrier fluids in circular conduits.

While viscous shear emerges as the dominant transport parameter, its role was by no means obvious at the outset. Other correlation quantities tested include vertical and lateral components of shear rates and stresses, axial velocity, apparent viscosity, and Stokes product. These correlated somewhat well, particularly at low inclinations, but shear stress almost *always* worked. Consider apparent viscosity as an example. Whereas Figure 6 of Becker et al. (1989) shows significant wide-band scatter, listing *rotational viscometer* values ranging from 1 to 50 cp, our *exact* computations gave good correlations with *actual* apparent viscosities ranging up to 300 cp. Computed viscosities expectedly showed no meaningful connection to the apparent viscosities given by the University of Tulsa investigators, because the latter were inferred from unrealistic Fann dial readings. This point is illustrated quantitatively later.

We emphasize that Figure 5.55(a–c) are based on unweighted muds. On a separate note, the effect of "pure changes in fluid density" should not alter computed shear stresses, at least theoretically, since the convective terms in the steady governing equations vanish. In practice, however, oilfield weighting materials are likely to alter n and K; thus, some change in stress level might be anticipated. The effects of buoyancy, not treated here, will of course help without regard to changes in shear.

We have shown how cuttings concentration correlates in a satisfactory manner with *the mean viscous shear stress averaged over the lower half of the annulus*. Thus, impending hole-cleaning problems can be alleviated by first determining the existing average stress level and then by adjusting n, K, and gpm values in the actual drilling fluid to increase that stress. Once this danger zone is past, additives can be used to reduce shear stress and thus mud pump pressure requirements. Simply increasing gpm may also help, although the effect of rheology on stress is probably more significant.

Interestingly, Seeberger et al. (1989) described an important field study where extremely high velocities together with very high yield points did not alleviate hole-cleaning problems. They suggested that extrapolated YP values may not be useful indicators of transport efficiency. Also, the authors pointed to the importance of elevated stress levels at low shear rates in cleaning large-diameter holes at high angles. They experimentally showed how oil- and water-based muds having like rheograms, despite their obvious textural or "look and feel" differences, will clean with like efficiencies. This implies that a knowledge of n and K alone suffices in characterizing real muds.

The procedure just described requires minimal change to field operations. Standard viscometer readings still represent required input information, but they should be used to determine actual downhole properties through computer analysis. Yield point and plastic viscosity, arising from older Bingham models, play no direct role in the present methodology, although these parameters sometimes offer useful correlations.

Cuttings transport database

The viscometer properties and cuttings concentrations data for the 15 muds (at all angles and flow rates), together with exact computed results for shear rate, stress, apparent viscosity, annular speed, and Stokes product, have been assembled into a comparative database for continuing study. These detailed results are available from the author upon request. Tables 5.7, 5.8, and 5.9 summarize bottom-averaged results for the eccentric hole used in the Tulsa experiments. Computations show that the bottom of the hole supports a low shear rate flow, ranging from 10 to 20 reciprocal seconds. These values are consistent with the authors' low shear rate conclusions, established by trial and error from the experimental data. However, their rule of thumb is not universally correct; for example, the same muds and flow rates gave high shear rate results for several different downhole geometries.

Shear rates *can* vary substantially depending on eccentricity and diameter ratio. Direct computational analysis is the only legitimate and final arbiter. These tables also give calculated apparent viscosities along with values extrapolated from rotating viscometer data (shown in parentheses). Comparison shows that no correlation between the two exists, a result not unexpected, since the

Table 5.7 Bottom-Averaged Fluid Properties at 1.91 ft/sec

Mud	n	K lbf sec^n/in.^{2}	Shear Rate 1/sec	Shear Stress (psi)	Apparent-Viscosity (cp)
1	1.00	0.15E−6	9.1	0.13E−5	1 (1)
2	0.74	0.72E−5	8.1	0.29E−4	27 (8)
3	0.59	0.13E−4	7.8	0.34E−4	35 (5)
4	0.74	0.14E−4	8.3	0.59E−4	54 (15)
5	0.59	0.25E−4	7.6	0.67E−4	71 (9)
6	0.42	0.57E−4	7.4	0.95E−4	116 (6)
7	0.74	0.24E−4	8.1	0.97E−4	89 (25)
8	0.59	0.43E−4	7.6	0.11E−3	118 (15)
9	0.42	0.94E−4	7.5	0.16E−3	191 (10)
10	0.74	0.38E−4	8.2	0.16E−3	143 (40)
11	0.59	0.68E−4	7.7	0.18E−3	190 (24)
12	0.42	0.15E−3	7.5	0.25E−3	307 (16)
13	0.74	0.48E−4	8.0	0.19E−3	180 (50)
14	0.59	0.85E−4	7.6	0.22E−3	237 (30)
15	0.42	0.19E−3	7.4	0.32E−3	388 (20)

Table 5.8 Bottom-Averaged Fluid Properties at 2.86 ft/sec

Mud	n	K lbf sec^n/in.^{2}	Shear Rate 1/sec	Shear Stress (psi)	Apparent-Viscosity (cp)
1	1.00	0.15E−6	14	0.20E−5	1 (1)
2	0.74	0.72E−5	12	0.39E−4	24 (8)
3	0.59	0.13E−4	11	0.42E−4	30 (5)
4	0.74	0.14E−4	12	0.78E−4	49 (15)
5	0.59	0.25E−4	11	0.84E−4	60 (9)
6	0.42	0.57E−4	11	0.11E−3	91 (6)
7	0.74	0.24E−4	12	0.13E−3	80 (25)
8	0.59	0.43E−4	11	0.14E−3	100 (15)
9	0.42	0.94E−4	11	0.19E−3	152 (10)
10	0.74	0.38E−4	12	0.21E−3	129 (40)
11	0.59	0.68E−4	11	0.23E−3	161 (24)
12	0.42	0.15E−3	11	0.30E−3	242 (16)
13	0.74	0.48E−4	12	0.26E−3	161 (50)
14	0.59	0.85E−4	11	0.28E−3	199 (30)
15	0.42	0.19E−3	11	0.38E−3	305 (20)

measurements bear little relation to the downhole flow. On the other hand, calculated apparent viscosities correlated well with cuttings concentration, although not as well as did viscous stress. This correlation was possible because bottom-averaged shear rates did not vary appreciably from mud to mud at any given flow speed. This effect may be fortuitous.

Table 5.9 Bottom-Averaged Fluid Properties at 3.82 ft/sec

Mud	n	K lbf sec^n/in.^{2}	Shear Rate 1/sec	Shear Stress (psi)	Apparent Viscosity (cp)
1	1.00	0.15E−6	18	0.27E−5	1 (1)
2	0.74	0.72E−5	16	0.49E−4	22 (8)
3	0.59	0.13E−4	15	0.50E−4	27 (5)
4	0.74	0.14E−4	17	0.98E−4	45 (15)
5	0.59	0.25E−4	15	0.10E−3	53 (9)
6	0.42	0.57E−4	15	0.13E−3	78 (6)
7	0.74	0.24E−4	16	0.16E−3	74 (25)
8	0.59	0.43E−4	15	0.17E−3	88 (15)
9	0.42	0.94E−4	15	0.21E−3	128 (10)
10	0.74	0.38E−4	16	0.26E−3	119 (40)
11	0.59	0.68E−4	15	0.27E−3	142 (24)
12	0.42	0.15E−3	15	0.34E−3	205 (16)
13	0.74	0.48E−4	17	0.33E−3	148 (50)
14	0.59	0.85E−4	15	0.33E−3	177 (30
15	0.42	0.19E−3	15	0.43E−3	258 (20)

Invert emulsions versus "all oil" muds

Conoco's early Jolliet project successfully drilled a number of deviated wells, ranging 30° to 60° from vertical, in the deepwater Green Canyon Block 184 using a new "all oil" mud. Compared with wells previously drilled in the area with conventional invert emulsion fluids, the oil mud proved vastly superior with respect to cuttings transport and overall hole cleaning (Fraser, 1990a,b,c). High levels of cleaning efficiency were maintained consistently throughout the drilling program. In this section we explain, using the fully predictive, steady eccentric annular flow model, *why* the particular oil mud employed by Conoco performed well in comparison with the invert emulsion.

Given the success of the correlations developed in the prior discussion, it is natural to test our "stress hypothesis" under more realistic and difficult field conditions. Conoco's Green Canyon experience is ideal in this respect. Unlike the unweighted, bentonite-polymer water-based muds used in the University of Tulsa experiments, the drilling fluids employed by Conoco were "invert emulsion" and "all oil" muds. Again, Seeberger et al. (1989) have demonstrated how oil-based and water-based muds having similar rheograms, despite obvious textural differences, will clean holes with like efficiencies. This experimental observation implies that a knowledge of n and K alone suffices in characterizing the carrying capacity of water, oil-based, or emulsion-based drilling fluids. Thus, the use of a Power law annular flow model as the basis for comparison for the two Conoco muds is completely warranted.

We assumed for simplicity a 2-in.-radius drill pipe centered halfway down a 5-in.-radius borehole. This eccentricity is consistent with the 30° to 60° inclinations reported by Conoco. The n and K values we required were calculated from Figure 2 of Fraser (1990b), using Fann dial readings at 13 and 50 rpm. For the invert emulsion, we obtained n = 0.55 and K = 0.0001 lbf sec^n/in.^{2}; the values n = 0.21 and K = 0.00055 lbf sec^n/in.^{2} were found for the "all oil" mud. Our annular geometry is identical to that used in the previous discussion and in Becker et al. (1989). It was chosen so

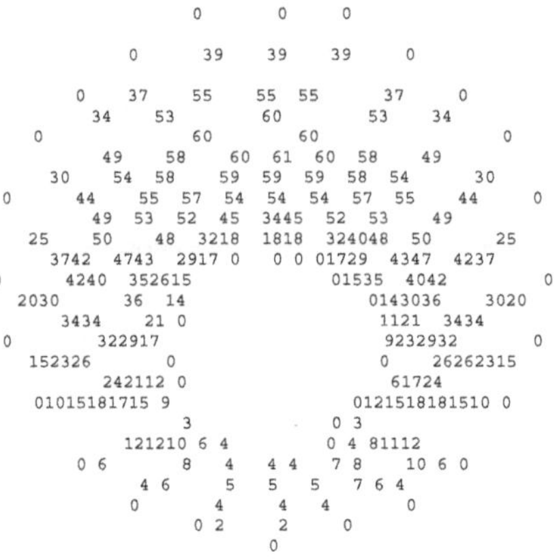

```
                     0    0   0
                 0   39   39  39   0
             0  37  55  55 55    37   0
               34  53     60     53  34
            0                          0
               49  58  60 61 60 58   49
            30  54 58 59 59 59 58 54    30
            0  44   55 57 54 54 54 57 55   44   0
               49 53 52 45 3445 52 53    49
            25  50   48 3218 1818 324048 50    25
              3742 4743 2917 0  0 0 01729 4347 4237
            0  4240 352615          01535 4042   0
            2030    36 14            0143036  3020
              3434   21 0            1121 3434
            0   322917               9232932    0
              152326    0             0  26262315
                242112 0             61724
              01015181715 9          0121518181510 0
                   3                   0 3
                 121210 6 4           0 4 81112
               0 6      8   4   4 4  7 8   10 6 0
                  4 6     5   5  5  7 6 4
                   0      4   4  4      0
                      0 2    2      0
                             0
```

FIGURE 5.57

Annular velocity, invert emulsion.

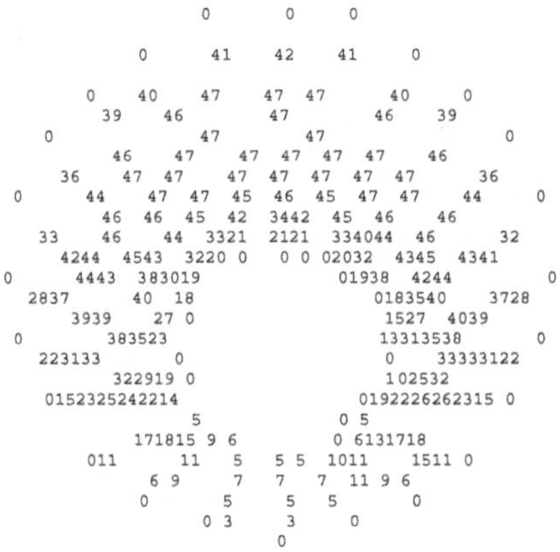

```
                     0    0   0
                 0   41   42  41   0
             0  40  47  47 47    40   0
               39  46     47     46  39
            0                          0
               46  47  47 47 47 47   46
            36  47 47 47 47 47 47 47    36
            0  44   47 47 45 46 45 47 47   44   0
               46 46 45 42 3442 45 46    46
            33  46   44 3321 2121 334044 46    32
              4244 4543 3220 0  0 0 02032 4345 4341
            0  4443 383019          01938 4244   0
            2837    40 18            0183540  3728
              3939   27 0            1527 4039
            0   383523               13313538    0
              223133    0             0  33333122
                322919 0             102532
              0152325242214          0192226262315 0
                   5                   0 5
                 171815 9 6           0 6131718
               011      11   5   5 5  1011   1511 0
                  6 9     7   7  7  11 9 6
                   0      5   5  5      0
                      0 3    3      0
                             0
```

FIGURE 5.58

Annular velocity, all oil mud.

that the shear stress results obtained for the Tulsa water-based muds (shown earlier in Figure 5.55) can be directly compared with those found for the weighted invert emulsion and oil fluids considered.

For comparative purposes, the two runs described here were fixed at 500 gpm. To maintain this flow rate, the invert emulsion required a local axial pressure gradient of 0.010 psi/ft; Conoco's all oil mud, by contrast, required 0.029 psi/ft. Figures 5.57 and 5.58, for invert emulsion and all oil muds, give calculated results for axial velocity in in./sec. Again, note how all no-slip conditions are identically satisfied.

Figures 5.59 and 5.60 display the absolute values of the *vertical* component of viscous shear stress; the leading significant digits are shown, corresponding to magnitudes that are typically $O(10^{-3})$ to

```
                6       6       6
            5       7   7   7       5
          4   6   4   4   4     6   4
            4   3     2       3   4
          3       2     2       3
            2   1   1 1 1 1   2
          3   0 1     3 3 3   1 0     3
        1     1   1 3 4   4 4 3 1   1   1
            0 3   4 6   8 6   4   3     0
        1   2     4   7 9 10 9   7 5 4   2         1
          0 1   3 4   6 810 1110 9 8 6   4 3   1 0
        0     3 3   4 4 7         7 7 4   3 3         0
          0 1       3   5       5 5 3 3     1 0
        2       2 3     3 2           2 3   3 2
              3 2 1               0 2 2 3       2
        2 2 3         0           0     3 3 2 2
            2 1 0 1                 1 0 2
        3 3 3 3 2 0 1               3 0 0 3 3 3 3 3
                  3                 4 3
              3 2 0 2 3         4 3 1 1 3
          4 4         0   2   3 2   1 0       3 4 4
              4 3       1   2   1   2 3 4
            4         3       2   3       4
              4 3               3       4
                      3
```

FIGURE 5.59

Viscous stress, invert emulsion.

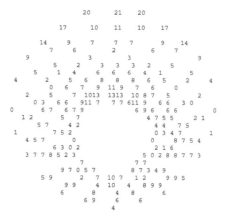

FIGURE 5.60

Viscous stress, all oil mud.

$O(10^{-4})$ psi. This shear stress is obtained as the product of local apparent viscosity and shear rate, both of which vary throughout the cross section. That is, the viscous stress is obtained *exactly* as "apparent viscosity (x, y) $\times$ dU(x, y)/dx."

Figure 5.57 shows that the invert emulsion yields maximum velocities near 61 in./sec on the high side of the annulus; the maximums on the low side, approximately 5 in./sec, are less than ten times this value. By comparison, the "all oil" results in Figure 5.58 demonstrate how a smaller n tends to redistribute velocity more uniformly. Still, the contrast is high, being 47 in./sec to 7 in./sec. The difference between the low-side maximum velocities of 5 and 7 in./sec are not significant and certainly do *not* explain observed large differences in cleaning efficiency.

Our earlier results in the first discussion provided experimental evidence suggesting that mean viscous shear stress is the correct correlation parameter for hole-cleaning efficiency. This is, importantly, the case here. First note how Figure 5.59 gives a bottom radial stress distribution of "3-2-2-3-3" for the invert emulsion mud. In the case of Conoco's "all oil" mud, Figure 5.60 shows that these values significantly increase to "10-10-4-6-4." We calculated mean shear stress values averaged over the lower half of the annulus. These values, for oil-based and invert-emulsion muds, respectively, were 0.00061 and 0.00027 psi. Their ratio, a sizable 2.3, substantiates the positive claims made in Fraser (1990b). Calculated shear stress averages for the University of Tulsa experiments in no case exceeded 0.0004 psi.

Similarly averaged apparent viscosities also correlated well, leading to a large ratio of 2.2. (The "apparent viscosities" in Becker et al. (1989) did not correlate at all, because nonmeaningful rotational viscometer readings were used.) Bottom-averaged shear rates, for oil-based and invert-emulsion muds, were calculated as 12.7 and 9.6/sec, respectively; at least in this case, we have again justified the "6 rpm (or 10/sec) recommendation" offered by many drilling practitioners. In general, however, shear rates will vary widely; they can be substantial depending on the particular geometry and drilling fluid.

The present results and the detailed findings of the first discussion, together with the recommendations of Seeberger et al. (1989), strongly suggest that "bottom-averaged" viscous shear stress correlates well with cuttings carrying capacity. Thus, as before, a driller suspecting cleaning problems should first determine his current downhole stress level; then he should alter n, K, and gpm to increase that stress. Once the danger is past, he can lower overall stress levels to reduce mud pump pressure requirements. This "stress hypothesis" for hole cleaning, first proposed by the author in 1991 in *Borehole Flow Modeling*, has been adopted for internal use at several oil companies.

Issues in cuttings transport

The empirical cuttings transport literature contains confusing observations and recommendations that, in light of the foregoing results, can be easily resolved. We will address several questions commonly raised by drillers. First and foremost is "Which parameters control transport efficiency?" In vertical wells, the drag or uplift force on small isolated chips can be obtained from lubrication theory via Stokes's or Oseen's low Reynolds number equations. This force is proportional to the product between local viscosity and the first power of relative velocity between chip and fluid. The so-called "Stokes product" correlates well in vertical holes.

In deviated and horizontal holes with eccentric annular geometries, cuttings beds invariably form on the low side. These beds consist of well-defined mechanical structures with nonzero yield stresses; to remove or erode them, viscous fluid stresses must be sufficiently strong to overcome their resilience. The stresses computed on a laminar basis are sufficient for practical purposes, because low-side, low-velocity flows are almost always laminar. In this sense, any turbulence in the high-side flow is unimportant, since it plays no direct role in bed removal (the high-side flow does convect debris that are uplifted by rotation). This observation is reiterated by Fraser (1990c). In his paper, Fraser correctly points out that too much significance is often attached to velocity criteria and fluid turbulence in deviated wells.

A second question concerns drillpipe rotation. With rotation, centrifugal effects throw cuttings circumferentially upward, where they are convected uphole by the high-side flow; then they fall downward. In the first part of this cycle, the cuttings are subject to drag forces not unlike those found in vertical wells. Turbulence can be important, determining the amount of axial throw traversed before the cuttings are redeposited into the bed. In addition to "throwing" estimates, the capabilities offered by the transient simulator now permit calculations for rotation-induced stress effects on cuttings bed erosion.

Other effects of rotation are subtly tied to the rheology of the background fluid. Conflicting observations and recommendations are often made regarding drillpipe rotation for concentric annuli. To resolve them, we need to reiterate some of the theoretical results from Example 5.6. There we demonstrated that axial and circumferential speeds completely decouple for laminar Newtonian flows despite the nonlinearity of the Navier-Stokes equations. This is so because the convective terms exactly vanish, allowing us to "naively" (but correctly) superpose the two orthogonal velocity fields.

This fact was, apparently, first deduced by Savins et al. (1966), who noted that no coupling between the discharge rate, axial pressure gradient, relative rotation, and torque could be found through the viscosity coefficient for Newtonian flows. The decoupling implies that experimental findings obtained using Newtonian drilling fluids (primarily water and air) cannot be extrapolated to more general Power law or Bingham plastic rheologies. Likewise, rules of thumb deduced using real drilling muds will not be consistent with those found for water. Newtonian (e.g., brines) and "real" muds behave differently in the presence of pipe rotation. In a Newtonian fluid, rotation will not affect the axial flow, although "centrifugal throwing" is still important. Note that in an initially steady, concentric non-Newtonian flow, where the mud pump is operating at constant pressure, a momentary increase in rpm leads to a temporary surge in flow rate and thus improved hole cleaning.

The decoupling just discussed applies to Newtonian flows in concentric annuli only. The coupling between axial and circumferential velocities reappears, even for Newtonian flows, when the rotating motion occurs in an eccentric annulus. This is so because the nonlinear convective terms will not identically vanish. This coupling is amply demonstrated in our transient calculations for flows with rotation. In general, concentric flow loop tests using Newtonian fluids provide little benefit or information in terms of field usefulness. In fact, their results will be subject to misinterpretation.

And the role of fluid rheology? We have demonstrated how bottom-averaged shear stress can be used as a meaningful correlation parameter for cuttings transport in eccentric deviated holes. This mean viscous stress can be computed using steady methods for flows without rotation or transient models for flows with rotation. The arguments given in several discussions are sound on physical grounds; in cuttings transport, rheology is a significant player by way of its effect on fluid stress.

We have *not* modeled, nor do we suggest computational studies focusing on the dynamics of single chips or ensembles of cuttings chips, for reasons cited in Chapter 1. As we have demonstrated, it suffices to use viscous stress as a correlation parameter. Modeling the dynamics of aggregates of chips involves mathematics so complicated that it is difficult to anticipate any practical significance, even in the long term.

Finally, we comment on the role of increasing fluid density to improve hole cleaning. For steady flows without rotation, fluid density completely drops out of the governing equations, but these equations were written without body forces. In this sense, density is not important. However, when more complete equations are considered, it is clear that higher densities will always increase buoyancy effects and that the consequent uplift is beneficial as far as helical "throwing" is concerned; it is therefore helpful in hole cleaning.

MODEL 5.8

Cuttings Bed Growth as an Unstable Flow Process

In vertical wells in which drilled cuttings move unimpeded, cuttings transport and hole-cleaning efficiency vary directly as the product between mud viscosity and "relative particle and annular velocity." For inclined wells, bed formation introduces a new physical source for clogging. Often, this means that rules of thumb developed for vertical holes are not entirely applicable to deviated wells. For example, Seeberger et al. (1989) pointed out that substantial increases in both yield point and annular velocity did not help in alleviating their hole problems. They suggested that high shear stresses at low shear rates would be desirable and that stress could be a useful indicator of cleaning efficiency in deviated wells. We have given compelling evidence for this hypothesis.

Using the steady eccentric annular flow model, we have demonstrated that "cuttings concentration" correlates *linearly* with mean shear stress—that is, *the viscous stress averaged over the lower half of the annulus*—for a wide range of oncoming flow speeds and well inclinations. Apparently, this empirical correlation holds for invert emulsions and oil-based muds as well.

Having established that shear stress is an important parameter in bed formation, it is natural to ask whether cuttings bed growth itself helps or hinders further growth; that is, does bed buildup constitute a self-sustaining, destabilizing process? The classic "ball on top of the hill," for instance, continually falls once it is displaced from its equilibrium position. In contrast, the "ball in the valley" consistently returns to its origin, demonstrating "absolute stability." In this section, we will consider the effect of cuttings bed thickness.

If cuttings bed growth itself induces further growth, the cleaning process will be unstable in the foregoing sense. This instability will underline, in field applications, the importance of controlling downhole rheology so as to increase stress levels at the onset of danger. Field site flow simulation can play an important role in operations—that is, in determining existing stress levels with a view towards optimizing fluid rheology in order to increase them. In this section, calculations are described that suggest that instability is possible.

In the nonrotating eccentric flow calculations that follow, we assume a 2-in.-radius nonrotating drill pipe, displaced 1.5 in. downward in a 5-in.-radius borehole. This annular geometry is the same as the experimental setup reported in Becker et al. (1989). For purposes of evaluation, we arbitrarily selected "Mud No. 10," which was used by the University of Tulsa team. It has a Power law exponent of 0.736 and a consistency factor of 0.0000383 lbf $\sec^n$/in.^{2} The total annular volumetric flow rate was fixed for all of our runs, corresponding to usual operating conditions. The average linear speed was held to 1.91 ft/sec or 22.9 in./sec. In the reported experiments, this speed yielded laminar flow at all inclination angles.

Four case studies were performed, the first containing no cuttings bed; the next three assuming flat cuttings beds successively increasing in thickness. The level surfaces of the "small," "medium," and "large" beds were located at 0.4, 0.8, and 1.0 in., respectively, from the bottom of the annulus. Required pressure drops varied from 0.0054 to 0.0055 psi/ft. The "Steady 2D" simulator offers highly visual text output that directly overlays computed quantities on the cross-sectional geometry, thus facilitating physical interpretation and correlation with annular position. Computed results for axial velocity in in./sec are shown in Figures 5.61 through 5.64.

All four velocity distributions satisfy the no-slip condition exactly; the text plotter used, we note, does not always show 0's at solid boundaries because of character spacing issues. The "no bed" flow given in Figure 5.61 demonstrates very clearly how velocity can vary rapidly about the annulus. For example, it has maximums of 51 and 5 in./sec above and below the pipe, a ten-fold difference. Figures 5.62 through 5.64 show that this factor increases—that is, worsens—as the cuttings bed increases in thickness.

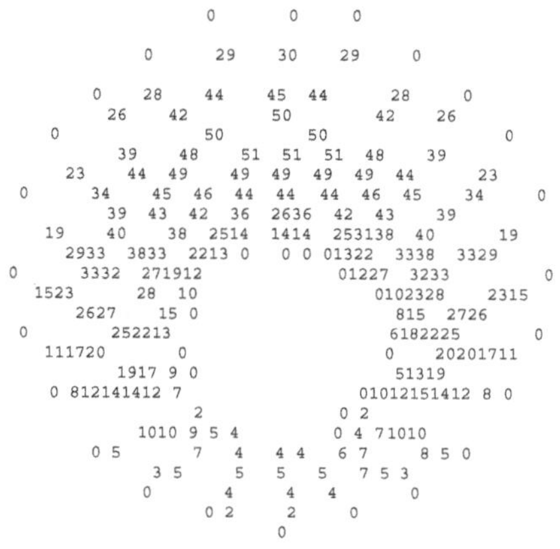

```
                    0     0     0
                0   29    30    29    0
            0   28   44   45 44      28    0
              26   42      50      42   26
          0            50            50          0
              39   48   51 51 51 48   39
            23   44 49   49 49 49 49 44      23
          0   34   45 46 44 44 44 46 45   34    0
              39 43 42 36 2636 42 43     39
          19   40   38 2514 1414 253138 40      19
          2933 3833 2213 0   0 0 01322 3338 3329
        0   3332 271912          01227 3233        0
        1523     28 10             0102328     2315
          2627    15 0              815 2726
        0      252213                6182225      0
        111720     0                0  20201711
          1917 9 0                  51319
        0 812141412 7              01012151412 8 0
                2                      0 2
            1010 9 5 4              0 4 71010
          0 5       7   4        4 4   6 7      8 5 0
              3 5        5      5   5   7 5 3
                0        4        4   4        0
                    0 2      2        0
                        0
```

FIGURE 5.61

Annular velocity, "no bed."

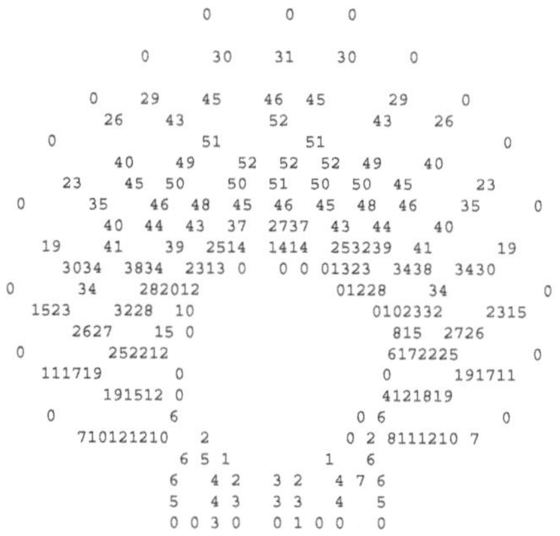

```
                    0     0     0
                0   30    31    30    0
            0   29   45   46 45      29    0
              26   43      52      43   26
          0            51            51          0
              40   49   52 52 52 49   40
            23   45 50   50 51 50 50 45      23
          0   35   46 48 45 46 45 48 46   35    0
              40 44 43 37 2737 43 44     40
          19   41   39 2514 1414 253239 41      19
          3034 3834 2313 0   0 0 01323 3438 3430
        0   34    282012          01228    34        0
        1523   3228 10             0102332     2315
          2627    15 0              815 2726
        0      252212                6172225      0
        111719     0                0    191711
          191512 0                 4121819
        0      6                    0 6          0
        710121210    2            0 2 8111210 7
              6 5 1                1 6
              6   4 2     3 2    4 7 6
              5   4 3     3 3    4   5
              0 0 3 0     0 1 0 0    0
```

FIGURE 5.62

Annular velocity, "small bed."

Figures 5.65 through 5.68 give computed results for the vertical component of the shear stress—that is, "apparent viscosity (x, y) × strain rate dU/dx," where "x" increases downward. Results for the stress related to "dU/dy," not shown because of space limitations, behaved similarly. For clarity, only the absolute values are displayed; the actual values, which are separately available in tabulated form, vary from $O(10^{-4})$ to $O(10^{-3})$ psi.

Note how the bottom viscous stresses decrease in magnitude as the cuttings bed builds in thickness. This decrease, which is accompanied by decreases in throughput area, further compounds cuttings transport problems and decreases cleaning efficiency. Thus, hole clogging is a self-sustaining, destabilizing process. Unless the mud rheology itself is changed in the direction of increasing stress, differential sticking and stuck pipe are possible. This decrease of viscous stress with increasing bed thickness is also supported experimentally. Quigley et al. (1990) measured "unexpected" decreases in

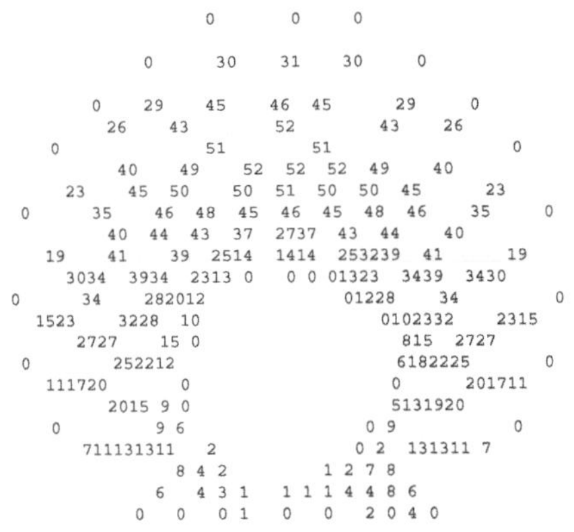

```
                      0      0    0

                0    30   31   30    0

            0   29   45   46  45      29    0
              26    43        52      43   26
          0           51        51            0
              40   49    52  52  52  49     40
          23    45  50    50  51  50  50  45      23
        0    35    46  48  45  46  45  48  46    35    0
              40  44  43  37  2737  43  44     40
          19    41    39  2514  1414  253239  41      19
            3034  3934  2313 0    0 0 01323  3439  3430
        0    34    282012              01228    34        0
        1523    3228  10              0102332    2315
          2727    15 0              815  2727
        0      252212              6182225        0
        111720      0              0    201711
            2015 9 0              5131920
        0      9 6              0 9        0
          711131311    2            0 2  131311 7
                  8 4 2          1 2 7 8
                6    4 3 1    1 1 1 4  4 8 6
                0    0    0 1    0    0    2 0 4 0
```

FIGURE 5.63

Annular velocity, "medium bed."

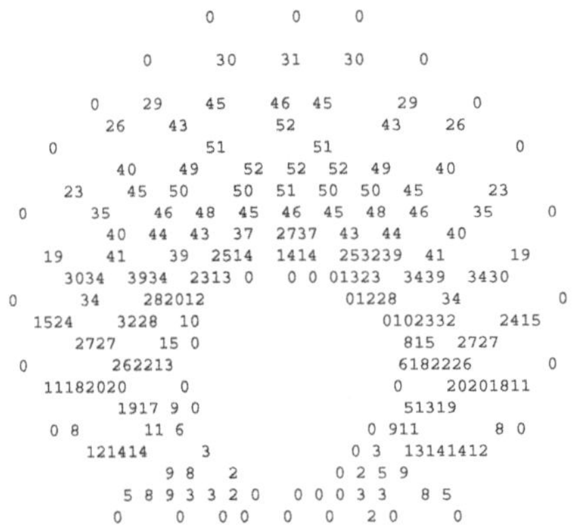

```
                      0      0    0

                0    30   31   30    0

            0   29   45   46  45      29    0
              26    43        52      43   26
          0           51        51            0
              40   49    52  52  52  49     40
          23    45  50    50  51  50  50  45      23
        0    35    46  48  45  46  45  48  46    35    0
              40  44  43  37  2737  43  44     40
          19    41    39  2514  1414  253239  41      19
            3034  3934  2313 0    0 0 01323  3439  3430
        0    34    282012              01228    34        0
        1524    3228  10              0102332    2415
          2727    15 0              815  2727
        0      262213              6182226        0
        11182020      0              0    20201811
            1917 9 0              51319
        0 8      11 6              0 911        8 0
          121414      3            0 3  13141412
                  9 8    2          0 2 5 9
                5 8 9 3 3 2 0    0 0 0 3 3    8 5
                0    0    0 0    0    0    2 0      0
```

FIGURE 5.64

Annular velocity, "large bed."

fluid (as opposed to mechanical) friction in a carefully controlled flow loop where cuttings beds were allowed to grow. While concluding that "cuttings beds can reduce friction," the authors clearly do not recommend its application in the field, since it increases the possibility of differential sticking.

Numerical results such as those shown in Figures 5.65 through 5.68 provide a quantitative means for comparing cleaning capabilities between different muds at different flow rates. "Should I use the "high-tech" mud offered by Company A when the simpler drilling fluid of Company B, run at a different speed, will suffice?" With numerical simulation, these and related questions are readily answered. The present results indicate that the smaller the throughput height, the smaller the viscous stresses will be.

This is intuitively clear, since narrow gaps impose limits on the peak bottom velocity and thus the maximum stress. We caution that this result applies only to the present calculations and may not hold in general. The physical importance of cuttings beds indicates that they should be modeled in any serious well-planning activity. This necessity also limits the potential of recently developed

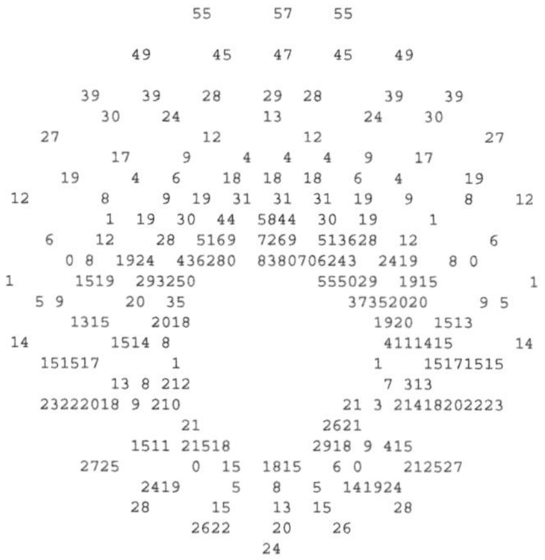

```
                55      57     55

           49      45     47     45      49

      39     39    28    29  28      39     39
         30     24        13      24    30
   27             12           12               27
      17      9     4    4   4     9    17
   19     4    6    18  18  18    6    4       19
12        8    9   19  31  31  31  19    9      8     12
      1  19   30   44  5844   30  19     1
    6    12     28  5169  7269  513628  12        6
    0  8  1924  436280  8380706243  2419    8 0
1      1519  293250           555029  1915         1
   5 9      20   35           37352020      9 5
     1315     2018             1920  1513
14         1514 8             4111415         14
 151517      1             1   15171515
       13 8 212             7 313
23222018 9 210           21  3 21418202223
         21               2621
      1511 21518         2918 9 415
  2725      0  15  1815   6 0      212527
       2419      5    8    5 141924
         28      15    13 15      28
           2622     20    26
                    24
```

FIGURE 5.65

Viscous stress, "no bed."

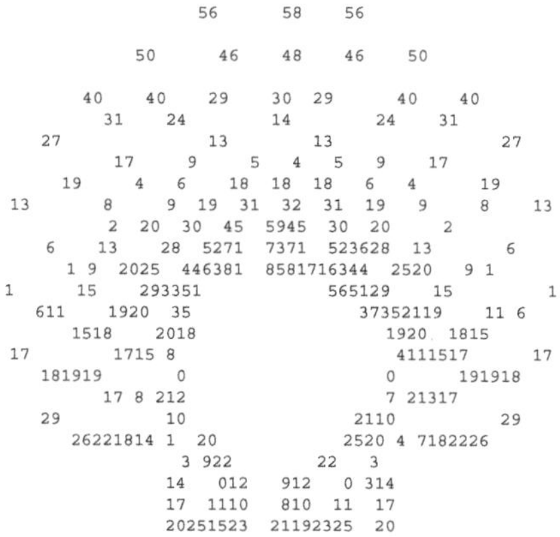

```
                56      58     56

           50      46     48     46      50

      40     40    29    30  29      40     40
         31     24        14      24    31
   27             13           13               27
      17      9     5    4   5     9    17
    19     4    6    18  18  18    6    4       19
13        8    9   19  31  32  31  19    9      8     13
      2   20   30   45  5945   30  20     2
    6    13     28  5271  7371  523628  13        6
    1 9  2025  446381  8581716344  2520   9 1
1      15  293351           565129   15         1
  611    1920  35           37352119      11 6
     1518     2018             1920  1815
17       1715 8             4111517        17
 181919      0             0   191918
       17 8 212             7 21317
29          10             2110             29
 26221814 1  20           2520 4 7182226
        3 922             22  3
          14     012  912    0 314
          17  1110  810   11  17
          20251523  21192325  20
```

FIGURE 5.66

Viscous stress, "small bed."

bipolar coordinate annular flow models. These handle circular eccentric annular geometries well, but they cannot be generalized to handle more difficult holes with cuttings beds. Computationally, though, even when dealing with purely eccentric circles, bipolar methods demand larger resources because many transcendental functions need to be evaluated repeatedly.

MODEL 5.9

Spotting Fluid Evaluation for Stuck Pipe and Jarring Applications

Stuck pipe due to differential pressure between the mud column and the formation often results in costly time delays. The mechanics governing differential sticking are well known (see, for example,

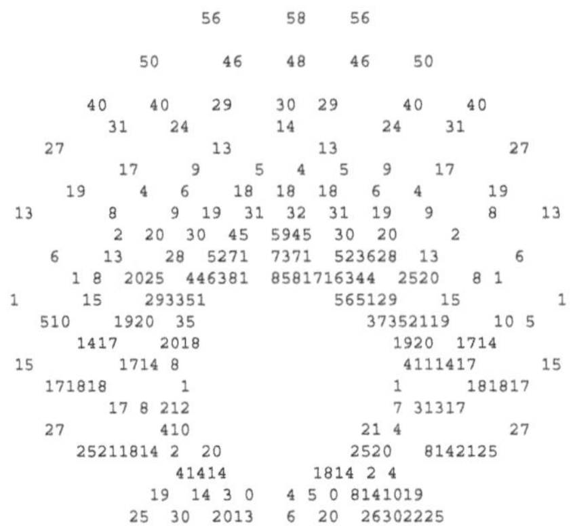

```
                      56      58    56

                50      46    48    46     50

          40    40    29    30  29      40    40
             31     24          14        24    31
       27            13            13                27
             17     9       5    4   5     9    17
       19      4    6     18   18  18    6    4           19
   13       8      9   19   31   32  31   19    9       8     13
         2   20   30   45   5945  30   20       2
      6     13     28   5271  7371  523628  13          6
        1 8   2025   446381  8581716344  2520    8 1
   1       15      293351            565129    15           1
     510     1920  35                 37352119     10 5
       1417      2018                   1920   1714
   15          1714 8                    4111417          15
     171818         1                    1       181817
           17 8 212                      7 31317
       27          410                  21 4              27
     25211814 2   20                    2520    8142125
               41414             1814 2 4
             19   14 3 0     4 5 0 8141019
             25   30   2013    6   20   26302225
```

FIGURE 5.67

Viscous stress, "medium bed."

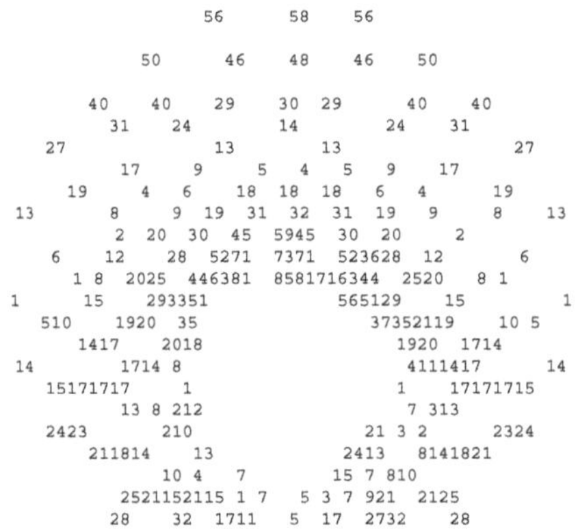

```
                      56      58    56

                50      46    48    46     50

          40    40    29    30  29      40    40
             31     24          14        24    31
       27            13            13                27
             17     9       5    4   5     9    17
       19      4    6     18   18  18    6    4           19
   13       8      9   19   31   32  31   19    9       8     13
         2   20   30   45   5945  30   20       2
      6     12     28   5271  7371  523628  12          6
        1 8   2025   446381  8581716344  2520    8 1
   1       15      293351            565129    15           1
     510     1920  35                 37352119     10 5
       1417      2018                   1920   1714
   14          1714 8                    4111417          14
     15171717        1                   1       17171715
           13 8 212                      7 313
     2423         210                   21 3 2          2324
       211814       13                  2413    8141821
              10 4    7                 15 7 810
            2521152115 1 7     5 3 7  921   2125
              28      32   1711    5   17   2732      28
```

FIGURE 5.68

Viscous stress, "large bed."

Outmans (1958)). In the past, diesel oil, mineral oil, and mixtures of these with surfactants, clays, and asphalts were usually spotted to facilitate the release of the drillstring. However, the use of these conventional spotting fluids is now stringently controlled by government regulation; environmentally safe alternatives must be found.

Halliday and Clapper (1989) described the development of a successful, nontoxic, water-based system. Their spotting fluid, identified using simple laboratory screening procedures, was used to free 1,000 feet of stuck pipe in a 39° hole, from a sand section in the Gulf of Mexico. Since water-based spotting fluids have seldom been studied in the literature, it is natural to ask whether or not they really work and, if so, how. This section calculates, on an exact eccentric flow basis, three important mechanical properties: the apparent viscosity, shear stress, and shear rate of the drilling mud, with and without the spot additive. Then we provide a complete physical explanation for the reported success. The spotting fluid essentially works by mechanically reducing overall apparent viscosity; this enables the resultant fluid to perform its chemical functions better.

The eccentric borehole annular model for steady nonrotating flow was used. While we have successfully applied it to hole cleaning before the occurrence of stuck pipe, it is of interest to apply it to other drilling problems—for example, determining the effectiveness of spotting fluids in freeing stuck pipe. Which mechanical properties are relevant to spotted fluids? What should their orders of magnitude be? We examined the water-based system described in Halliday and Clapper (1989) because such systems are becoming increasingly important. Why they work is not yet thoroughly understood. But it suffices to explain how the water-based spotting fluid behaves, insofar as mechanical fluid properties are concerned, on a single-phase, miscible flow basis. Conventional capillary pressure and multiphase considerations for "oil on aqueous filter cake" effects do not apply here, since we are dealing with "water on water" flows.

We performed our calculations for a 7.75-in.-diameter drill collar located eccentrically within a 12.5-in.-diameter borehole. This corresponds to the bottomhole assembly reported by the authors. A small bottom annular clearance of 0.25 in. was selected for evaluation purposes. This almost closed gap is consistent with the impending stuck pipe conditions characteristic of typical deviated holes.

In Hallliday and Clapper, Table 11 gives Fann 600- and 300-rpm dial readings for the water-based mud used, before and after spot addition; both fluids, incidentally, were equal in density. In the former case, these values were 46 and 28; in the latter, 41 and 24. These properties were measured at 120°F. The calculated n and K Power law coefficients are, respectively, 0.70 and 0.000025 lbf $sec^n/in.^2$ for the original mud; for the spotted mud, we obtained 0.77 and 0.0000137 lbf $sec^n/in.^2$.

Halliday and Clapper (1989) reported that attempts to free the pipe by jarring down, with the original drilling fluid in place, were unsuccessful. At that point, the decision to spot the experimental nonoil fluid was made. Since jarring operations are more impulsive, rather than constant pressure drop processes, we calculated our flow properties for a wide range of applied pressure gradients. Note that the unsteady, convective term in the governing momentum equation has the same physical dimensions as pressure gradient. It was in this approximate engineering sense that our exact simulator was used.

The highest-pressure gradients shown correspond to volumetric flow rates near 1,100 gpm. Computed results for several parameters averaged over the lower half of the annulus are shown in Tables 5.10 and 5.11. We emphasize that calculated averages are sensitive to annular geometry; thus, the results shown in the tables may not apply to other borehole configurations.

In general, any required numerical quantities should be recomputed with the exact downhole geometry. The results for averaged shear stress are "almost" Newtonian in the sense that stress increases linearly with applied pressure gradient. This unexpected outcome is not generally true of non-Newtonian flows. Both treated and untreated muds, in fact, show exactly the same shear stress values. However, shear rate and volumetric flow rate results for the two muds vary differently, and certainly nonlinearly with pressure gradient. The most interesting results, those concerned with spotting properties, are related to apparent viscosity.

The foregoing calculations importantly show how the apparent viscosity for the spotted mud, which varies spatially over the annular cross section, has a nearly constant "bottom average" close to 0.000010 lbf $sec/in.^2$ over the entire range of flow rates. This value is approximately 69 cp, far in excess of the viscosities inferred from rotational viscometer readings, but still *two to three times*

Table 5.10 Fluid Properties, Original Mud

Pressure Gradient (psi/ft)	Flow Rate (gpm)	Apparent Viscosity (lbf sec/in.^{2})	Shear Rate (sec^{-1})	Viscous Stress (psi)
0.0010	69	0.000036	0.4	0.000011
0.0020	185	0.000027	1.2	0.000022
0.0030	329	0.000022	2.1	0.000033
0.0035	410	0.000021	2.6	0.000038
0.0040	497	0.000020	3.2	0.000044
0.0050	683	0.000018	4.3	0.000055
0.0060	886	0.000017	5.6	0.000066
0.0070	1105	0.000016	7.0	0.000077

Table 5.11 Fluid Properties, Spotted Mud

Pressure Gradient (psi/ft)	Flow Rate (gpm)	Apparent Viscosity (lbf sec/in.^{2})	Shear Rate (sec^{-1})	Viscous Stress (psi)
0.0010	140	0.000014	1.0	0.000011
0.0020	344	0.000012	2.4	0.000022
0.0023	412	0.000011	2.8	0.000025
0.0030	582	0.000010	4.0	0.000033
0.0035	711	0.000010	4.9	0.000039
0.0040	846	0.000010	5.8	0.000044
0.0050	1130	0.000009	7.8	0.000055

less than those of the original untreated mud. The importance of "low viscosity" in spotting fluids is emphasized in several mud company publications the author is aware of. Whether the apparent viscosity is high or low, of course, cannot be determined independently of hole geometry and pressure gradient.

The apparent viscosity is relevant because it is related to the lubricity factor conventionally used to evaluate spotting fluids. Importantly, it is calculated on a true eccentric flow basis, rather than determined from (unrelated) rotational viscometer measurements. As in cuttings transport, viscometer measurements are only valid to the extent that they provide accurate information for determining n and K over a limited range of shear rates.

That the treated fluid exhibits much lower viscosities over a range of applied pressures is consistent with its ability to penetrate the pipe and mudcake interface. This lubricates and separates the contact surfaces over a several-hour period; thus, it enables the spotting to perform its chemical functions efficiently, thereby freeing the stuck drillstring. The effectiveness of any spotting fluid, of course, must be determined on a case by case basis.

While computed averages for apparent viscosity are almost constant over a range of pressure gradients, we emphasize that exact cross-sectional values for each flow property can be quite variable. For example, consider the annular flow for the spotted mud under a pressure gradient of

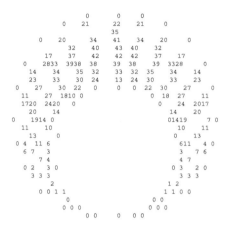

FIGURE 5.69

Annular velocity.

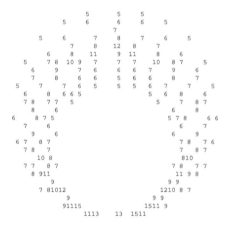

FIGURE 5.70

Apparent viscosity.

0.002 psi/ft, with a corresponding flow rate of 344 gpm. The velocity solutions in in./sec, using the visual text output format discussed previously, are shown in Figure 5.69; note, again, how no-slip conditions are rigorously enforced at all solid surfaces.

Figure 5.70 gives results for *exact apparent viscosity*, which varies with spatial position plotted over the eccentric geometry itself. Although the text plotter does not provide visual resolution at the very bottom, tabulated solutions indicates pipe surface values of 13, increasing to 29 at the mid-section, finally decreasing to 13×10^{-6} lbf sec/in.^{2} at the borehole wall. The flatness of the cuttings bed, or the extent to which it modifies annular bottom geometry, will also be an important factor as far as lubricity is concerned. Any field-oriented hydraulics simulation should also account for such bed effects.

We demonstrated earlier that modeling can be used to correlate laboratory and field cuttings transport efficiency data against actual (computed) downhole flow properties. Bottom-averaged viscous shear stress importantly emerged as the physically significant correlation parameter. This section indicates that annular flow modeling can also be used to evaluate spotting fluid effectiveness in freeing stuck pipe. The important correlation parameter is average apparent viscosity, a fact mechanical engineers might have anticipated. This is directly related to the lubricity factor usually obtained in laboratory measurements.

MODEL 5.10

Newtonian Flow in Rectangular Ducts

In this model, we provide solutions for Newtonian flow in rectangular ducts. These solutions and methods were used in our research to validate more general algorithms for non-Newtonian flow in complicated cross sections. Of course, they are useful in their own right. We will observe that, even with our restriction to the simplest fluid, very different mathematical techniques are needed even for a "simple" change in duct shape. From an engineering point of view, this is impractical: A more "robust" approach applicable to large classes of problems is needed and motivated, particularly by the discussion that follows.

Exact analytical series solution

Here, a closed-form solution for unidirectional, laminar, steady Newtonian viscous flow in a rectangular duct is obtained. Unlike $d^2u(r)/dr^2 + 1/r \, du/dr = 1/\mu \, dp/dz$, which applies to Newtonian flow in circular pipes and takes the form of an *ordinary* differential equation requiring data only at two points in space, we now have the *partial* differential equation

$$\partial^2 u/\partial x^2 + \partial^2 u/\partial y^2 = (1/\mu) \, dp/dz \qquad (5.115)$$

Its solution is obtained, subject to the "no-slip" velocity boundary conditions

$$u(-1/2 \, b < y < +1/2 \, b, x = 0) = 0 \qquad (5.116a)$$

$$u(-1/2 \, b < y < +1/2 \, b, x = c) = 0 \qquad (5.116b)$$

$$u(y = -1/2 \, b, 0 < x < c) = 0 \qquad (5.116c)$$

$$u(y = +1/2 \, b, 0 < x < c) = 0 \qquad (5.116d)$$

where "b" and "c" denote the lengths of the sides of the rectangular duct shown in Figure 5.71.

The solution is obtained by taking u(y, x) as the sum of "particular" and "complementary" solutions—that is, $u = u_p(x) + u_c(y, x)$. To simplify the analysis, we allow $u_p(x)$ to vanish at $x = 0$ and c, while satisfying $d^2u/dx^2 = (1/\mu) \, dp/dz$, where dp/dz is a prescribed constant. Then the particular solution is obtained as $u_p(x) = dp/dz \, x^2/(2\mu) + C_1 x + C_2$, where the constants of integration can be evaluated to give $u_p(x) = -dp/dz \, (xc - x^2)/(2\mu)$. This involves no loss of generality, since the complementary solution $u_c(y, x)$ has not yet been determined and will be expressed as a

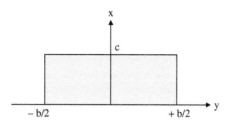

FIGURE 5.71

Rectangular duct cross section.

function of $u_p(x)$. With this choice, the partial differential equation for $u_c(x)$ reduces to the classical Laplace equation

$$\partial^2 u_c/\partial x^2 + \partial^2 u_c/\partial y^2 = 0 \tag{5.117}$$

Now, since $u = 0$ and $u_p(x) = 0$ along the upper and lower edges of the rectangle in Figure 5.71, it follows that $u_c(x) = 0$ there also, since $u_c = u - u_p(x)$. By separating variables in the conventional manner, it is possible to show that product representations of $u_c(y, x)$ involve combinations of trigonometric and exponential functions. In particular, we are led to the combination

$$u_c = \sum_{n=1}^{\infty} A_n \, \cosh\,(n\pi y/c) \, \sin\,(n\pi x/c) \tag{5.118}$$

The factor "sin $(n\pi x/c)$" allows $u_c(y, x)$ to vanish at the lower and upper boundaries $x = 0$ and c. The linear combination of exponentials "cosh $(n\pi y/c)$" is selected because the velocity distribution must be symmetric with respect to the vertical line $y = 0$. Specific products cannot be disallowed, so the infinite summation accounts for the maximum number permitted. The coefficient A_n must be determined in such a way that side wall no-slip conditions are satisfied. To do this, we reconstruct the complete solution as

$$\begin{aligned} u = u_p(x) + u_c(y, x) &= -\, dp/dz \, (xc - x^2)/(2\mu) \\ &+ \sum A_n \, \cosh\,(n\pi y/c) \, \sin\,(n\pi x/c) \end{aligned} \tag{5.119}$$

and apply the boundary conditions given by Equations 5.116a and 5.116b. The coefficients of the resulting Fourier series can be used, together with the orthogonality properties of the trigonometric sine function, to show that

$$A_n = dp/dz/(\mu c)c^3[2 - \{2\cos(n\pi) + n\pi \sin\,(n\pi)\}/[(n\pi)^3 \cosh\,\{n\pi b/(2c)\}] \tag{5.120}$$

With A_n defined, the solution to u_c, and thus to Equations 5.115 and 5.116, is determined. The shear rates $\partial u/\partial x$ and $\partial u/\partial y$, and the viscous stresses $\mu \, \partial u/\partial x$ and $\mu \, \partial u/\partial y$, can be obtained by differentiating Equation 5.119. Again, analytical methods suffer limitations (e.g., the superpositions in "$u = u_p + u_c$" and "Σ" are not valid when the equation for "u" is nonlinear, as for non-Newtonian rheologies. Also, while there are no "log" function or "centerline" problems, as for radial formulations, it is clear that even if "y and x" coordinates are found for general ducts, it will not be

```
C       SERIES.FOR
C       INPUTS (Observation point (Y,X) assumed)
        B  = 1.
        C  = 1.
        Y  = 0.
        X  = 0.5
        VISC = 0.0000211/144.
        PGRAD = 0.001/12.
C       SOLUTION (Consider 100 terms in series)
        PI = 3.14159
        C2  = C**2.
        SUM = 0.
        DO 100  N=1,100
        TEMP = 2.*(C**3) - (C**3)*(2.*COS(N*PI) +N*PI*SIN(N*PI))
        TEMP = TEMP/((N**3.)*(PI**3.))
        A = PGRAD*TEMP/(VISC*C)
        A = A/COSH(N*PI*B/(2.*C))
        SUM = SUM + A*COSH(N*PI*Y/C)*SIN(N*PI*X/C)
100     CONTINUE
        UC = SUM
        UP = -PGRAD*(C*X-X**2.)/(2.*VISC)
        U = UC + UP
        WRITE(*,200) U
200     FORMAT(1X,' Velocity = ',E10.4,' in/sec')
        STOP
        END
```

FIGURE 5.72

Fortran code, series solution for rectangular duct.

possible to find the analogous "sin" and "cosh" functions. In general, for arbitrarily clogged ducts, there will be no lines of symmetry to help in defining solution products.

Classical techniques are labor intensive in this sense: Each problem requires its own special solution strategy. The Fortran code required to implement Equations 5.119 and 5.120 is shown in Figure 5.72. The input units will be explained later. Note that large values of the summation index "n" will lead to register overflow; thus, the apparent generality behind Equation 5.118 is limited by practical machine restrictions.

Finite difference solution

Now we obtain the solution for flow in a rectangular duct by purely numerical means. For circular ducts governed by the ordinary differential equation $d^2u(r)/dr^2 + 1/r \, du/dr = 1/\mu \, dp/dz$, the complete solution can be obtained in a single pass using a tridiagonal equation solver if the equation is discretized implicitly using second-order central differences. For problems in two independent variables, iterative methods are generally required to obtain practical solutions. For linear problems, say Newtonian flows, it is possible to obtain the solution in a single pass using "direct solvers." However, these are not practical for complicated geometries because numerous meshes are required to characterize the defining contours. In the analysis that follows, we will illustrate the use of iterative methods, since these are used in the solution of our governing grid generation and transformed flow equations.

We now turn to Equation 5.115 and consider it in its entirety, without resolving the dependent variable into particular and complementary parts. That is, we address $\partial^2 u/\partial x^2 + \partial^2 u/\partial y^2 = (1/\mu) \, dp/dz$ directly. Now, from elementary numerical analysis, it can be shown that the central difference formula

$$d^2u(r_i)/dr^2 = (u_{i-1} - 2u_i + u_{i+1})/(\Delta r)^2 \tag{5.121a}$$

holds to second-order accuracy. Thus, we can similarly write

$$\partial^2 u(y_i)/\partial y^2 = (u_{i-1} - 2u_i + u_{i+1})/(\Delta y)^2 \tag{5.121b}$$

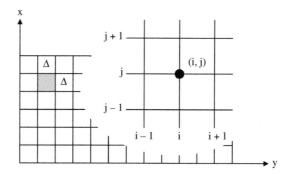

FIGURE 5.73

Rectangular finite difference grid.

for second derivatives in the "y" direction. In the present problem, we have an additional "x" direction, as shown in Figure 5.73. The grid depicted there overlays the cross section of Figure 5.71. Since "y, x" requires two indexes, we extend Equation 5.121b in the obvious manner. For example, for a fixed j, the second derivative is

$$\partial^2 u(y_i, x_j)/\partial y^2 = (u_{i-1,j} - 2u_{i,j} + u_{i+1,j})/(\Delta y)^2 \tag{5.121c}$$

Similarly,

$$\partial^2 u(y_i, x_j)/\partial x^2 = (u_{i,j-1} - 2u_{i,j} + u_{i,j+1})/(\Delta x)^2 \tag{5.121d}$$

Thus, at the "observation point" (i,j), Equation 5.115 becomes

$$(u_{i-1,j} - 2u_{i,j} + u_{i+1,j})/(\Delta y)^2 + (u_{i,j-1} - 2u_{i,j} + u_{i,j+1})/(\Delta x)^2 = 1/\mu \ dp/dz \tag{5.122}$$

We can proceed to develop a rectangular duct solver allowing arbitrarily different Δx and Δy values. However, that is not our purpose. For simplicity, we will therefore assume constant meshes $\Delta x = \Delta y = \Delta$, which allows us to rewrite Equation 5.122 in the form

$$u_{i,j} = 1/4 \ (u_{i-1,j} + u_{i+1,j} + u_{i,j-1} + u_{i,j+1}) - \Delta^2/(4\mu) \ dp/dz \tag{5.123}$$

Equation 5.123 is a central difference approximation to governing Equation 5.115, which is second-order accurate. Interestingly, it can be used as a "recursion formula" that iteratively produces improved numerical solutions. For example, suppose that some approximate solution for u(i, j) is available. Then an improved (left-side) solution can be generated by evaluating the right side of Equation 5.123 with it. It can be shown that, if this method converges, it will tend to the correct physical solution whatever the starting guess. Thus, if an initial approximation were not available, a trivial "zero solution" for u would be perfectly acceptable! Such methods are known as "relaxation methods." Since we have calculated improvements point by point (e.g., as opposed to an entire line of points at a time), the method used is a "point relaxation" method.

The Fortran source code implementing Equation 5.123 and the boundary conditions in Equations 5.116a through 5.116d is given in Figure 5.74. The units used are identical to those of

```
C      SQFDM.FOR (SQUARE DUCT, FINITE DIFFERENCE METHOD)
       DIMENSION U(11,11)
C      SQUARE IS 1" BY 1" AND THERE ARE 10 GRIDS
       DEL = 1./10.
       VISC = 0.0000211/144.
       PGRAD = 0.001/12.
       DO 100  I=1,11
       DO 100  J=1,11
       U(I,J) = 0.
100    CONTINUE
       DO 300  N=1,100
       DO 200  I=2,10
       DO 200  J=2,10
       U(I,J) =  (U(I-1,J) + U(I+1,J) + U(I,J-1) + U(I,J+1))/4.
      1           - PGRAD*(DEL**2)/(4.*VISC)
200    CONTINUE
300    CONTINUE
       Q = 0.
       DO 400  I=2,11
       DO 400  J=2,11
       Q = Q + U(I-1,J-1)*(DEL**2)
400    CONTINUE
       Q = Q*0.2597
       Q = -Q
       WRITE(*,500) Q
500    FORMAT(' Volume flow rate = ',E10.4,' gal/min')
       WRITE(*,510) U(6,6)
510    FORMAT(' Umax = ',E10.4,' in/sec')
       STOP
       END
```

FIGURE 5.74

Finite difference code, rectangular ducts.

the previous example, but here, a square duct having one-inch sides is considered. A mesh width of 0.1 in. is assumed, so that ten grids are taken along each side of the square. Loop 100 initializes the "starting guess" for U(I, J) to zero, also setting vanishing velocities along the duct walls I = 1 and 11, and J = 1 and 11. Loop 300 updates U(I, J) in the internal flow domain bounded by I = 2, ... , 10 and J = 2, ... , 10. One hundred iterations are taken, which more than converges the calculation; in a more refined implementation, suitable convergence criteria would be defined. "Q" provides the volumetric flow rate in gallons per minute, while "U" is calculated in inches per second. For the Fortran code shown, computations are completed in less than one second on standard personal computers.

Example calculation

Here a pressure gradient with dp/dz = 0.001 psi/ft is assumed, and a square duct with 1-in. sides is taken; also, we consider a unit centipoise viscosity fluid, with $\mu = 0.0000211$ lbf sec/ft^2. Units of "in., sec, and lbf" are used in the source listing. The program breaks each side of the square into 10 equal increments, with $\Delta x = \Delta y = 0.1$ in. This is done for comparative purposes with radial flow results. For the finite difference method, the maximum velocity is found at the center of the duct— that is, y = 0 and x = 0.5 in., and it is given by the value u(6, 6) = −0.4157E +02 in./sec. The code in Figure 5.72 gives the exact series solution at the center as −0.4190E +02, so that the difference method incurs less than 1 percent error.

Again, this accuracy is achieved with a coarse "10 × 10" constant mesh. We note in closing this chapter that the rectangular duct solution, while not bearing directly on petroleum applications, was developed in part to validate the singly connected duct flow curvilinear grid scheme used in our solids deposition and wellbore "hole enlargement" applications.

Transient, Two-Dimensional, Single-Phase Flow Modeling

In Chapter 2, a broad foundation for annular flow problems with arbitrary cross-sectional geometries was developed, and we showed how for nonrotating drillpipe and casing, the initial formulation must be presented in rectangular or Cartesian variables in order to implement the boundary-conforming, curvilinear grid procedure. New subtleties accompanying the treatment of inner body rotation were discussed. There we demonstrated how, for rotating flow applications, it was necessary to start with a host formulation in circular cylindrical coordinates so that tangential surface speeds can be adequately described, and then to progress to Cartesian coordinates, and only then to recast the formulation in general curvilinear coordinates. Of course, in software development, the geometric transformations do not end with these mappings: There are also screen transforms and pixel mappings to contend with. However, the latter two are not discussed in this book.

With the underlying formulation issues addressed in Chapter 2, the work in Chapter 3 focused on the details of the curvilinear grid mappings—for example, the manner in which branch cut boundary conditions are developed for doubly connected annular flow domains (and how they contrast with singly connected geometries), the exact relaxation or iteration schemes needed to solve the nonlinear mesh generation equations in a fast and numerically stable manner, and the way in which velocity gradients are expressed in both physical and mapped coordinates.

We also addressed the modeling of yield stress fluids and explained how plug zone size, shape, and location can be calculated naturally using a new extended Herschel-Bulkley relationship. The foundation and techniques that were developed in these two early chapters are now presumed to be understood by the reader; therefore, in the present chapter dealing with transient, two-dimensional, single-phase flow modeling, we will only outline the broad strategy and omit lower-level details associated with the mathematics and computational analysis.

SECTION 6.1

Governing Equations for Transient Flow

The transient formulation handles, of course, axial reciprocation and general pump schedules (ramp-up, ramp-down, changes in flow rate, and so on). But the added complexity inherent in our use of circular cylindrical coordinates was basically driven by the need to model pipe or casing rotation, an effect that cannot be modeled by steady flow formulations with absolute numerical stability.

We emphasize that the model developed in this chapter, and amply illustrated with calculations in Chapter 7, handles general axial reciprocation, arbitrary pipe rotation, and pump schedules taken in any transient form. These unsteady actions may be constant, linearly varying in time, sinusoidal, or any combination thereof. While the menus in Figure 1.28 might appear restrictive, we emphasize that very general simultaneous actions can be modeled by modifying just several lines of Fortran source code.

As noted earlier, we formulate the problem first in circular cylindrical coordinates, recognizing that these also apply (although not conveniently) to arbitrary annular cross-sectional geometries. The momentum equations in the "r," "θ," and "z" directions were discussed previously and are, respectively,

$$\rho(\partial v_r/\partial t + v_r\,\partial v_r/\partial r + v_\theta/r\,\partial v_r/\partial\theta - v_\theta^2/r + v_z\,\partial v_r/\partial z)$$
$$= F_r - \partial P/\partial r + 1/r\,\partial(r\,S_{rr})/\partial r + 1/r\,\partial S_{r\theta}/\partial\theta - 1/r\,S_{\theta\theta} + \partial S_{rz}/\partial z \tag{6.1}$$

$$\rho(\partial v_\theta/\partial t + v_r\,\partial v_\theta/\partial r + v_\theta/r\,\partial v_\theta/\partial\theta + v_r v_\theta/r + v_z\,\partial v_\theta/\partial z)$$
$$= F_\theta - 1/r\,\partial P/\partial\theta + 1/r^2\,\partial(r^2\,S_{\theta r})/\partial r + 1/r\,\partial S_{\theta\theta}/\partial\theta + \partial S_{\theta z}/\partial z \tag{6.2}$$

$$\rho(\partial v_z/\partial t + v_r\,\partial v_z/\partial r + v_\theta/r\,\partial v_z/\partial\theta + v_z\,\partial v_z/\partial z)$$
$$= F_z - \partial P/\partial z + 1/r\,\partial(r\,S_{zr})/\partial r + 1/r\,\partial S_{z\theta}/\partial\theta + \partial S_{zz}/\partial z \tag{6.3}$$

where F denotes body forces, while mass conservation takes the form

$$\partial v_r/\partial r + v_r/r + 1/r\,\partial v_\theta/\partial\theta + \partial v_z/\partial z = 0 \tag{6.4}$$

In the preceding, v_r, v_θ, and v_z are radial, azimuthal, and axial velocity components, respectively. Again, we have

$$\underline{S} = 2N(\Gamma)\underline{\underline{D}} \tag{6.5}$$

denoting the deviatoric stress tensor, $N(\Gamma)$; the apparent viscosity function, Γ; the shear rate; and now the deformation tensor $\underline{\underline{D}}$ whose elements are defined by

$$D_{rr} = \partial v_r/\partial r \tag{6.6a}$$

$$D_{\theta\theta} = 1/r\,\partial v_r/\partial\theta + v_r/r \tag{6.6b}$$

$$D_{zz} = \partial v_z/\partial z \tag{6.6c}$$

$$D_{r\theta} = D_{\theta r} = 1/2\,[r\,\partial(v_\theta/r)/\partial r + 1/r\,\partial v_r/\partial\theta] \tag{6.6d}$$

$$D_{\theta z} = D_{z\theta} = 1/2\,(\partial v_\theta/\partial z + 1/r\,\partial v_z/\partial\theta) \tag{6.6e}$$

$$D_{rz} = D_{zr} = 1/2\,(\partial v_r/\partial z + \partial v_\theta/\partial\theta) \tag{6.6f}$$

The solution process proceeds as follows and the steps are described qualitatively but sequentially next.

1. For most annular flow applications, formation influx and outflux are permitted as described in Chapter 4; however, we will assume that local effects are not large. Thus, v_r can be neglected in comparison to v_z and v_θ. This simplification eliminates one partial differential equation.
2. The work of Escudier et al. (2000) suggests that the azimuthal velocity solution in problems with rotation is dominated by a "dragging" mechanism due to pipe or casing shear. This

observation is acceptable physically and provides the basis for a second simplification allowing us to neglect the azimuthal pressure gradient $\partial P/\partial\theta$ in the formulation. What remains is the usual $\partial P/\partial z$ axial driver.

3. The extended Herschel-Bulkley constitutive law introduced is now invoked so that plug zones can be calculated naturally, with deep plug and interfacial transition boundaries appearing as part of any numerical solution. Thus, a single computational domain applies, and there is no need to consider multiple flow domains.

4. The circular cylindrical coordinate-based momentum equations are rewritten in Cartesian coordinates and then reexpressed in terms of general boundary-conforming, curvilinear coordinates. These coordinates and all transformation metrics are created using the computational scheme outlined in Chapter 3. The mappings for steady and unsteady flows are identical.

5. The resulting unsteady, nonlinear partial differential equation system is extremely complicated, containing variable coefficients, mixed derivatives, first- and second-order terms, and so on. While high-order accurate "approximate factorization" and "alternating-direction-implicit" (ADI) schemes are now available in the literature for their solution, implementation requires substantial research. Thus, a faster explicit time integration method (see Press et al. (1992)) is used. Consistent with this usage is the neglect of partial derivatives of apparent viscosity, although, of course, the basic $N(\Gamma)$ function with its nonlinearity and variability across the flow domain is retained to leading order. This treatment is also appropriate to the three-dimensional multiphase calculations pursued in Chapters 8 and 9, where the use of concentration functions to describe miscible mixing, an empirical procedure, introduces errors consistently of the same order of magnitude.

6. Before the transient partial differential equations can be integrated in time, initial conditions must be defined. We offer two options. First, we permit a quiescent state in which the annular fluid is completely at rest. Second, we allow the borehole fluid to move at steady state under general eccentric, non-Newtonian, nonrotating conditions.

7. The coupled nonlinear equations are integrated step by step in time, and at the end of each time step the apparent viscosity function is updated with the latest available velocity gradients. Although we have dealt only with the conventional models used in petroleum engineering, we emphasize that "memory" fluids are easily handled by a simple Fortran subroutine change.

8. When the time integration is completed, velocities are processed for display using the color graphics tools discussed earlier. In addition, shear rates, apparent viscosities, and viscous shear stresses are computed by postprocessing available velocity results and prepared for on-demand static displays. For the present transient calculations, axial velocity results are also collected at different user-selected time intervals and assembled into a movie available for playback after the calculations terminate.

SECTION 6.2

Rotation Paradox

Developing a transient algorithm, at least for the present annular flow research, involved much more than the "mechanical" steps just outlined. Numerous observation inconsistencies have been reported in the literature. Prior to 1990, drillers observed that the effect of pipe rotation (under a constant pressure gradient) is an increase in flow rate; equivalently, a fixed flow rate finds a pressure gradient that is less steep. The explanation was simple: Shear thinning of non-Newtonian fluids leads to reduced viscosities that increase throughput. Mathematicians during this period reproduced these results analytically and numerically. Observation and theory were consistent.

In the 1990s and beyond, empirical observations were completely opposite: For the same flow rate, rotation leads to a steepening of axial pressure gradients. Trends related to pressure, of course, warrant more than academic curiosity because of their application in ultra-deepwater drilling. Because drillers must navigate narrow pressure windows, errors related to pressures at the drillbit can lead to dangerous situations that gravely affect safety. Any transient solver claiming to model rotation, in this practical scenario, must explain the apparent inconsistency before its predictions can be credible.

The present research indicates that no contradictions or inconsistencies exist. Prior to 1990, all reported empirical observations and mathematical models dealt with the concentric annuli encountered in vertical well drilling. For this application, the only manner in which rheology appears in flow formulations is through the apparent viscosity function in viscous terms; that is, the only physical effect is shear thinning. After 1990, most published works dealt with deviated and horizontal wells, for which relevant borehole annulus cross sections are eccentric. Thus, certain terms in the momentum equations that dropped out by virtue of symmetry in concentric applications remain.

The work developed here shows that these new terms effectively modify the applied pressure gradient in a manner that varies throughout the annulus. Computations using the integrated procedure outlined above gave stable results that were, significantly, consistent with both pre- and post-1990s field and laboratory observations. Again, the effects of eccentricity are subtle. While shear thinning nonetheless exists, these effects typically lead to flow rate decreases (for fixed pressure gradient) as rotation rate increases (see Figure 2.2 and the accompanying explanations).

It is important to emphasize strong axial and azimuthal velocity coupling and flow nonlinearity in general. In our approach, both are permitted without compromise. Time integrations are used that minimize artificial viscosity and lead to steady solutions that agree with known analytical solutions for concentric annuli. However, nonlinearity is often treated using ad hoc "recipes" in the literature.

One paper gives procedures for combining axial velocities across annular gaps (with no rotation), tangential velocities (with no axial flow), and so on. This represents some type of linearization about baseline no-flow and no-rotation conditions. Sometimes the method did not work, and "modeling efforts improved substantially when a nonlinear model for decay of shear rate across the annular gap was used. . . ." These reports only emphasize the value of addressing full nonlinearity at the outset so that empirical procedures can be avoided.

SECTION 6.3

Operational Consequences for the Transient Rotation Algorithm

Managed pressure drilling enhances the driller's ability to control pressures within the borehole with greater precision. The literature describes three methods for active pressure control. First, mud pump flow rates can be changed to affect dynamic friction. This is easily accomplished, although, of course, care must be exercised; flow transients induced by sudden pumping changes by positive displacement pumps can be dangerous. Second, mud rheologies and weights can be altered. This process, however, is time consuming and will not allow rig hands to rapidly respond to dangerous situations. Third, changing the overall system pressure level by adjusting the surface choke is a possibility that is effective and simple to carry out.

A fourth, newer, method is proposed in the present work. Our research has shown that drillstring rotation (or the lack of it) can affect borehole pressures significantly, and theoretical and numerical results are consistent with field and laboratory observation. Thus, drillstring rotation provides an important and rapid means for downhole pressure control. Moreover, the effects can be quantified by computer simulations such as those illustrated in Chapter 7. Because the calculations can be performed quickly using inexpensive computer resources, their role in job planning is enhanced: Prior to drilling, a range of "what-if" options and responses can be prepared for different contingencies.

Another important applications area is hole cleaning. While we successfully demonstrated the application of our "stress hypothesis" to hole cleaning using the University of Tulsa data in Chapter 5, much remains in the way of understanding the role of rotation in removing debris. It is well known that hole-cleaning problems are heightened in large-diameter wells and that rotation effects can be subtle. The mathematical reasons are simple. For the larger diameters, axial velocities are lower. In other words, both axial and azimuthal velocities are comparable in magnitude, so their nonlinear coupling is not small. Thus, the outcome of any drilling program or simulation, for that matter, is not obvious.

Computer models provide tools that help to verify and refute possible explanations for hole cleaning. Does rotation affect the stress field significantly and thus change bed erosion characteristics? Or is the primary effect an "upward throwing" of debris where cuttings chips are consequently transported by turbulent convection in the wide side of the annulus? These and other questions can be answered using computed results as correlative tools, in much the same way that we used them in more elementary applications in Chapter 5. We pursue detailed calculations designed to validate the method against available solutions and to demonstrate potential applications of the new transient model. Some comments on pressure gradients are offered next.

SECTION 6.4

Transient Pressure Gradient and Volume Flow Rate

One of the most important relationships in non-Newtonian flow is that between pressure gradient and volume flow rate. For steady flows, the "Steady 2D" simulator provides an option that automatically calculates and plots the nonlinear curve once annular geometry and borehole fluid are specified. For unsteady flows, an analogous relationship is desirable for modeling purposes, but typically is not available. Integration of the axial momentum equation in time requires an explicit $\partial P / \partial z$ input value (see Equation 6.3). But very often, it is the volume flow rate Q that is specified in a practical problem, and its expression in terms of $\partial P / \partial z$ is not generally possible.

However, the $\partial P / \partial z$ value needed to produce a given Q(t) *can* be explicitly written for certain problems when properties for a baseline steady flow are known—say via calculation using "Steady 2D." Again we turn to fundamentals. The axial velocity u satisfies

$$\rho \partial u / \partial t + \{ \ldots \} = -\partial P / \partial z + < \ldots > \tag{6.7}$$

where {...} represents smaller nonlinear convection terms and rotation effects that we will ignore or that are nonexistent, while $<...>$ denotes viscous terms that we keep in their entirety. We multiply Equation 6.7 by the infinitesimal area element $J\Delta\xi\Delta\eta$ throughout, where J is the transformation Jacobian. This leads to the equation

$$\partial(\rho u\ J\ \Delta\xi\Delta\eta)/\partial t = -\partial P/\partial z\ J\ \Delta\xi\Delta\eta + <...> J\ \Delta\xi\Delta\eta \tag{6.8}$$

Note that the cross-sectional annular area A and the volume flow rate Q(t) through it satisfy

$$A = \int\int J\ \Delta\xi\Delta\eta \tag{6.9}$$

$$Q(t) = \int\int u\ J\ \Delta\xi\Delta\eta \tag{6.10}$$

Thus, we have

$$\rho\partial Q/\partial t = -A\ \partial P/\partial z + \int\int <...> J\ \Delta\xi\Delta\eta \tag{6.11}$$

Suppose we have a steady flow with $\partial/\partial t = 0$ and Q* specified. The pressure gradient for this problem can be computed from the "Steady 2D" solver in the "flow rate specified" mode and denoted as $(\partial P/\partial z)^*$. Therefore, we have the equality $\int\int <...> J\ \Delta\xi\Delta\eta = A(\partial P/\partial z)^*$ for Q*. Using this to approximate the integral above, we find that $\rho\partial Q/\partial t = -A\partial P/\partial z + A(\partial P/\partial z)^*$ or

$$\partial P/\partial z = (\partial P/\partial z)^* - (\rho/A)\ \partial Q/\partial t \tag{6.12}$$

Equation 6.12 can be used to evaluate the pressure gradient input in the "Transient 2D" solver if a baseline steady-state flow with $(\partial P/\partial z)^*$ and Q* is available. If mudpump flow rate increases so that $\partial Q/\partial t > 0$, then the axial pressure gradient $\partial P/\partial z$ will become more negative, as expected, by the amount shown. If flow rate is slowing so that $\partial Q/\partial t < 0$, then the pressure gradient will be less negative. In either case, time variations in pressure gradient are proportional to mud weight, as would be anticipated on physical grounds. We emphasize that Equation 6.12 provides an estimate for pressure gradient only.

Transient Applications
Drillpipe or Casing Reciprocation and Rotation

We will demonstrate how non-Newtonian flows in highly eccentric annuli can be computed under general transient conditions with the drillstring or casing undergoing arbitrary coupled reciprocation and rotation together with flow changes in the mud pump. Here the annulus can be highly eccentric, with the drillpipe almost resting on the formation, thus risking economic losses due to stuck pipe. Aside from its role in calculating pressure losses and velocity fields, our simulator capability is important in jarring applications, with the effects of flow rate ramp-up and ramp-down accounted for.

An important use of the algorithm described here does not include unsteady flow as an end effect at all. In Chapter 4, we noted that the steady flow formulation with nonzero rotation cannot presently be solved on an unconditionally stable numerical basis; in fact, the method destabilizes as specific gravity and rpm values approach those used in field practice. An application developed in Chapter 5 applies the transient method to steady-state swab-surge where pipe rotation is significant. Because rotation significantly affects pressure fields, the algorithm described here is extremely important, and all the more so because it is the only available method serving this function.

In the examples that follow, we first design examples to validate the integration method, in particular by seeking agreement with known analytical solutions. The effects of rotation are studied for Newtonian flows (which, because of constant viscosities, never exhibit shear thinning) as well as non-Newtonian fluids. Foams as well as heavy muds are considered. Then the separate effects of transient pipe reciprocation, unsteady pipe rotation, and general mud pump flow variation are considered, and finally, all three simulation modes are permitted. Importantly, the analytical and numerical formulations are constructed in such a manner that the least and most complicated applications require almost identical computing times. More general transient capabilities can be developed by simply modifying modularized Fortran subroutines.

EXAMPLE 7.1

Validation Runs: Three Different Approaches to Steady, Nonrotating Concentric Annular Power Law Flow

Before studying transient effects in detail, we explore the accuracy of three different methods we have developed in the limit of steady, nonrotating, concentric, non-Newtonian Power law flow. Specifically, we consider an inner radius of 2 in., an outer radius of 4 in., n = 0.8,

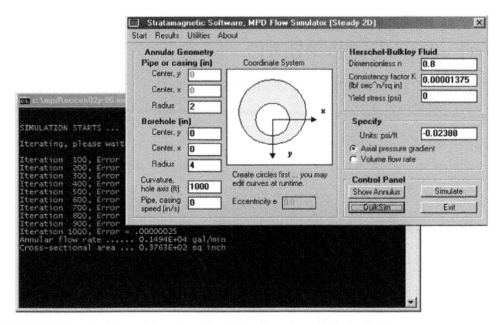

FIGURE 7.1

Finite difference curvilinear grid simulator result.

K = 0.00001375 lbf sec^n/in.^{2}, and a pressure gradient of −0.02388 psi/ft. In the first case, we run the finite difference−based simulator based on curvilinear meshes in QuikSim fine-mesh mode to find a flow rate of 1,494 gpm, as shown in Figure 7.1. (This simulator *does* allow for pipe or casing axial movement and yield stress modeling.)

Next we consider the simulator used for rotating Power law flow, noting that it does not allow axial pipe movement. Recall that approximations were employed to facilitate closed-form analytical solutions; the nature of the math used does not allow "0 rpm" to be entered directly, so a value of "1" is used instead (this simulator also will not model Newtonian flows with n = 1). The software produces a solution of 1,491 gpm, as shown in Figure 7.2. Finally, we use the *exact* Herschel-Bulkley solver, which assumes a completely immobile inner pipe, running it in the limit of vanishing yield stress; this gives a solution of 1,523 gpm, as shown in Figure 7.3. The difference between the largest and smallest predictions is about 2 percent, which is reassuring given that the three models are derived from completely different assumptions and methods.

EXAMPLE 7.2

Validation Run for Transient, Newtonian, Nonrotating, Concentric Annular Flow

The excellent agreement obtained in Example 7.1 between three completely different steady flow models should provide a strong degree of user confidence. In the present validation example, we consider Newtonian, nonrotating, concentric annular flow, for which an exact, closed-form, steady

Example 7.2   275

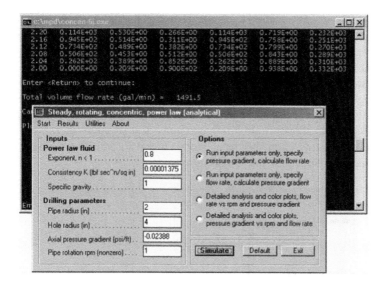

FIGURE 7.2

Rotating Power law approximate flow result.

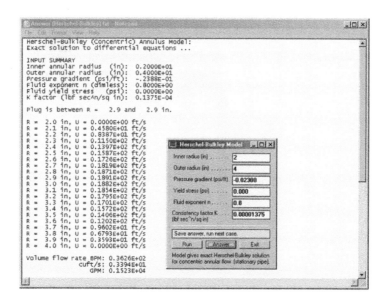

FIGURE 7.3

Herschel-Bulkley simulator, *exact* results.

solution of the Navier-Stokes equation is available using the simulator in Figure 7.4. Here, inner and outer radii are 3 and 6 in., respectively, and a viscosity of 1 cp and a pressure gradient of −0.0001 psi/ft are assumed. This simulator also allows axial pipe movement, but we disallow it in the validation that follows. Figure 7.4 shows that the flow rate is 947.1 gpm.

Now we use the finite difference–based, curvilinear grid, transient simulator in Figure 7.5 to show how the large-time solution of a transient problem is consistent with the steady-state solution obtained previously from an analytical method. We have assumed a very small specific gravity of 0.01. As will be seen, this allows finite difference numerical solutions to achieve steady state rapidly, since in the nonrotating case, the dependence on density vanishes—small densities, in fact, imply small mechanical inertias for fast equilibration.

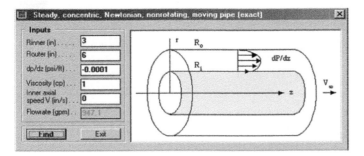

FIGURE 7.4

Exact, steady, Newtonian flow solution.

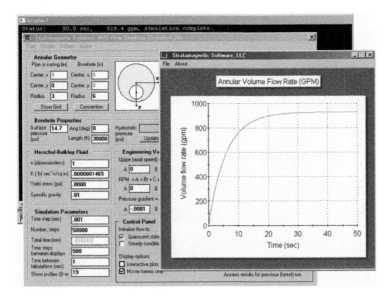

FIGURE 7.5

Low-specific-gravity transient solution.

Example 7.3   **277**

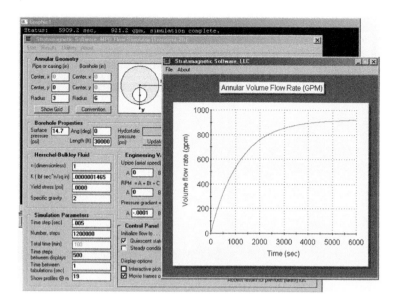

FIGURE 7.6

High-specific-gravity transient solution.

The asymptotic flow rate is 928.4 gpm for a small 2 percent error. Importantly, the unsteady model shows that the physical time scale required to achieve this steady condition, starting from a quiescent state, is about 30 sec (the computation requires about 15 seconds). The reader should note the inputs used. Also, the "engineering variables" hidden by the graph are identically zero.

In Figure 7.6, we rerun the foregoing simulation with all inputs unchanged except that the specific gravity is increased to 2.0, corresponding to a heavy 16.7-lbm/gal mud, and the time step is increased to 0.005 sec. The same asymptotic flow rate of 928.4 gpm is achieved, and the time scale to attain steady state from quiescent conditions is about one hour (the simulation itself, for 1,200,000 time steps, requires about four minutes of computing). The transient simulator illustrates the role of inertia in establishing steady conditions.

We have demonstrated that our transient finite difference results are consistent with the exact analytical steady solution. (We used our curvilinear grid approach and considered both low and high specific gravity runs.) Importantly, if transient analysis is used to find steady flows, at least in nonrotating problems, then low-specific-gravity fluids should be assumed because low mechanical inertias lead to very rapid physical equilibration.

EXAMPLE 7.3

Validation Run for Transient, Newtonian, Nonrotating, Eccentric Annular Flow

In this consistency check, we examine *eccentric* annular flows, for which no analytical or exact solutions are available. We assume a Newtonian fluid with a viscosity of 10 cp and that the pipe is

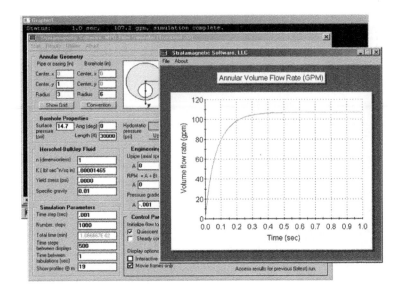

FIGURE 7.7

Transient, Newtonian, nonrotating flow solution.

not rotating or moving axially. The transient solution in Figure 7.7 requires about five seconds of computing time and yields an asymptotic steady-state flow rate of 107.2 gpm.

The complementary steady flow computation in Figure 7.8, using the same 25 × 11 mesh, yields an identical 107.2 gpm, which is much better than this author had anticipated. This is all the more remarkable because the steady solver uses an iterative, implicit, successive line over relaxation (SLOR) method, whereas the transient method uses an explicit time integration procedure. Note that the QuikSim fine-mesh solution yields 109.2 gpm, for less than a 2 percent difference. That the two final results for unsteady and steady flow are consistent bodes well for our transient and steady solvers. Again, we emphasize that the steady flow solver handles constant axial pipe speed motion without rotation, while the unsteady solver handles coupled axial and rotary movement, both under general transient conditions.

EXAMPLE 7.4

Effect of Steady Rotation for Laminar Power Law Flows in Concentric Annuli

In this example, we use our closed-form analytical solution developed for steady, rotating Power law fluids in *concentric annuli* to explore pressure gradient and flow rate relationships in a non-Newtonian application. The user interface is shown in Figure 7.9, where the third option is selected. Using automatically defined internal parameters, this simulation plots flow rate (gpm) on the vertical axis and pressure gradient (dp/dz) and rotational rate (rpm) on the two horizontal axes, as shown in Figure 7.10. It is clear from this figure that as the (absolute value of) pressure gradient increases for fixed rpm, flow rate increases, as would be expected. Interestingly, as the rotational

Example 7.4   **279**

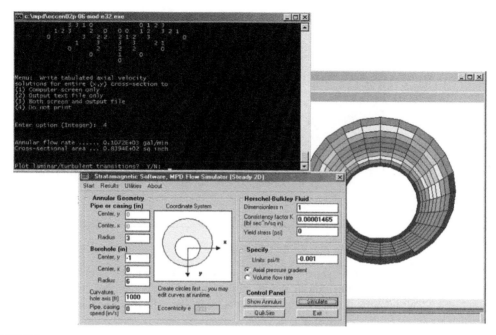

FIGURE 7.8

Steady flow computation on identical mesh.

FIGURE 7.9

Steady, rotating Power law simulator.

rate increases at fixed dp/dz, the flow rate also increases. This is explained by the reduction in apparent viscosity induced by rotation due to shear thinning.

This result also appears in several related and well-known investigations external to the petroleum industry. Significantly, it is consistent with the results of classical studies reported in the well-regarded book *Dynamics of Polymeric Liquids* by Bird, Armstrong, and Hassager (1987).

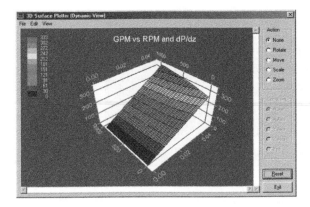

FIGURE 7.10

GPM versus RPM and dp/dz.

Their Example 4-2-5 conclusion "shows that the flow in the axial direction is enhanced because of the imposed shearing in the tangential direction, since this additional shearing causes the viscosity to be lowered." The numerical analysis by the respected authors Savins and Wallick (1966) also supports our findings. From their abstract,

> *The most interesting consequence of the coupling effect is that the axial flow resistance is lowered in a helical flow, with the result, for example, that for a given applied axial pressure gradient the axial discharge rate in a helical flow field is higher than in a purely annular flow field.*

In their analysis, the authors observe that

> *It is seen that the effect of a helical flow produced by impressing a relative rotation on the z directed annular flow is to increase the axial discharge rate. This result is not unexpected. The preceding viscosity profile analyses showed that the shear-dependent viscosity is lowered, hence the axial flow resistance is lowered.*

Finally, from their summary,

> *In contrast, if the fluid were Newtonian, the superimposed laminar flows would be noninterfering in that there would be no coupling among the discharge rate, axial pressure gradient, relative rotation, and torque through the viscosity coefficient.*

Recall that we have proven this latter observation directly from the governing Navier-Stokes equations. Several subsequent theoretical and experimental petroleum publications also support the foregoing results.

It is important to emphasize that, in all of the just mentioned works and in the present example, laminar, *concentric* annular flows are considered. For concentric flows, the nonlinear inertia (or convective) terms in the governing momentum equations vanish identically, and velocity coupling is possible only through changes to apparent viscosity or shear thinning. Fortuitously, early publications focused on this limit—from the mathematical perspective for simplicity and from the drilling perspective by the vertical well applications prior to 1990. In the past two decades, with deviated

Example 7.5   **281**

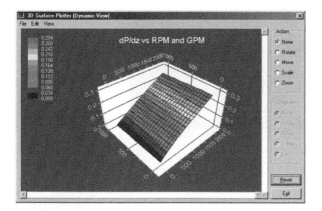

FIGURE 7.11

dp/dz versus RPM and GPM.

and horizontal wells becoming predominant in exploration, conflicting relationships between pressure gradient and flow rate have been reported.

These conflicts arise because of annular eccentricity. In general non-Newtonian flows, shear thinning is always present; however, when eccentricity exists, the applied pressure gradient is effectively modified by a spatially dependent convective term that is proportional to fluid density and rotational rate. The complicated interplay among flow rate, applied pressure gradient, fluid rheology, rotational rate, and annular geometry cannot be described by casual "rules of thumb"; however, it can be obtained as the solution to coupled nonlinear partial differential equations, as described in this book and particularly in this chapter.

Let us return for now to concentric annular flow analysis. Figure 7.11 offers a different view of the results from that provided in Figure 7.10. It is obtained by selecting the last option in Figure 7.9. Note that each figure uses hundreds of solution points, and both are produced because analytical solutions are used in less than one second of computing time. Again, the increase in flow rate (for a fixed pressure gradient) obtained when the rotational rate increases is well accepted in the older literature, but confusion and inconsistencies have arisen in recent studies, a point we address in several examples next.

EXAMPLE 7.5

Effect of Steady-State Rotation for Newtonian Fluid Flow in Eccentric Annuli

Here we consider the effects of annular eccentricity. To isolate rheological effects, we assume a Newtonian fluid with constant viscosity so shear thinning is impossible. The eccentricity is 0.333. As a validation point, we first obtain the flow rate under nonrotating conditions using the steady-state, curvilinear grid flow solver in Figure 7.12. For the assumptions shown, the flow rate is 109.2 gpm. Next we run the transient simulator for the same nonrotating flow conditions, as shown in Figure 7.13, where the engineering variables not shown are identically zero, to obtain a nearly identical flow rate of 107.2 (the difference is less than 2 percent). The agreement is excellent.

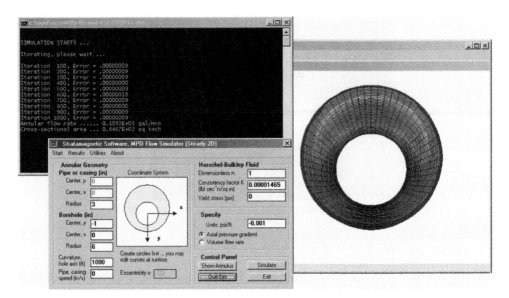

FIGURE 7.12

Steady-state solution without rotation.

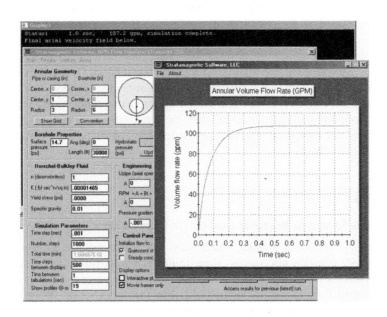

FIGURE 7.13

Transient Newtonian solution without rotation.

Example 7.5   **283**

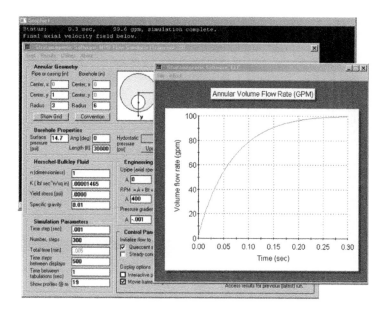

FIGURE 7.14

Transient rotating solution from quiescent state (the curve actually peaks at 100 and then asymptotes to 99.4 gpm).

Now it is important to ask, "What if the drillpipe or casing is rotated? Does the flow rate increase or decrease, assuming the same pressure gradient?" In Figure 7.14, we assume a somewhat high 400 rpm to demonstrate numerical stability, but also the fact that the asymptotic steady flow rate *decreases* to 99.4 gpm, or about 8 percent. Thus, in the complementary problem where flow rate is specified and pressure gradient is to be determined, we can expect to see similar order-of-magnitude increases to pressure drop. These changes are significant to drilling safety in managed pressure drilling.

The exact decrease or increase depends on rheological and geometric parameters, and will vary from run to run. Differences as high as 50 percent have been observed. But why did flow rate increase in Example 7.4 but decrease here? The explanation is simple. In the previous example, the gpm increase was due to a decrease in non-Newtonian apparent viscosity arising from rotation; also, for concentric annuli the inertia terms in the axial momentum equations vanish identically. In this example, the viscosity is constant and does not change.

A nonvanishing "$\rho v/r\ \partial U/\partial\theta$" inertia term is new. The azimuthal velocity v is proportional to rpm, while $\partial U/\partial\theta$" is related to eccentricity. The term acts as a spatially variable pressure gradient modifier. These factors are subtle but clear when we examine the governing partial differential equations. We chose Newtonian fluids in this example to isolate rheological effects in order to ascertain the importance of the rotating flow inertia terms alone.

In the control panel of Figure 7.13, we checked "Initialize flow to quiescent state." This assumes vanishing initial flow. We now check "Steady conditions" for our starting point. The simulator first calculates a steady nonrotating flow, and then at t = 0 uses this flow to initialize time

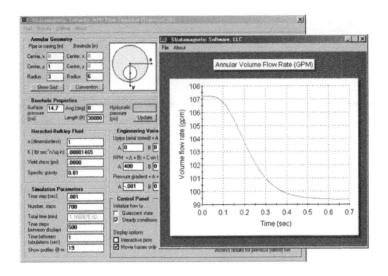

FIGURE 7.15

Transient rotating solution from flowing state.

integrations. This corresponds to a nonrotating pipe with flow that is suddenly rotated. Figure 7.15 shows how the flow rate decreases suddenly from 107.2 to 99.4 gpm, highlighting the effects of rotation (computing time is about one second). Even for this high rotational rate, the transient algorithm for coupled axial and azimuthal movement is fast and stable. Our results demonstrate the usefulness of numerical simulation in drilling safety and operations.

EXAMPLE 7.6

Effect of Steady Rotation for Power Law Flows in Highly Eccentric Annuli at Low Densities (Foams)

The annulus in Figure 7.12, while not concentric, is not highly eccentric. In this example, we examine a cross section with high eccentricity and also allow for nonlinear Power law fluid motion. Here, the eccentricity is 0.5. Results for a nonrotating pipe are given in Figure 7.16, where a steady flow rate of 1,052 gpm is indicated. The time required to achieve steady state is approximately one second. What happens if we rotate the drillpipe at 300 rpm? Figure 7.17 shows that with rotation the time to reach steady conditions is reduced; also, the flow rate decreases to 905.8 gpm. This suggests that in the complementary problem when volumetric flow rate is fixed, the effect of rotation is to increase (the absolute value of) pressure gradient.

Consistent with the previous example, the decrease in flow rate occurs because of inertia effects. We emphasize that the flow rate reduction due to rotation seen here is a sizeable 16 percent. Finally, in Figure 7.18, we rerun the simulation with the initial fluid assumed to be nonrotating and flowing. The results show an equilibration time of one second between steady states so that flow changes are sudden and dangerous. The steady-state flow rate is again about 900 gpm. There is a "bump" in the gpm versus time curve, one seen repeatedly in many such simulations. Whether this

Example 7.6   285

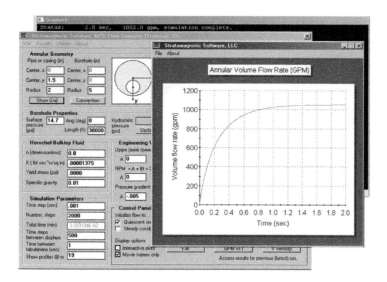

FIGURE 7.16

Power law flow with nonrotating pipe.

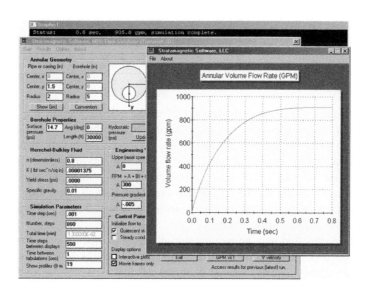

FIGURE 7.17

Power law flow with rotating pipe (zero starting conditions).

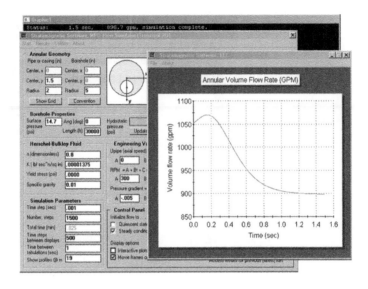

FIGURE 7.18

Power law flow with rotating pipe (from flowing conditions).

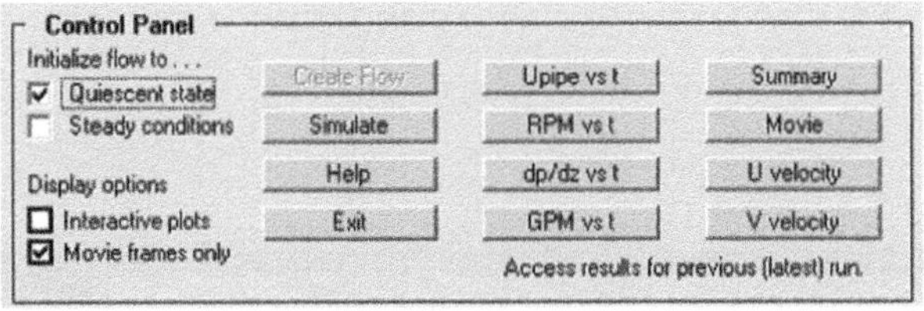

FIGURE 7.19

Assumed quiescent, stagnant-flow initial conditions.

effect is real will require laboratory observation. All of the calculations for this example were performed stably, as our line graphs show, and required only two to three seconds of computing time.

It is important to point out some significant software details associated with flow initialization. For steady flow formulations, the initial state of flow does not appear as a parameter because there is no variation in time. (Actually, it does in a numerical sense, since initial solution guesses are taken, although internally to the software.) For transient formulations, the initial state must be specified. If quiescent stagnant-flow conditions are selected, the box shown in Figure 7.19 is checked and "Simulate" can be clicked immediately.

On the other hand, the fluid may be moving initially at t = 0, and *then* the transient flow specifications shown in the user interface are applied. If the initial flow is not rotating, we know that its solution does not depend on density; we can therefore calculate it assuming a very small value of ρ together with large time steps. If we wish to initialize to a nonrotating steady flow, the message box in Figure 7.20 appears, reminding the user to click "Create Flow" to start this process. Once this is completed, the "Simulate" button can be clicked to perform the required transient analysis.

Example 7.8 **287**

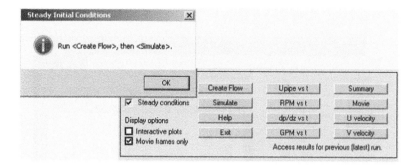

FIGURE 7.20

Creating a nonrotating, steady initial flow.

If the starting flow is rotating, its solution does depend on density, and time steps will need to be very small to ensure convergence. This initialization is not supported at the present time because the solution procedure cannot be made as robust or automatic as desired by the author, but continuing research is being pursued in this area.

EXAMPLE 7.7

Effect of Steady Rotation for Power Law Flows in Highly Eccentric Annuli at High Densities (Heavy Muds)

We emphasized earlier that for nonrotating flows the effects of density vanish at large times. Thus, in computing nonrotating steady-state flows with the transient algorithm, it is advantageous to use as small a fluid density as possible in order to quickly converge the calculations. Here we wish to evaluate the effects of mud weight under rotating conditions. For the non-Newtonian eccentric flow in Figure 7.21, a very low specific gravity of 0.01 leads to a flow rate of 898.5 gpm.

Next we wish to consider the opposite extreme—for example, a heavy mud or cement with a specific gravity of 2. Because the unstable convective term never vanishes when the pipe rotates (its magnitude is proportional to fluid density and pipe rpm), we decrease the time step to 0.0001 sec and increase the number of time steps simulated (Figure 7.22). The resulting flow rate is a much lower 135.1 gpm. Computation times for the two runs are approximately five seconds and two minutes. Finally, we reduce the specific gravity to 1.0 (i.e., an unweighted mud). Will the flow rate vary linearly with density—that is, fall midway between 135.1 and 898.5 gpm? Figure 7.23 shows that the flow rate is, in fact, 160.1 gpm. This unpredictability shows why computer models are important to real-world field job planning.

EXAMPLE 7.8

Effect of Mud Pump Ramp-Up and Ramp-Down Flow Rate under Nonrotating and Rotating Conditions

In Figure 7.24, we consider a Power law fluid in an eccentric annulus under a constant imposed pressure gradient of -0.005 psi/ft with the drillpipe completely stationary. This is seen to produce

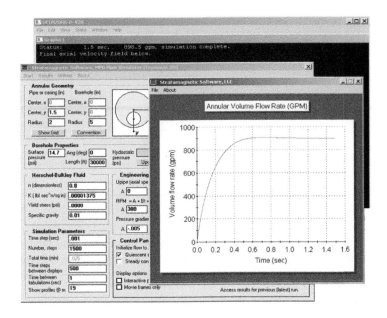

FIGURE 7.21

Very-low-density fluid (e.g., foam) at high rpm.

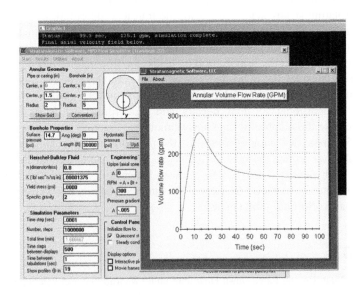

FIGURE 7.22

Very-high-density fluid (e.g., heavy mud or cement) at high rpm.

Example 7.8   289

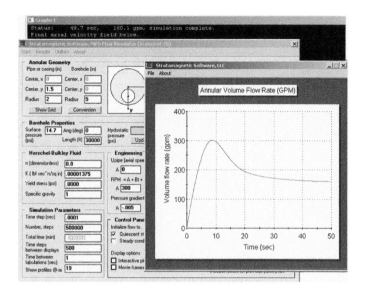

FIGURE 7.23

Unweighted fluid (e.g., water or brine) at high rpm.

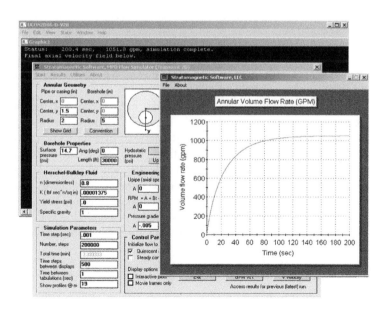

FIGURE 7.24

Constant pressure gradient calculation.

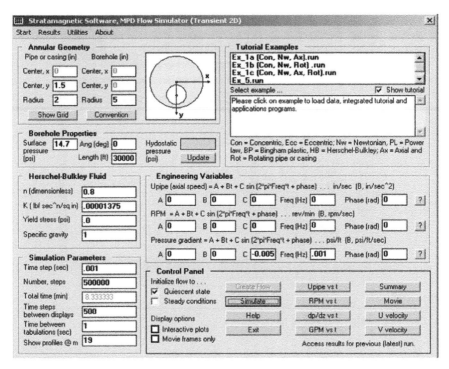

FIGURE 7.25

Mud pump ramp-up and ramp-down.

a steady-state flow rate of 1,051.8 gpm as shown. In practice, the mud pump starts and stops, and transient effects are associated with ramp-up and ramp-down. We ask, "How are pressure gradient and flow rate transient properties related?"

To answer this question, we modify several menu entries in Figure 7.24 so that the pressure gradient is no longer constant. The assumption shown in Figure 7.25 allows a sinusoidal ramp-up from quiet conditions to our previous value of −0.005 psi/ft, followed by a full ramp-down. This is accompanied by time mesh refinement plus the use of additional time steps. Clicking on the "?" to the far right of the pressure gradient menu produces the left-side diagram of Figure 7.26 showing pressure assumptions. The right-side diagram gives the computed volumetric flow rate as a function of time.

Next we determine the effect of drillstring rotation. We simply change the zero rotation input in Figure 7.25 to allow for a 100-rpm rotational rate as shown in Figure 7.27. For the same pressure gradient variation as before, the flow rate is now substantially reduced, as shown in Figure 7.28.

EXAMPLE 7.9

Effect of Rotational and Azimuthal Start-up

In this example, we study the effects of drillstring rotational start-up on the baseline nonrotating problem defined in Figure 7.29 for a Power law fluid in an eccentric annulus. Figure 7.30 shows that after 100 sec, the (almost) steady flow rate is 1,024.0 gpm.

Example 7.10   291

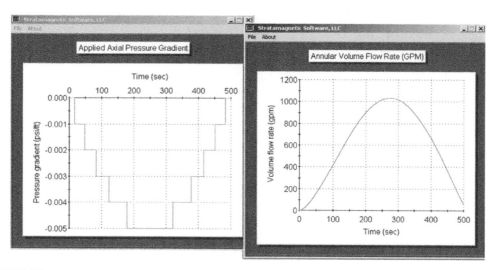

FIGURE 7.26

Assumed pressure gradient and calculated flow rate.

What happens when the drillstring is rotating at a fixed constant 100 rpm for the duration of the start-up process? This new flow is easily obtained by changing the constant rpm input in Figure 7.29 to that in Figure 7.31, to produce the flow rate history shown in Figure 7.32. After 100 sec, the flow has fully equilibrated at the reduced rate of 221.1 gpm. There is a flow rate "overshoot" near 350 gpm early on that we have observed in all rotational flow calculations.

We next determine the effects of rotational start-up. In Figure 7.33, we choose the "Bt" input option for rpm definition, typing "1" into that box for the time step information assumed. In Figure 7.34, we show at the left how the same 100 rpm is achieved as before, but at the end of the 100-sec period. The right-side diagram shows a flow rate returning to the 200-gpm range; however, the flow rate overshoot is now near 600 gpm.

EXAMPLE 7.10

Effect of Axial Drillstring Movement

In this *nonrotating* drillstring example, we study the effects of axial movement on the baseline problem defined in Figure 7.29 for a Power law fluid in an eccentric annulus. Again, Figure 7.30 shows that after 100 sec, the (almost) steady flow rate is 1,024.0 gpm, assuming stationary pipe. If a constant +20 in./sec is modeled instead, we have an increased 1,132.6 gpm, whereas if −20 in./sec is taken, we find a reduced 912.6 gpm. Computer screens for these simple constant-speed dragging calculations are not shown.

In field applications, the drillstring is often reciprocated axially to facilitate jarring operations or cuttings removal while the mud pump acts under an almost constant pressure gradient

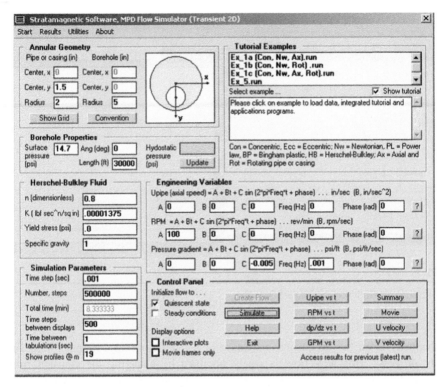

FIGURE 7.27

Increasing rotational rate to 100 rpm.

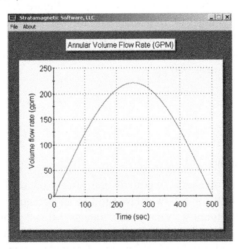

FIGURE 7.28

Significantly reduced volumetric flow rate with rotation.

Example 7.10   **293**

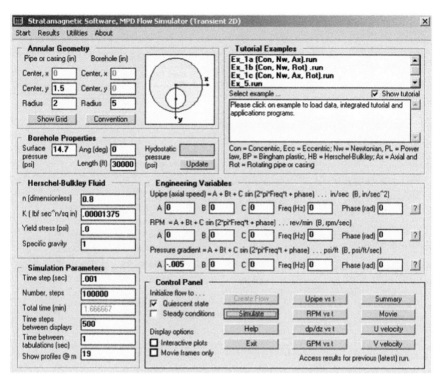

FIGURE 7.29

Nonrotating flow.

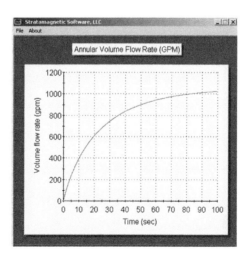

FIGURE 7.30

Nonrotating flow.

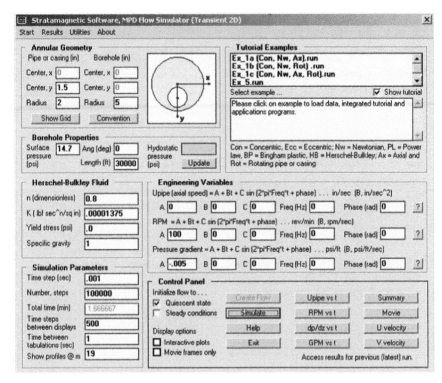

FIGURE 7.31

Constant 100 rpm throughout.

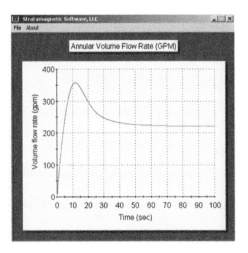

FIGURE 7.32

Constant 100 rpm throughout.

Example 7.10   **295**

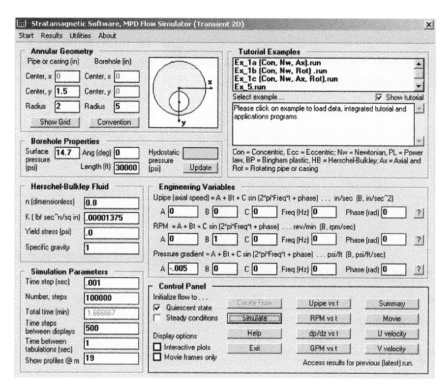

FIGURE 7.33

Linearly increasing rpm with time.

condition. One might ask what the effects on flow rate, apparent viscosity, shear rate, and viscous stress are, with the answers sure to assist the engineer in interpreting the physical consequences of his or her actions. For example, increases in bottomhole stress may improve hole cleaning, while reductions in apparent viscosity may lubricate the drillstring. In Figure 7.35, we alter the "U_{pipe}" input to allow sinusoidal drillstring reciprocation with a peak-to-peak amplitude of 20 in./sec and a frequency of 0.1 Hz. Clicking on the "?" at the far right will produce the pipe displacement speed history at the left of Figure 7.36. At the right is the stably computed oscillatory flow rate.

The "Results" menu in Figure 7.37 provides additional postprocessed results useful for correlation purposes. For instance, "Color plots" provides displays of the physical quantities appearing in the list, several of which are shown in Figure 7.38. Notice in Figure 7.35 that we elected to save "movie frames" showing the axial velocity distribution evolving in time. (The "interactive plot" option would produce line graph results during simulation.) Playing the "Axial velocity—Movie" option produces a movie, which can be viewed continuously or frame by frame. Typical movie frames (with time increasing to the right) are shown in Figure 7.39. All of the postprocessing options described here are also available for rotating flow problems.

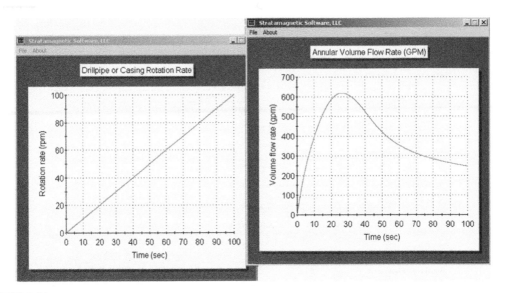

FIGURE 7.34

Linearly increasing rpm with time.

EXAMPLE 7.11

Combined Rotation and Sinusoidal Reciprocation

In this example, again for transient, nonlinear, non-Newtonian Power law flow in an eccentric annulus, we combine two previous calculations and demonstrate the ease with which combined sinusoidal axially reciprocating pipe motion and drillstring rotation can be modeled, literally by filling in input boxes and clicking. The assumptions are given in Figure 7.40, assumed pipe displacement histories are displayed in Figure 7.41, and the computed volumetric flow rate is provided in Figure 7.42. Note from this curve the pronounced overshoots and flow rate fluctuations. We have modeled the mud pump as a constant pressure gradient source in our work that leads to variable flow rate.

In reality, the pump may act more as a constant-rate source that leads to time-dependent pressure gradients. This latter model is much more complicated mathematically and cannot be solved within a reasonable time. However, the percent fluctuations seen from flow rate curves such as the one in Figure 7.42 represent those for pressure gradient and can be used meaningfully for managed pressure job planning.

EXAMPLE 7.12

Combined Rotation and Sinusoidal Reciprocation in the Presence of Mud Pump
Flow Rate Ramp-Up for Yield Stress Fluid

This comprehensive example illustrates the high level of simulation complexity offered by our math model. Here we again consider an eccentric annulus, now containing a Herschel-Bulkley yield

Example 7.12   **297**

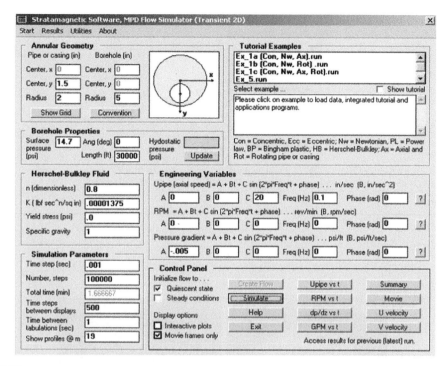

FIGURE 7.35

Sinusoidal drillstring reciprocation.

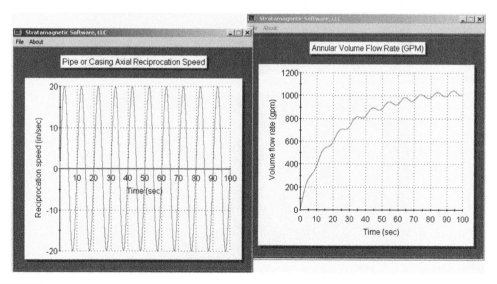

FIGURE 7.36

Pipe displacement history and computed flow rate.

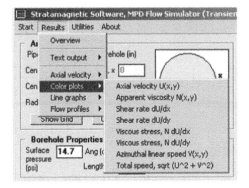

FIGURE 7.37

Example color output.

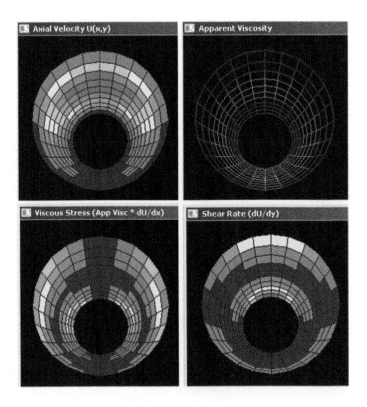

FIGURE 7.38

Example color output for several physical quantities.

Example 7.12   **299**

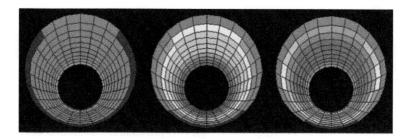

FIGURE 7.39

Frames from axial velocity movie (time increasing).

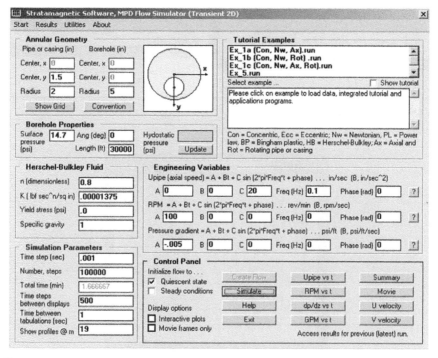

FIGURE 7.40

Combined transient reciprocation and rotation.

stress fluid. The drillpipe is allowed to axially reciprocate sinusoidally in time, while rotational rate increases linearly with time. The mud pump pressure gradient is allowed to steepen with time from start-up to describe increased pumping action.

All of these effects are coupled nonlinearly. They can be computed quickly and stably, and if numerical instabilities are encountered, they can be remedied by decreasing time step size. To accommodate this possibility, the algorithm is efficiently coded to make optimal use of memory resources and will allow up to 10,000,000 time steps, for which calculations may require 15 minutes

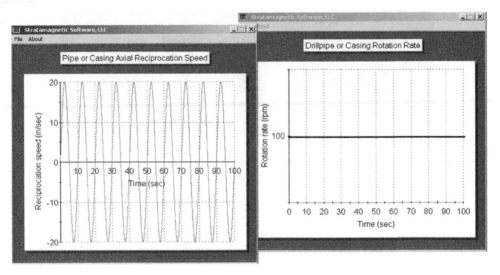

FIGURE 7.41

Pipe displacement history display.

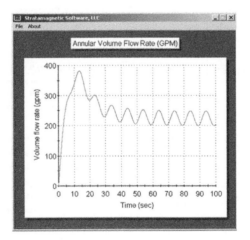

FIGURE 7.42

Computed volumetric flow rate.

or more. The assumptions are shown in Figure 7.43, while detailed pipe displacement histories, applied pressure gradients, and computed volumetric flow rate are given in Figure 7.44.

In this chapter, we demonstrated how the most general transient single-phase, constant-density non-Newtonian Hershel-Bulkley fluid with yield stress can be studied as it flows through an eccentric annulus in the presence of coupled and arbitrary drillpipe axial reciprocation, unsteady rotation, and time-varying axial pressure gradient. The algorithm and its strengths and limitations

Example 7.12   **301**

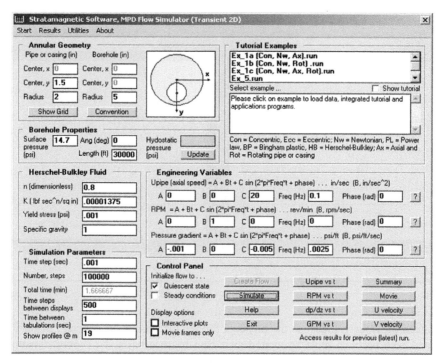

FIGURE 7.43

Basic assumptions, comprehensive example.

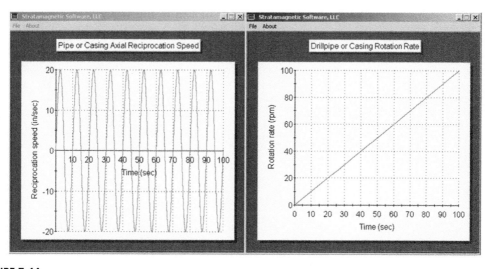

FIGURE 7.44

Additional assumptions and computed flow rate with time.

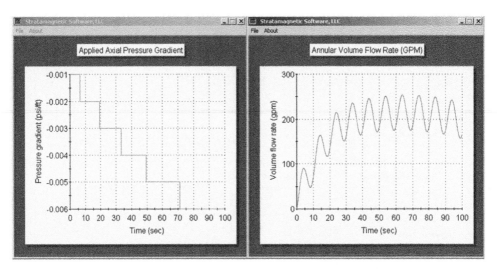

FIGURE 7.44

(Continued).

were explained previously. Because the physical problem is nonlinear, general conclusions are not available and each problem must be treated on a case-by-case basis. To support this endeavor, all efforts have been made to render the method simple to use, with all text output, report generation, and color graphics completely automated. There is no requirement on the part of the user for any special skills in fluid dynamics, advanced mathematics, or computer modeling. The model is new, and, certainly, as more becomes known about its properties and the consequences of general borehole flows, we will update our exposition accordingly.

Cement and Mud Multiphase Transient Displacements

In Chapters 1 through 7 we dealt with steady and transient single-phase non-Newtonian flows in two dimensions, allowing arbitrary borehole eccentricity, significant yield stress (with plug zone size and shape computed as part of the solution), and combined drillpipe or casing reciprocation and rotation in the most general limit. Such "single-phase" flows may in fact be two-phase (e.g., typical fluid-solid mixtures in which barite is the weighting agent or liquid-gas foams used in underbalanced drilling). However, they are single-phase in that a set of Herschel-Bulkley parameters "n," K," and "τ_0" can characterize the mixture as if it were a unique homogeneous fluid. In this chapter and in Chapter 9, we consider two-phase flows, by which we imply displacements of one such single-phase flow by another in a general borehole annulus. Such problems are necessarily transient and three-dimensional and consequently involve more computation.

These are not academic exercises but real flows arising in different important applications. In drilling, multiple non-Newtonian fluids are pumped down the drillpipe according to a prescribed rate schedule (e.g., fluid A for t_A min at flow rate GPM_A, fluid B for t_B min at GPM_B, and so on). This transient fluid column then flows downward through the drillbit and finally up the eccentric borehole annulus. For managed pressure drilling, the simulation objective is the pressure profile along the borehole, and particularly at the bit, as a function of time. Drillers wish to control pressures at the drillbit, perhaps targeting constant values for safety considerations. This can be accomplished by various means—for example, changing pump rates to affect dynamic friction, altering mud type and weight, directly controlling surface choke pressures, or, as we have demonstrated, varying drillstring rotational rates.

Simulation methods assist engineers in these objectives by "assigning numbers" to these steps to quantify their effectiveness. In cementing, the same overall pump scenario applies. However, flow details at the diffusing interfaces of different fluids are of interest: the extent to which they mix at different positions in the annular cross section, the time needed to establish target concentration levels, the role of rheology and diffusion in achieving displacement objectives, and so on.

While the objectives are simple, the formulations and calculations are not. They must be scientifically rigorous and numerically stable. Many strategies are available depending on the research objectives. For instance, equations for "microscopic" particle interactions between fluid components with adjustable parameters are often postulated together with closure hypotheses to model kinematic and dynamic processes. Such models can require numerically intensive computation. A simpler and well-known "macroscopic" method offered by physicists is described in Landau and Lifschitz (1959) and is easily explained. We explain the method in the next section and extend it for increasing levels of flow and geometric sophistication in the following discussions.

DISCUSSION 8.1

Unsteady Three-Dimensional Newtonian Flows with Miscible Mixing in Long Eccentric Annular Ducts

The simplest differential equation for transient, single-phase axial flow is given by the two-dimensional model $\rho \; \partial u/\partial t = -\partial p/\partial z + \mu(\partial^2 u/\partial x^2 + \partial^2 u/\partial y^2)$, where rotation is neglected. In dealing with flows where miscible mixing is permitted with general initial species variations, purely two-dimensional flow is not possible, since concentrations must vary in the direction of fluid convection. Thus, an additional "$\partial^2 u/\partial z^2$" term at the right side is expected. We now interpret u(x, y, z, t) as the axial speed for an evolving heterogeneous mixture characterized by a mass density function $\rho(C)$ and a Newtonian viscosity function $\mu(C)$, where the local fluid concentration C(x, y, z, t) must satisfy an isotropic convection-diffusion equation.

The functions $\rho(C)$ and $\mu(C)$ for the two-fluid system are determined from simpler experiments, say steady-state concentric rheometer measurements. Once obtained, our boundary value problem formulations predict their effects in real physical problems with complicated geometries acting under general boundary and initial conditions. The general axial velocity equation now takes the form

$$\rho(C) \, (\partial u/\partial t + u \, \partial u/\partial z) = -\partial p/\partial z + \mu(C)(\partial^2 u/\partial x^2 + \partial^2 u/\partial y^2 + \partial^2 u/\partial z^2) \qquad (8.1)$$

Equation 8.1 arises from an intuitive ad hoc argument. Note that the viscous terms in Equation 8.1 actually derive from the divergence form "$\partial S_{zx}/\partial x + \partial S_{zy}/\partial y + \partial S_{zz}/\partial z$," so that the product terms "$\partial u/\partial x \; \partial \mu/\partial x + \partial u/\partial y \; \partial \mu/\partial y + \partial u/\partial z \; \partial \mu/\partial z$"—that is, $d\mu(C)/dC\{(\partial u/\partial x)^2 + (\partial u/\partial y)^2 + (\partial u/\partial z)^2\}$—have been omitted in comparison to the retained terms shown. These nonlinear terms represent higher-order effects comparable to those neglected by our use, owing to the complexity of the resulting equation system, of simpler explicit time integration methods. The important second-derivative terms, retained here, are required to impose no-slip boundary conditions on solid boundary surfaces. Equation 8.1 for the axial velocity u(x, y, z, t) is nonlinearly coupled to the convection-diffusion equation

$$\partial C/\partial t + u \, \partial C/\partial z = \varepsilon(\partial^2 C/\partial x^2 + \partial^2 C/\partial y^2 + \partial^2 C/\partial z^2) \qquad (8.2)$$

where $\varepsilon(C) > 0$ is an isotropic diffusion coefficient, possibly dependent on C, determined experimentally. Equation 8.1 is solved together with:

- $\partial u/\partial z = 0$ far upstream and downstream
- No-slip velocity conditions at solid surfaces (e.g., $u = u_{speed}(t)$ on the pipe surface if the pipe moves with speed $u_{speed}(t)$)
- u(x, y, z, 0) = 0 for the assumed initial quiescent state
- Note that the additional convective term "$u \, \partial u/\partial z$" (from the more complete acceleration "$\partial u/\partial t + u \, \partial u/\partial z$") has been retained, although it is usually negligible for ducts that are uniform in the axial direction.

Equation 8.2 is solved with boundary conditions C(x, y, z_{left}, t) = C_{left} and C(x, y, z_{right}, t) = $C_{right}(t)$, plus the initial condition C(x, y, z, 0) = C_{left} for z < $z_{interface}$ and C(x, y, z, 0) = C_{right} for z > $z_{interface}$, where z = $z_{interface}$ is the initial position of the flat interface separating the two fluids. Also, the normal partial derivative of C to a solid wall vanishes because there is no diffusion into a solid. Other boundary and initial conditions are, of course, possible.

For the concentration equation, the u ∂C/∂z term must be retained in comparison with all of the others, since it is this term that provides the required convection. The foregoing formulation applies to borehole flows with arbitrary annular cross sections. However, the use of rectangular coordinates that do not fit inner and outer pipe and hole contours leads to inaccuracies associated with numerical noise.

As noted earlier, the use of classical coordinate systems (e.g., Cartesian or cylindrical-radial) is not appropriate for the eccentric annuli domains encountered in petroleum engineering. Borehole contours, for instance, must also describe nonideal washout and cuttings bed boundaries. The mapping method devised in Chapter 3 is ideal for hosting transformed versions of Equations 8.1 and 8.2. When these equations are combined with Equation 3.59, we obtain the transformed model

$$\rho(C)(\partial u/\partial t + u\,\partial u/\partial z) = -\partial p/\partial z + \mu(C)\partial^2 u/\partial z^2 + \mu(C)(\alpha u_{\xi\xi} - 2\beta u_{\xi\eta} + \gamma u_{\eta\eta})/J^2 \qquad (8.3a)$$

$$\partial C/\partial t + u\,\partial C/\partial z = \varepsilon(C)\,\partial^2 C/\partial z^2 + \varepsilon(C)(\alpha C_{\xi\xi} - 2\beta C_{\xi\eta} + \gamma C_{\eta\eta})/J^2 \qquad (8.3b)$$

where α, β, γ, and J are metrics of the mapping transformations available numerically from the solutions for $x(\xi, \eta)$ and $y(\xi, \eta)$. This is the problem solved numerically for this first illustrative example. The preceding model does not allow rotation, but it does allow transient axial reciprocation.

The general transient three-dimensional algorithm was checked in the steady two-dimensional concentric limit where the exact closed-form velocity solution for concentric annuli given in Chapter 1 is used. This numerical check is extremely important because different discretization procedures for partial derivatives in parabolic equations can yield additive diffusion to those diffusive effects indicated explicitly in the equation. This is well known in numerical modeling (see the discussions on artificial viscosity and von Neumann analysis in Press et al. (1992)). In fact, various methods attempted at first gave total volumetric flow rates that were noticeably inconsistent with the exact solution. The validation method used also served to narrow the range of numerical fixes needed to stabilize the rotating flow version of Equation 8.1. In this sense, the differencing of partial differential equations is as much an art as it is a science.

DISCUSSION 8.2

Transient, Single-Phase, Two-Dimensional Non-Newtonian Flow with Inner Pipe Rotation in Eccentric Annuli

Here we review a different limit of the general flow equations: steady single-phase non-Newtonian flow with inner pipe rotation in constant eccentric annuli. The flow is driven by an axial pressure gradient so that rotational and axial velocities both exist. However, because the flow is identical in all cross sections, the problem is two-dimensional mathematically. This initial discussion is necessary because it explains the detailed mathematical strategy used to model flows with rotation in eccentric annuli, and it represents the basis for extension when we address complicated multiphase flows.

In our initial mapping discussion, we started with Cartesian coordinates in order to derive curvilinear grid transformations for the general cross section. This approach is suitable so long as the inner circle does not rotate. When it does rotate, circular cylindrical formulations are needed at first to accommodate the no-slip velocity condition at the pipe or casing surface. The formulation is next reexpressed in rectangular coordinates so that the mapping procedure explained in Chapter 3 can be applied. Thus, rotating flow problems require several sets of coordinate transformations, not to mention those that project physical plane properties onto computer screen displays.

The general non-Newtonian rheological equations for unsteady single-fluid flow with and without yield stress are given in three dimensions by

$$
\begin{aligned}
\rho(\partial v_r/\partial t + v_r\, \partial v_r/\partial r + v_\theta/r\, \partial v_r/\partial \theta - v_\theta^2/r + v_z \partial v_r/\partial z) \\
= F_r - \partial p/\partial r + 1/r\, \partial(r\, S_{rr})/\partial r + 1/r\, \partial S_{r\theta}/\partial \theta - 1/r S_{\theta\theta} + \partial S_{rz}/\partial z
\end{aligned}
\tag{8.4a}
$$

$$
\begin{aligned}
\rho(\partial v_\theta/\partial t + v_r\, \partial v_\theta/\partial r + v_\theta/r\, \partial v_\theta/\partial \theta + v_r v_\theta/r + v_z \partial v_\theta/\partial z) \\
= F_\theta - 1/r\, \partial p/\partial \theta + 1/r^2\, \partial(r^2\, S_{\theta r})/\partial r + 1/r\, \partial S_{\theta\theta}/\partial \theta + \partial S_{\theta z}/\partial z
\end{aligned}
\tag{8.4b}
$$

$$
\begin{aligned}
\rho(\partial v_z/\partial t + v_r\, \partial v_z/\partial r + v_\theta/r\, \partial v_z/\partial \theta + v_z \partial v_z/\partial z) \\
= F_z - \partial p/\partial z + 1/r\, \partial(r\, S_{zr})/\partial r + 1/r\, \partial S_{z\theta}/\partial \theta + \partial S_{zz}/\partial z
\end{aligned}
\tag{8.4c}
$$

$$
\partial v_r/\partial r + v_r/r + 1/r\, \partial v_\theta/\partial \theta + \partial v_z/\partial z = 0
\tag{8.4d}
$$

where v_r, v_θ, and v_z are radial, azimuthal, and axial velocity components, respectively, with

$$
\underline{\underline{S}} = 2\, N(\Gamma)\underline{\underline{D}}
\tag{8.4e}
$$

denoting the deviatoric stress tensor, $N(\Gamma)$; the apparent viscosity function, Γ; the shear rate; and $\underline{\underline{D}}$, the deformation tensor whose elements are defined by

$$
D_{rr} = \partial v_r/\partial r
\tag{8.4f}
$$

$$
D_{\theta\theta} = 1/r\, \partial v_\theta/\partial \theta + v_r/r
\tag{8.4g}
$$

$$
D_{zz} = \partial v_z/dz
\tag{8.4h}
$$

$$
D_{r\theta} = D_{\theta r} = 1/2[r\, \partial(v_\theta/r)/\partial r + 1/r\, \partial v_r/\partial \theta]
\tag{8.4i}
$$

$$
D_{\theta z} = D_{z\theta} = 1/2(\partial v_\theta/\partial z + 1/r\, \partial v_z/\partial \theta)
\tag{8.4j}
$$

$$
D_{rz} = D_{zr} = 1/2(\partial v_r/\partial z + \partial v_z/\partial r)
\tag{8.4k}
$$

We ignore body forces in this discussion. For steady-state flow ($\partial/\partial t = 0$) under fully developed ($\partial/\partial z = 0$) conditions, with the *further* assumption of Newtonian flow, Equations 8.4a through 8.4c possess a remarkable property. Equations 8.4a and 8.4b are independent of v_z and z, so they can be solved first (we have assumed that the azimuthal $\partial p/\partial \theta$ driver is insignificant compared to the dragging effect offered by the rotating surface). Once solutions for v_r and v_θ are available, they are used to evaluate the left-side convective terms of Equation 8.4c for the solution of axial velocity.

This decoupling is possible because the Newtonian assumption removes v_z from Equations 8.4a and 8.4b by rendering the apparent viscosity a simple constant. That is, rotation affects axial flow but not conversely. This conclusion is borne out by the more detailed calculations of Escudier et al. (2000) for eccentric annuli comprising off-centered circles. In flows without rotation, the axial velocity field is symmetric with respect to a line passing through the centers of both circles.

Escudier et al. showed that inner circle rotation removes this symmetry and displaces the location of maximum axial velocity azimuthally.

The steady analysis of Escudier et al. is overly simple for our purposes in several respects. First, the Newtonian fluid assumption decouples the rotating flow from the axial flow. The resulting linear second-derivative operators present at all right sides of the differential equations simplify numerical analysis but preclude any modeling of non-Newtonian shear-thinning effects. In general, the apparent viscosity function depends on all velocity components and strongly couples all velocity fields. Our transient formulation, in contrast, is posed in terms of a general stress tensor valid for fluids with and without yield stresses where the extended Herschel-Bulkley flow model is used. Also, Escudier et al. use bipolar coordinate systems to model eccentered circles. Our formulation is developed using general curvilinear coordinates whose outer boundaries may conform to washout and cuttings beds contours.

Again, while the approach just discussed may be acceptable for Newtonian flows, which allow v_z to be efficiently solved once a complicated cross-flow is obtained, it is not practical for non-Newtonian fluids because velocities are dynamically coupled through the apparent viscosity. In our approach, we use both physical and mathematical arguments to facilitate a fast algorithm applicable to general fluids on arbitrary curvilinear meshes.

We begin with the flow equations in cylindrical radial coordinates, as shown in Equations 8.4a through 8.4d, which, we emphasize, apply to annuli with circular as well as noncircular boundaries. It is clear that, in these coordinates, barring the possibility of strong formation influx, the radial velocity v_r is much smaller than both the azimuthal speed v_θ and the axial speed v_z (the latter two may be large and comparable). More precisely, "v_r is small" because it vanishes at the inner and outer boundaries, by virtue of zero velocity slip, and also between the two boundaries, because strong waviness in the contours is absent.

When these conditions are fulfilled, the rotating fluid in the cross-plane is defined by v_θ alone, solved together with $\partial p/\partial\theta \approx 0$. Note that $\partial p/\partial\theta$ vanishes identically in concentric problems even when rotations are extremely rapid. For eccentric problems, we are justified in neglecting this azimuthal driver because the primary source of rotating flow is the dragging effect of the boundary. In other words, the problem with fully coupled rotating and axial flow for non-Newtonian fluids in eccentric annuli with washouts and cuttings beds can be determined by just two nonlinearly coupled equations for v_θ and v_z.

In steady, concentric two-dimensional problems, the left side of Equation 8.4b is identically zero because $\partial/\partial t = 0$, $\partial/\partial z = 0$, $v_r = 0$, and $\partial/\partial\theta = 0$. Also, $\partial p/\partial\theta$ vanishes by virtue of symmetry, and what remains is $1/r^2 \ \partial(r^2 S_{\theta r})/\partial r + 1/r \ \partial S_{\theta\theta}/\partial\theta = 0$. If the fluid is Newtonian, this reduces to a simple linear ordinary differential equation for v_θ. Once v_θ is solved subject to "rpm" constraints at the pipe and no-slip boundary conditions at the borehole wall, Equation 8.4a is used to calculate the radial pressure gradient $\partial p/\partial r$, therefore resulting in the classical "$-v_\theta^2/r$" centrifugal force effect.

In our approach to transient non-Newtonian fluids with rotating axial flow coupling in eccentric annuli, the simple "$1/r^2 \ \partial(r^2 S_{\theta r})/\partial r + 1/r \ \partial S_{\theta\theta}/\partial\theta = 0$" relationship is replaced by the more general azimuthal approximation

$$\rho(\partial v_\theta/\partial t + v_\theta/r \ \partial v_\theta/\partial\theta) \approx 1/r^2 \ \partial(r^2 S_{\theta r})/\partial r + 1/r \ \partial S_{\theta\theta}/\partial\theta \approx 0 \qquad (8.4l)$$

where a more general apparent viscosity function, $N(\Gamma)$, is solved together with Equation 8.4c. Consistently assuming $\partial/\partial t = 0$, $\partial/\partial z = 0$, and $v_r \approx 0$, we have

$$\rho(\partial v_z/\partial t + v_\theta/r \; \partial v_z/\partial \theta) \approx -\partial p/\partial z + 1/r \; \partial(r S_{zr})/\partial r + 1/r \; \partial S_{z\theta}/\partial \theta + \partial S_{zz}/\partial z \qquad (8.4m)$$

When the viscous stress terms of Equations 8.4l and 8.4m are expanded, the second-order linear partial derivative operator $\Lambda = \partial^2/\partial r^2 + 1/r \; \partial/\partial r + 1/r^2 \; \partial^2/\partial \theta^2 - 1/r^2$ appears (additional nonlinear terms are present that we do not neglect here). In order to use the curvilinear mapping method developed in Chapter 3, we first recognize that "$\partial^2/\partial r^2 + 1/r \; \partial/\partial r + 1/r^2 \; \partial^2/\partial \theta^2$" is equivalent to "$\partial^2/\partial x^2 + \partial^2/\partial y^2$" and that "$r^2 = x^2 + y^2$." Thus, we have $\Lambda = \partial^2/\partial x^2 + \partial^2/\partial y^2 - 1/(x^2 + y^2)$. In curvilinear coordinates, this becomes

$$\Lambda = (\alpha \; \partial^2/\partial \xi^2 - 2\beta \; \partial^2/\partial \xi \partial \eta + \gamma \; \partial^2/\partial \eta^2)/J^2 - 1/(x(\xi, \eta)^2 + y(\xi, \eta)^2) \qquad (8.4n)$$

where the mapping functions $x(\xi, \eta)$, $y(\xi, \eta)$, α, β, and γ are known. Stable approximation methods for "$(\alpha \; \partial^2/\partial \xi^2 - 2\beta \; \partial^2/\partial \xi \partial \eta + \gamma \partial^2/\partial \eta^2)/J^2$" were discussed earlier in this book and are used here. The negative nature of the additional term "$-1/(x^2 + y^2)$" increases the numerical stability of the schemes by increasing diagonal dominance.

All spatial derivatives are approximated by second-order accurate central difference formulas. The convective term $\rho v_\theta/r \; \partial v_z/\partial \theta$ in Equation 8.4m is also evaluated by central differences, although several intermediate transformations are required. For example in the cylindrical coordinate derivation underlying Equation 8.4m, we can reexpress $\partial/\partial \theta$ as $\partial/\partial \theta = -y \; \partial/\partial x + x \; \partial/\partial y$. Thus, in general curvilinear coordinates, it follows that

$$\begin{aligned} \partial v_z/\partial \theta &= -y \; \partial v_z/\partial x + x \; \partial v_z/\partial y \\ &= -y(\xi, \eta)\{\xi_x \; \partial v_z/\partial \xi + \eta_x \; \partial v_z/\partial \eta\} + x(\xi, \eta)\{\xi_y \; \partial v_z/\partial \xi + \eta_y \; \partial v_z/\partial \eta\} \end{aligned} \qquad (8.4o)$$

In rotating flows, the combination "$\partial p/\partial z + \rho v_\theta/r \; \partial v_z/\partial \theta$" serves as an "effective axial pressure gradient" that depends on the azimuthal coordinate θ by way of solutions to the equation for v_θ.

Computations for steady rotating flows show that the location of maximum axial velocity at the wide side of eccentric annuli is displaced azimuthally, with displacements increasing with rotational speed in a manner consistent with Escudier et al. (2000). Again, our method applies to transient non-Newtonian flow in arbitrary eccentric annuli. The unsteady boundary conditions possible with this extension include general axial reciprocation coupled with arbitrary transient rotation. It is thus possible to simulate the effects of drillstring axial vibration and torsional stick-slip on annular fluid flow characteristics.

DISCUSSION 8.3

Transient, Three-Dimensional Non-Newtonian Flows with Miscible Mixing in Long Eccentric Annular Ducts with Pipe or Casing Rotation and Reciprocation

For presentation purposes, we introduced the general modeling ideas starting with unsteady, three-dimensional Newtonian flow, and miscible mixing problems without rotation, and then we considered transient, two-dimensional non-Newtonian single-phase problems with rotation. With these ideas developed, we now present their obvious combination in order to treat the complicated problem indicated in the discussion title. It is clear that we replace the axial flow momentum description in Equation 8.4c with the approximation

$$\rho(C)(\partial v_z/\partial t + v_\theta/r\ \partial v_z/\partial\theta) = -\partial p/\partial z + 1/r\ \partial(rS_{zr})/\partial r + 1/r\ \partial S_{z\theta}/\partial\theta + \partial S_{zz}/\partial z \qquad (8.5a)$$

while the azimuthal flow statements in Equations 8.4b and 8.4l are replaced with

$$\rho(C)(\partial v_\theta/\partial t + v_\theta/r\ \partial v_\theta/\partial\theta) \approx 1/r^2\ \partial(r^2 S_{\theta r})/\partial r + 1/r\ \partial S_{\theta\theta}/\partial\theta \approx 0 \qquad (8.5b)$$

where again C is the concentration function. However, the left operator "$\partial C/\partial t + u\ \partial C/\partial z$" in Equation 8.2, or "$\partial C/\partial t + v_z\ \partial C/\partial z$" in the present nomenclature, must be replaced by "$\partial C/\partial t + v_\theta/r\ \partial C/\partial\theta + v_z\ \partial C/\partial z$," so that

$$\partial C/\partial t + v_\theta/r\ \partial C/\partial\theta + v_z\ \partial C/\partial z = \varepsilon\ (\partial^2 C/\partial x^2 + \partial^2 C/\partial y^2 + \partial^2 C/\partial z^2) \qquad (8.5c)$$

The new term models the addition of fluid convection in the azimuthal direction. Boundary conditions for velocity and concentration were given previously.

The numerical strategy for solving Equations 8.5a through 8.5c is straightforward. As suggested previously, all three equations are first expressed in Cartesian x, y, and z coordinates. Then the curvilinear coordinate transformations derived in Chapter 3 are used to map the system to convenient computational (ξ, η, z) coordinates. Ideally, the transient equations are integrated implicitly, using newer alternating-direction-implicit equation methods allowing mixed partial derivatives. However, for simplicity an explicit scheme is used for which concentric flow validations indicate reasonable accuracy.

An important question that arises is the approximation for the stress tensor, $\underline{S}$, used. Again recall that the general rheology model involves the Herschel-Bulkley parameters n, K, and τ_0. It is tempting to try to fit each of these to linear functions of C—for example, $n(C) = (n_{right} - n_{left})$ C + n_{left} so that $n = n_{left}$ when C = 0 at the left inlet and $n = n_{right}$ when C = 1 at the right outlet. However, this approach is not desirable because the physical consequences of this curve fit are unclear. Instead, at each spatial node (for a fixed time step), the apparent viscosity function based on the extended Herschel-Bulkley formula is first calculated to give values for N_{left} and N_{right}.

Then a local weighted average of the viscosity, N, based on the Todd-Longstaff formula using these two inputs is taken in Equation 8.4e. This alternative approach, dealing with apparent viscosities directly, is physically satisfying. Consistently with the use of an explicit time integration scheme, we neglect spatial derivatives of apparent viscosity, although, of course, the apparent viscosity function used is definitely variable throughout the field of flow, barring a perfectly Newtonian fluid.

For the local mass density, ρ, appearing in both axial and azimuthal momentum equations, the linear interpolation $\rho(C) = (\rho_{right} - \rho_{left})$ C + ρ_{left} (so that $\rho = \rho_{left}$ when C = 0 at the left inlet and $\rho = \rho_{right}$ when C = 1 at the right outlet) is reasonable and thus used. We do emphasize that the apparent viscosity does not directly appear in the concentration equation, which shows an explicit diffusion coefficient $\varepsilon(C)$ at the right side. However, viscosity and density enter through the velocities in the convection terms "$v_\theta/r\ \partial C/\partial\theta + v_z\ \partial C/\partial z$," which are directly affected by the two parameters.

A second question that arises is the functional form of the shear rate expression needed to evaluate local apparent viscosities. In Chapter 2, we indicated that $\Gamma = \{(r\ d(v_\theta/r)/dr)^2 + (dv_z/dr)^2\}^{1/2} = \{(r\ d\Omega/dr)^2 + (dv_z/dr)^2\}^{1/2}$, where the usual rotational rate is defined by $\Omega = v_\theta/r$, applies when flow variations in the streamwise direction are small. This expression, which strictly applies to concentric flows, also holds eccentrically for most practical flows, as computations show. This usage significantly enhances numerical stability. As before, the expression in cylindrical coordinates must be recast in Cartesian coordinates for subsequent transformation to general curvilinear coordinates. To provide some indication of Fortran source code complexity, we have duplicated several lines from the apparent viscosity update carried out for each point at the end of an integration time step:

```
C
      UX = YETA(I,M,N)*((UTN(I,M+1,N)-UTN(I,M-1,N))/TDPSI)/GAKOB(I,M,N)
     1      -YPSI(I,M,N)*((UTN(I,M,N+1)-UTN(I,M,N-1))/TDETA)/GAKOB(I,M,N)
      UY = XPSI(I,M,N)*((UTN(I,M,N+1)-UTN(I,M,N-1))/TDETA)/GAKOB(I,M,N)
     1      -XETA(I,M,N)*((UTN(I,M+1,N)-UTN(I,M-1,N))/TDPSI)/GAKOB(I,M,N)
C
      OMEGAX = YETA(I,M,N)*(OMEGA(I,M+1,N)-OMEGA(I,M-1,N))
     1           -YPSI(I,M,N)*(OMEGA(I,M,N+1)-OMEGA(I,M,N-1))
      OMEGAY = XPSI(I,M,N)*(OMEGA(I,M,N+1)-OMEGA(I,M,N-1))
     1           -XETA(I,M,N)*(OMEGA(I,M+1,N)-OMEGA(I,M-1,N))
      OMEGAX = OMEGAX/(2.*GAKOB(I,M,N))
      OMEGAY = OMEGAY/(2.*GAKOB(I,M,N))
      RDWDR  = XACTUAL(I,M,N)*OMEGAX + YACTUAL(I,M,N)*OMEGAY
C
      ARG = UX**2 + UY**2 + RDWDR**2
```

The terms to the right of UX and UY (that is, $\partial U/\partial x$ and $\partial U/\partial y$) are expressions in the computation coordinates (ξ, η, z). The second code block provides results for $\partial\Omega/\partial x$ and $\partial\Omega/\partial y$. Finally, ARG captures the argument term in the shear rate function.

DISCUSSION 8.4

Subtleties in Non-Newtonian Convection Modeling

The axial momentum equation contains a pressure gradient term $\partial p/\partial z$ that drives the flow. In steady flow, one formulation specifies its value, computes the velocity field, and then integrates to obtain the total volumetric flow rate, Q. If Q is specified, the steady formulation is solved repeatedly using guessed values of $\partial p/\partial z$. These are refined iteratively using a half-step procedure until the target flow rate is achieved. Although the method is iterative, the final converged solution for the velocity field is exact.

For transient flows where Q(t) is specified, no such simple procedure exists. It is possible, in principle, to similarly determine $\partial p/\partial z$ within each time step in order to fulfill the target flow rate; however, such a procedure would be extremely computation intensive. This method is therefore unacceptable for practical use. The partial differential equation nonetheless requires some value of $\partial p/\partial z$ for numerical time marching to proceed; the question, of course, is *which* approximate pressure gradient value to use.

A Newtonian fallacy

Some investigators have approached the nonrotating problem with a seemingly clever solution, which turns out to be incorrect. We explain this technique so that others will not repeat it. The method first determines a pressure gradient associated with an equivalent Newtonian flow based, say, on average shear rate for the annular cross section or some other physical criterion for the assumed volumetric flow rate. The steady axial velocity field associated with this $\partial p/\partial z$ is then used in the concentration partial differential equation, and standard time integrations are performed. For constant Q, transient solutions for axial velocity generally vary with time, as expected—a "warm" feeling, although, of course, the correctness of the solutions cannot be verified. However, the space-time solution for concentration does not vary from run to run even when input viscosities are substantially changed.

No explanation or fix for this "bug" was ever obtained, nor could it be. The answer is clear from simple physical arguments. For steady, two-dimensional Newtonian flows, whether they apply to pipes or complicated annular geometries, the axial velocity field and the corresponding total volumetric flow rates can be generally written as $v_z(\xi, \eta) = 1/\mu \; \partial p/\partial z \; G(\xi, \eta)$ and $Q = 1/\mu \; \partial p/\partial z \; H$, where the function $G(\xi, \eta)$ and the constant H both depend exclusively on geometry.

Now, the two relationships can be combined to give $v_z(\xi, \eta) = (Q/H) \; G(\xi, \eta)$, which is completely independent of the viscosity. In other words, two problems (without rotation) having very different viscosities but the same flow rate, Q, and geometry will have identical velocity fields $v_z(\xi, \eta)$ and thus identical space-time histories for concentration. This is clear because setting $v_\theta = 0$ in Equation 8.5c leaves the left side unchanged and obviously incorrect physically. The reason is apparent: Steady Newtonian *approximations* must not be used locally to simplify the mathematics whatever the physical justification.

Correct physical solution

In determining total pressure drops for managed pressure drilling or cementing, the foregoing approach leads to grave mistakes. This method illustrates the dangers lurking in "engineering recipes" that may appear physically justifiable when they in fact lead to incorrect mathematics. For the general problem considered in Chapter 9, illustrated here as Figure 8.1, each "slug" containing non-Newtonian fluid at any particular time t_n with flow rate $Q(t_n)$ is represented by the exact pressure gradient obtained from our "Steady 2D" and "Transient 2D" solvers. (The latter solver is used for rotating flow.)

A fluid slug with higher average apparent viscosities will have a stronger pressure gradient than one with lower viscosities. When a very viscous fluid displaces a less viscous slug, neighboring pressure gradients can vary substantially in magnitude depending on rheology. This is the actual situation physically, and the challenge in solving for the complete miscible mixing field is a stable numerical solution in the presence of strong discontinuous axial pressure gradients. The results of Chapter 9 show that such stability can in fact be achieved in practical displacement problems in which contiguous fluids can differ substantially.

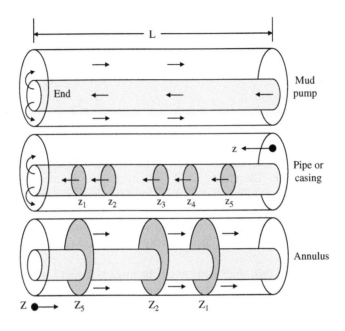

FIGURE 8.1

General pumping schedule with non-Newtonian flow.

DISCUSSION 8.5

Simple Models for Multiple Non-Newtonian Fluids with Mixing

The multifluid mixing problem in the drillpipe annulus represented in Figure 8.1 is in general difficult to formulate and solve with any degree of accuracy unless certain conditions (introduced later) are met. A theoretical discussion is given in Bird, Stewart, and Lightfoot (2002), where the equations of change for multicomponent systems are derived and their solutions outlined. Basically, one postulates a mixture velocity function applicable at a point that applies to a weighted average for all of the fluids. The momentum equation for this velocity is solved and used as the convection driver to solve multiple sets of concentration differential equations for C_m applicable to the different fluid species.

This process must be repeated in time and involves substantial numerical work. If this method can be accurately formulated and solved, then the space-time diffusion history at fluid interfaces can be determined. For instance, one can in principle determine diffusion zone thicknesses as they vary throughout the cross section and monitor their evolution in time. That this can be done with any precision, however, is unclear when fluid slug lengths are comparable to cross-sectional diameters because empirical mixing laws must be introduced to augment local details of the physical flow.

Although it is not explicitly stated, this approach is implicit in the recent multifluid model of Savery, Darbe, and Chin (2007). The uncertainties associated with such methods, fortunately, are

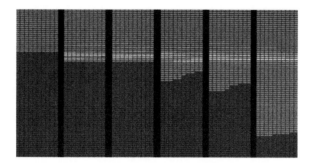

FIGURE 8.2

Propagating and diffusing front in time, constructed from movie frames for viscosity history using exaggerated diffusion.

not problematic when fluid slug lengths greatly exceed typical cross-sectional diameters, as is the case in all drilling and cementing problems. For such applications, the maximum diffusion thickness in the streamwise direction will always be much shorter than a typical slug length. Consequently, its effect on the overall pressure field is minimal. This observation allows us to solve for the pressure field first, ignoring diffusion effects associated with the coefficients ε_m.

As suggested earlier and demonstrated in Chapter 9, the positions of all fluid interfaces can be determined on a kinematic volumetric flow rate basis. The actual pressure gradient applicable to a particular fluid slug at any time t_n is obtained from the "Steady 2D" or "Transient 2D" solvers discussed in Chapters 1 through 7. Once the complete pressure profile is found on this basis, the mixing details between any two contiguous fluids can be obtained by solving a single concentration equation that supports two different pressure gradients without having to solve a complete system of coupled concentration equations.

In a sense, the method just outlined takes into account the disparate physical length scales present in the problem and substantially reduces the computational work needed to find practical solutions. At the same time, the results are obviously easier to interpret and do not require any analysis of highly mixed fields like those that result from methods outlined in Bird, Stewart, and Lightfoot (2002) or in Savery, Darbe, and Chin (2007) for cementing applications.

Our method, in a sense, draws upon "boundary layer-like" simplifications, allowing us to "zoom in" after the fact, not unlike classical boundary layer methods, which permit inviscid pressure determination over an airfoil first, followed by similar after-the-fact calculations for viscous drag effects that are confined to narrow zones near solid boundaries. As will be demonstrated in Chapter 9, time-dependent interfacial details associated with mixing zone growth and convection are given in Figure 9.9, reproduced as Figure 8.2, where the vertical mixing scale is greatly exaggerated.

Transient, Three-Dimensional, Multiphase Pipe and Annular Flow

In this chapter, we consider the general problems for managed pressure drilling and cementing flow simulation with the solutions posed as objectives in this book. All of the "building block" tools captured in the "Steady 2D" and "Transient 2D" simulators are brought to bear in the transient, three-dimensional, multiphase applications considered here. Again, we will address the computation of pressure profiles along the borehole and particularly at the drillbit for all times when a general pumping schedule is allowed at the mud pump. The problem models the complete system:

1. Surface pumping of general fluids with user-defined time schedules
2. Non-Newtonian flow down the drillpipe
3. Capture of pressure losses through the drillbit
4. Flow up the borehole annulus

For both pipe and annulus, fluid mixing is permitted via the introduction of coupled momentum and concentration equations. For the borehole, general annular eccentricity is allowed. Once the basic setup work is undertaken (i.e., defining fluid interface positions and completing the pressure gradient entries in Figure 1.40 using pressure solvers provided), the calculation of borehole pressure profiles at any instant in time requires just minutes of hand calculation (this process will be automated in the future).

For cementing applications, the degree to which contiguous fluids mix or do not mix is important to zonal isolation. Here, detailed calculations for interfacial mixing yield details related to diffusion zone geometry and time scales for mixing. These calculations, which are not required for managed pressure drilling applications, may require anywhere from minutes to an hour, depending on numerical stability requirements dictated by fluid density, apparent viscosity, and rotational rate parameters. (The controlling variable is $\rho\omega/\mu$, where ρ is density, ω is rotational rate, and μ represents an average apparent viscosity.) With these preliminaries taken care of, we now present detailed calculated validations and results.

DISCUSSION 9.1

Single Fluid in Pipe and Borehole System: Calculating Total Pressure Drops for General Non-Newtonian Fluids

The general problem considered is shown in Figure 9.1 (our analysis applies to open and closed systems). A positive displacement mud pump forces drilling fluid or cement into a drillpipe

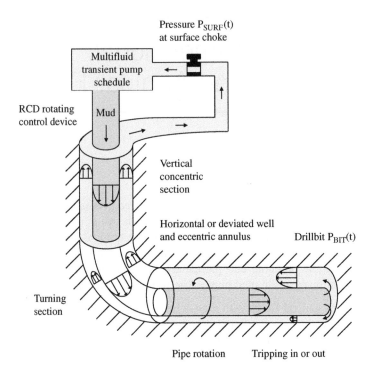

FIGURE 9.1

Managed pressure system simulation.

centralized in a concentric annulus. This vertical hole turns into a deviated or horizontal borehole with an eccentric annulus through an intermediate (possibly eccentric) section with radius of curvature, R. Note that Figure 9.1 is used to establish conventions and a frame of reference for discussion only. In fact, our "vertical concentric section" may represent another deviated or horizontal section with an eccentric cross section, and the turning section (however unlikely) may be concentric if desired. Length scales may be assigned arbitrarily, and out-of-plane sections are permissible. The geometry considered here is quite general.

We emphasize here "single" in the title of this discussion. When only a single fluid is considered, the problems are two-dimensional because the flows in the pipe and annulus are unchanged with time in the axial direction. Only the flow rate changes. For any given flow rate, a single calculation determines entirely what happens in the drillpipe and similarly for the annulus. When multiple fluids are introduced at the inlet with different slug lengths, a three-dimensional transient model is obviously required that supports moving interfaces. Net pressure drops will vary with time, since the fluid system is constantly changing. Significant complications arise that are studied in the remainder of this chapter.

DISCUSSION 9.2

Interface Tracking and Total Pressure Drop for Multiple Fluids Pumped in a Drillpipe and Eccentric Borehole System

In this example, we will consider a centered or eccentered drillpipe (with cross-sectional area A_{pipe}) located in a borehole annulus whose geometry is unchanged along its length. The annular area is $A_{annulus}$. Note that while pipe area is simply available from "πR_{pipe}^2," the same is not true for the annulus if the cross-sectional contours from two initially eccentered circles have been edited to incorporate washouts, cuttings beds, or fractures. If that is the case, the "Steady 2D" simulator automatically computes and displays total cross-sectional area by summing incremental trapezoidal areas constructed from the curvilinear grid.

Now mud progresses down the drillpipe and then out through the drillbit, and finally flows upward in the return annulus. At the outset $t = 0$, a single initial fluid with Herschel-Bulkley properties (n_0, K_0, $\tau_{0,0}$) is assumed to exist in the pipe and annular system (n is the fluid exponent, K is the consistency factor, and τ_0 is the yield stress). The initial fluid may be flowing or quiescent. At $t = 0+$, the mud pump starts to act according to a user-defined schedule with piecewise constant rates. At $t = t_0 = 0+$, fluid "1" with properties (n_1, K_1, $\tau_{0,1}$) is pumped into the pipe at the volumetric flow rate of Q_1; at $t = t_1$, fluid "2," with properties (n_2, K_2, $\tau_{0,2}$) is pumped at rate Q_2; and so on. In fact,

Fluid "1" pumps at rate Q_1: $t_0 \leq t < t_1$
Fluid "2" pumps at rate Q_2: $t_1 \leq t < t_2$
Fluid "3" pumps at rate Q_3: $t_2 \leq t < t_3$
Fluid "4" pumps at rate Q_4: $t_3 \leq t < t_4$
Fluid "5" pumps at rate Q_5: $t \geq t_4$

The overall pumping process is illustrated at the top of Figure 9.2. Here fluid introduced at the far right into the drillpipe travels to the left, then turns at the drillbit (not shown), and finally progresses to the very far right. The middle diagram shows five interfaces (starting at t_0, t_1, t_2, t_3, and t_4) associated with the onset of each pump action. The location "z_1" (using the "little z" left-pointing coordinate system shown) describes the interface separating the initial fluid ahead of it with fluid "1" just behind it. Similarly, "z_2" separates fluid "1" ahead of it and fluid "2" behind it. Fluid "5" is a single fluid that is pumped continuously without stoppage with flow rate Q_5 for $t \geq t_4$. While more interfaces are easily programmed, a limit of five (which model six fluid slugs) to enable rapid modeling and job prototyping, was assumed, since it suffices for most rig site planning purposes.

Once the first interface reaches the end of the drillpipe, shown with length L—that is, $z_1 = L$—it turns into the borehole annulus and travels to the right. Similar descriptions apply to the remaining interfaces. Annular interfaces are described by the "big (as in capital) Z" right-pointing coordinate system at the bottom in Figure 9.2. When $Z_1 = L$, the first fluid pumped will have reached the surface.

Figure 9.2 provides a "snapshot" obtained for a given instant in time. At different times, the locations of the interfaces will be different, as will pressure profiles along the borehole (and thus at the drillbit). Also, while our discussion focuses on drilling applications with distinct mud interfaces, it is clear that all of our results apply to cement-spacer-mud systems.

Now we wish to determine the locations of $z_{1,2,3,4,5}$ and $Z_{1,2,3,4,5}$ as functions of time. In general, this is a difficult problem if the fluids are compressible, if significant mixing is found at fluid

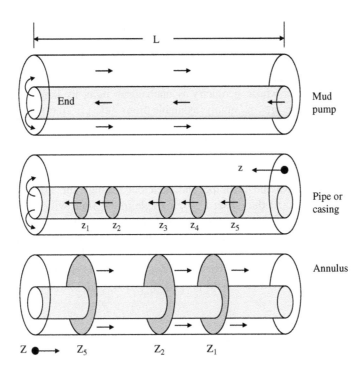

FIGURE 9.2

General pumping schedule.

interfaces, or both. However, if the the fluid slugs are long compared to the annular diameter (so that mixing zones are not dynamically significant), and, further, if the pump acts instantaneously and transient fluid effects reach equilibrium quickly, interface tracking can be accomplished kinematically. Once the locations of all interfaces are known for any instant in time, pressure drop calculations (for each fluid slug) proceed using the non-Newtonian flow models developed previously.

Two output tables are provided by our "interface tracker." The calculations are performed almost instantaneously by the software model. The two are, respectively, "Drillpipe Fluid Interfaces vs. Time" and "Annular Fluid Interfaces vs. Time," as shown in Figures 9.3 and 9.4. The numbers assumed for these tables are obviously not realistic, and for this reason the units shown in the headings should be ignored for now. They were chosen so that all results fit on the printed page, with all values allowing convenient visual checking and understanding of the computer output.

Note that 0's at early times along a z column indicate absence of the particular fluid in the drillpipe. Also, once the interface has reached the position "100" (the end of the borehole in this illustration), the subsequent 0's are no longer meaningful and are used only to populate the table. Note as well that the very small annular area of $A_{annulus}$ selected later was designed only so that we can "watch fluid move" in the table of Figure 9.4.

ELAPSED TIME		FLOW	Drillpipe Fluid Interface (feet)				
Minutes	Hours	GPMs	z(1)	z(2)	z(3)	z(4)	z(5)
0	0.	1	0	0	0	0	0
1	0.	1	1	0	0	0	0
2	0.	1	2	0	0	0	0
3	0.	1	3	0	0	0	0
4	0.	1	4	0	0	0	0
5	0.	2	5	0	0	0	0
6	0.	2	7	2	0	0	0
7	0.	2	9	4	0	0	0
8	0.	2	11	6	0	0	0
9	0.	2	13	8	0	0	0
10	0.	3	15	10	0	0	0
11	0.	3	18	13	3	0	0
12	0.	3	21	16	6	0	0
13	0.	3	24	19	9	0	0
14	0.	3	27	22	12	0	0
15	0.	4	30	25	15	0	0
16	0.	4	34	29	19	4	0
17	0.	4	38	33	23	8	0
18	0.	4	42	37	27	12	0
19	0.	4	46	41	31	16	0
20	0.	5	50	45	35	20	0
21	0.	5	55	50	40	25	5
22	0.	5	60	55	45	30	10
23	0.	5	65	60	50	35	15
24	0.	5	70	65	55	40	20
25	0.	5	75	70	60	45	25
26	**0.**	**5**	**80**	**75**	**65**	**50**	**30**
27	0.	5	85	80	70	55	35
28	0.	5	90	85	75	60	40
29	0.	5	95	90	80	65	45
30	0.	5	100	95	85	70	50
31	1.	5	0	100	90	75	55
32	1.	5	0	0	95	80	60
33	1.	5	0	0	100	85	65
34	1.	5	0	0	0	90	70
35	1.	5	0	0	0	95	75
36	1.	5	0	0	0	100	80
37	1.	5	0	0	0	0	85
38	1.	5	0	0	0	0	90
39	1.	5	0	0	0	0	95
40	1.	5	0	0	0	0	100
41	1.	5	0	0	0	0	0

FIGURE 9.3

Drillpipe Fluid Interfaces vs. Time.

To facilitate visual interpretation, we have assumed that $A_{pipe} = 1$ and $A_{annulus} = 0.5$, so that the nominal linear displacement speeds in the pipe and annulus are $U_{pipe} = Q/A_{pipe}$ and $U_{annulus} = Q/A_{annulus}$. The borehole length is assumed for clarity to be 100. At the same time, we pump according to the schedule

Fluid "1" at a rate of $Q_1 = 1$: $0 = t_0 < t < t_1 = 5$
Fluid "2" at a rate of $Q_2 = 2$: $5 = t_1 < t < t_2 = 10$
Fluid "3" at a rate of $Q_3 = 3$: $10 = t_2 < t < t_3 = 15$
Fluid "4" at a rate of $Q_4 = 4$: $15 = t_3 < t < t_4 = 20$
Fluid "5" at a rate of $Q_5 = 5$: $t > t_4 = 20$

where our five interfaces originate at t_0, t_1, t_2, t_3, and t_4.

ELAPSED TIME Minutes	Hours	FLOW GPMs	Annular Fluid Interface (feet)				
			Z(5)	Z(4)	Z(3)	Z(2)	Z(1)
0	0.	1	0	0	0	0	0
1	0.	1	0	0	0	0	0
2	0.	1	0	0	0	0	0
3	0.	1	0	0	0	0	0
4	0.	1	0	0	0	0	0
5	0.	2	0	0	0	0	0
6	0.	2	0	0	0	0	0
7	0.	2	0	0	0	0	0
8	0.	2	0	0	0	0	0
9	0.	2	0	0	0	0	0
10	0.	3	0	0	0	0	0
11	0.	3	0	0	0	0	0
12	0.	3	0	0	0	0	0
13	0.	3	0	0	0	0	0
14	0.	3	0	0	0	0	0
15	0.	4	0	0	0	0	0
16	0.	4	0	0	0	0	0
17	0.	4	0	0	0	0	0
18	0.	4	0	0	0	0	0
19	0.	4	0	0	0	0	0
20	0.	5	0	0	0	0	0
21	0.	5	0	0	0	0	0
22	0.	5	0	0	0	0	0
23	0.	5	0	0	0	0	0
24	0.	5	0	0	0	0	0
25	0.	5	0	0	0	0	0
26	**0.**	**5**	**0**	**0**	**0**	**0**	**0**
27	0.	5	0	0	0	0	0
28	0.	5	0	0	0	0	0
29	0.	5	0	0	0	0	0
30	0.	5	0	0	0	0	0
31	1.	5	0	0	0	0	10
32	1.	5	0	0	0	10	20
33	1.	5	0	0	0	20	30
34	1.	5	0	0	10	30	40
35	1.	5	0	0	20	40	50
36	**1.**	**5**	**0**	**0**	**30**	**50**	**60**
37	1.	5	0	10	40	60	70
38	1.	5	0	20	50	70	80
39	1.	5	0	30	60	80	90
40	1.	5	0	40	70	90	100
41	1.	5	10	50	80	100	0
42	1.	5	20	60	90	0	0
43	1.	5	30	70	100	0	0
44	1.	5	40	80	0	0	0
45	1.	5	50	90	0	0	0
46	1.	5	60	100	0	0	0
47	1.	5	70	0	0	0	0
48	1.	5	80	0	0	0	0
49	1.	5	90	0	0	0	0
50	1.	5	100	0	0	0	0
51	1.	5	0	0	0	0	0

FIGURE 9.4

Annular Fluid Interfaces vs. Time.

We next explain Figure 9.3. The left column provides elapsed minutes, while the second provides elapsed hours. The volumetric flow rate is given in the third column. The corresponding drillpipe fluid interfaces $z_{1,2,3,4,5}$ are given in the five remaining columns. Also, each change in flow rate, (associated with a new interface) is separated by a single horizontal line spacing to enhance clarity. Consider the result for z_1. In the first time block with $U_{pipe} = 1/1 = 1$, the interface advances at a rate of "1." In the second block with $U_{pipe} = 2/1$, the interface advances at the rate "2." As time increases, the easily recognized rate increments are 3, 4, and 5 following the pump schedule shown previously.

The z_1 interface starts moving at t = 0. Now we turn to the second interface and study the column for z_2 results. At t = 5, the second interface starts moving. Because we are already in the second time block, the interface moves at the rate "2." Subsequent speeds are 3, 4, and 5. Similarly, z_3 starts at t = 10 with rate increments of 3, followed by 4 and 5, and so on. We described Figure 9.3 from the perspective of tracking individual fronts. However, the table is important for pressure calculations.

Let us consider the results obtained at t = 26 (shown in bold font for emphasis). In particular, we have

ELAPSED	TIME	FLOW	\multicolumn{5}{c}{Drillpipe Fluid Interface (feet)}
Minutes	Hours	GPMs	z(1)	z(2)	z(3)	z(4)	z(5)
26	0.	5	80	75	65	50	30

This printout indicates that at t = 26 the front z_1 is located at z = 80, while the last front z_5 is located at z = 30. The drillpipe thus contains six distinct fluid slugs at $100 > z > 80$, $80 > z > 75$, $75 > z > 65$, $65 > z > 50$, $50 > z > 30$, and $30 > z > 0$, where "100" refers to the assumed borehole length. In fact,

- $100 > z > 80$ contains "initial fluid" with properties (n_0, K_0, $\tau_{0,0}$)
- $80 > z > 75$ contains fluid "1" with properties (n_1, K_1, $\tau_{0,1}$)
- $75 > z > 65$ contains fluid "2" with properties (n_2, K_2, $\tau_{0,2}$)
- $65 > z > 50$ contains fluid "3" with properties (n_3, K_3, $\tau_{0,3}$)
- $50 > z > 30$ contains fluid "4" with properties (n_4, K_4, $\tau_{0,4}$)
- $30 > z > 0$ contains fluid "5" with properties (n_5, K_5, $\tau_{0,5}$)

If a non-Newtonian 2D flow model for a Herschel-Bulkley fluid in a circular pipe is available that gives the pressure gradient $(\partial P/\partial z)_{pipe,n}$ for any of the given fluid slugs "n" flowing a rate Q with a pipe radius $(A_{pipe}/\pi)^{1/2}$, then the total drillpipe pressure drop is simply calculated from $(100 - 80) \ (\partial P/\partial z)_{pipe,0} + (80 - 75)(\partial P/\partial z)_{pipe,1} + (75 - 65) \ (\partial P/\partial z)_{pipe,2} + (65 - 50) (\partial P/\partial z)_{pipe,3} + (50 - 30) (\partial P/\partial z)_{pipe,4} + (30 - 0) (\partial P/\partial z)_{pipe,5}$. The flow rate, Q, used is the one applicable at the time the snapshot was taken—in this case, Q = 5 at t = 26. (A single rate applies to all slugs at any instant in time.) Now, at time t = 26, Figure 9.4 shows, as indicated by "0's," that none of the pumped fluids have arrived in the annulus:

ELAPSED	TIME	FLOW	\multicolumn{5}{c}{Annular Fluid Interface (feet)}
Minutes	Hours	GPMs	Z(5)	Z(4)	Z(3)	Z(2)	Z(1)
26	0.	5	0	0	0	0	0

Thus, the only fluid residing in the annulus is the initial fluid. If the pressure gradient obtained from a 2D eccentric flow analysis is $(\partial P/\partial z)_{annulus,0}$, then the pressure drop in the annulus is just $(100 - 0) (\partial P/\partial z)_{annulus,0}$. If we further denote by Δ the pressure drop through the drillbit, then the total pressure drop through the entire pipe-bit-annulus system is obtained by summing the prior three results—that is, $(100-80)(\partial P/\partial z)_{pipe,0} + (80-75)(\partial P/\partial z)_{pipe,1} + (75-65)(\partial P/\partial z)_{pipe,2} + (65 - 50) (\partial P/\partial z)_{pipe,3} + (50 - 30) (\partial P/\partial z)_{pipe,4} + (30 - 0) \ (\partial P/\partial z)_{pipe,5} + \Delta + (100 - 0) \ (\partial P/\partial z)_{annulus,0}$, which is the pressure (additive to the surface choke pressure, P_{SURF}) required at the mud pump to support this multislug flow.

The software that creates Figure 9.3 also provides the times at which fluid interfaces in the drill-pipe enter the borehole annulus. These are obtained from the table in the figure by noting the "100" marker. In this case, we have

```
Borehole total length L, is:    100 ft.
Fluid "1" enters annulus at:    30 min.
Fluid "2" enters annulus at:    31 min.
Fluid "3" enters annulus at:    33 min.
Fluid "4" enters annulus at:    36 min.
Fluid "5" enters annulus at:    40 min.
```

We next consider another time frame, say $t = 36$, for which our drillpipe interfaces have entered the annulus, and explain how annular pressure drops are determined (see Figure 9.5). For this time frame, Figure 9.4 gives

ELAPSED	TIME	FLOW	Annular Fluid Interface (feet)				
Minutes	Hours	GPMs	Z(5)	Z(4)	Z(3)	Z(2)	Z(1)
36	1.	5	0	0	30	50	60

This indicates that three interfaces exist in the annulus, with Z_1 located at the far right $Z = 60$, followed by Z_2 at $Z = 50$ and Z_3 at $Z = 30$. Since the fluid ahead of Z_1 is the "initial fluid," the total annular pressure drop is calculated from the sum $(100 - 60) \; (\partial P/\partial z)_{annulus,0} + (60 - 50) \; (\partial P/\partial z)_{annulus,1} + (50 - 30) \; (\partial P/\partial z)_{annulus,2} + (30 - 0) \; (\partial P/\partial z)_{annulus,3}$, where subscripts denote fluid type for the annular model.

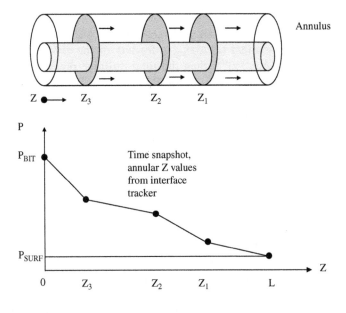

FIGURE 9.5

Example annular interface distribution.

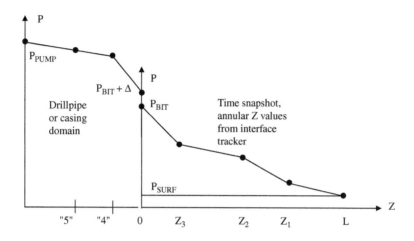

FIGURE 9.6

Complete drillpipe-drillbit-annulus system.

We note that the actual pressure P_{BIT} at the drillbit *in the formation* is obtained by adding the total annular pressure drop to the pressure P_{SURF} obtained at the surface choke. The value of P_{SURF} is in itself a "boundary condition," and, importantly, the pressure P_{BIT} at the bottom of the annulus *in the formation* does not depend on the pressure drop Δ through the drillbit. On the other hand, the pressure required at the pump to move the system includes pipe, bit, and annular losses, as shown in Figure 9.6 for one interface configuration.

Interface tracking and example

Here we describe the software module that has been implemented to track multiple fluid interfaces, leading to results such as those in Figures 9.3 and 9.4. For clarity, we did not work in physical units previously, choosing (unrealistic) numerical inputs whose results were simple to visualize and understand and parameters that allowed complete tables to fit on single printed pages. Here we return to physical units and work with a more realistic example. Our "Interface Tracker" is executed from the user screen in Figure 9.7, which shows default run parameters. Actual run parameters span many ranges and combinations of different numbers. For instance, pump rates will typically vary 100 to 1,500 gpm, and time schedules will vary up to days. Borehole lengths may vary from 5,000 to 30,000 ft. Pipe and annular areas are very different from run to run.

To provide meaningful tabulations that are reasonable in file size, time increments are therefore expressed in minutes. To execute this program, click "Run." When a blue status screen appears and instructs the user to click "Answer," results analogous to Figures 9.3 and 9.4 are provided. For completeness, we perform our calculations now and explain the outputs at selected instants in time. These calculations require approximately five seconds.

The output file reproduced that follows contains a summary of all input parameters. Again note that for interface tracking, provided that our fluid slugs are long compared to the annular diameter and interfacial mixing is confined to a small axial extent, the tracking process can be performed

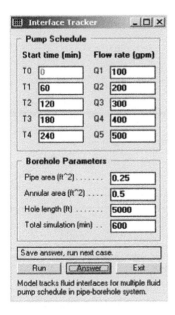

Interface Tracker with default inputs.

kinematically (using only the pumping schedule and overall geometric parameters) and does not depend on the fluid dynamics or rheologies. (These are used after the fact for pressure calculations as discussed previously.) We now explain selected entries at various times. At the present writing, interface positions must be inferred from tabular results; however, this process (together with integrated color graphics) will be automated in the near future.

```
Pump Schedule, Interface Tracking ...
100 gpm:        0 min < T <    60 min
200 gpm:       60 min < T <   120 min
300 gpm:      120 min < T <   180 min
400 gpm:      180 min < T <   240 min
500 gpm:     T >             240 min
Drillpipe area (ft^2):       0.250E +00
Annular area (ft^2):         0.500E +00
Borehole length (ft):        0.500E +04
Time simulation (min):       600
```

ELAPSED Minutes	TIME Hours	FLOW GPMs	Drillpipe Fluid Interface (feet)				
			z(1)	z(2)	z(3)	z(4)	z(5)
0	0.0	100	0	0	0	0	0
1	0.0	100	53	0	0	0	0
2	0.0	100	106	0	0	0	0
3	0.1	100	160	0	0	0	0
4	0.1	100	213	0	0	0	0
5	0.1	100	267	0	0	0	0

6	0.1	100	320	0	0	0	0
7	0.1	100	374	0	0	0	0
8	0.1	100	427	0	0	0	0
9	0.2	100	481	0	0	0	0
10	0.2	100	534	0	0	0	0
11	0.2	100	588	0	0	0	0
12	0.2	100	641	0	0	0	0
13	0.2	100	695	0	0	0	0
14	0.2	100	748	0	0	0	0
15	0.2	100	802	0	0	0	0
16	0.3	100	855	0	0	0	0
17	0.3	100	909	0	0	0	0
18	0.3	100	962	0	0	0	0
19	0.3	100	1015	0	0	0	0

The first table tracks fluid interfaces in the drillpipe or casing. At $t = 20$ min, the first interface is located at 1,069 ft (refer to the coordinate system in the middle diagram of Figure 9.2). By $t = 25$ min, it has traveled to 1,336 ft. No other fluid has entered the pipe. This means that the initial fluid is located in the range $5,000 > z > 1,336$, while the first fluid pumped is found in the range $1,336 > z > 0$.

20	0.3	100	1069	0	0	0	0
21	0.3	100	1122	0	0	0	0
22	0.4	100	1176	0	0	0	0
23	0.4	100	1229	0	0	0	0
24	0.4	100	1283	0	0	0	0
25	0.4	100	1336	0	0	0	0
26	0.4	100	1390	0	0	0	0
27	0.4	100	1443	0	0	0	0
28	0.5	100	1497	0	0	0	0
29	0.5	100	1550	0	0	0	0
30	0.5	100	1604	0	0	0	0
31	0.5	100	1657	0	0	0	0
32	0.5	100	1711	0	0	0	0
33	0.6	100	1764	0	0	0	0
34	0.6	100	1818	0	0	0	0
35	0.6	100	1871	0	0	0	0
36	0.6	100	1925	0	0	0	0
37	0.6	100	1978	0	0	0	0
38	0.6	100	2031	0	0	0	0
39	0.6	100	2085	0	0	0	0
40	0.7	100	2138	0	0	0	0
41	0.7	100	2192	0	0	0	0
42	0.7	100	2245	0	0	0	0
43	0.7	100	2299	0	0	0	0
44	0.7	100	2352	0	0	0	0
45	0.8	100	2406	0	0	0	0
46	0.8	100	2459	0	0	0	0
47	0.8	100	2513	0	0	0	0
48	0.8	100	2566	0	0	0	0
49	0.8	100	2620	0	0	0	0
50	0.8	100	2673	0	0	0	0
51	0.9	100	2727	0	0	0	0
52	0.9	100	2780	0	0	0	0
53	0.9	100	2834	0	0	0	0
54	0.9	100	2887	0	0	0	0
55	0.9	100	2940	0	0	0	0
56	0.9	100	2994	0	0	0	0
57	0.9	100	3047	0	0	0	0
58	1.0	100	3101	0	0	0	0
59	1.0	100	3154	0	0	0	0
60	1.0	200	3208	0	0	0	0
61	1.0	200	3315	106	0	0	0

62	1.0	200	3422	213	0	0	0
63	1.0	200	3529	320	0	0	0
64	1.1	200	3636	427	0	0	0
65	1.1	200	3743	534	0	0	0
66	1.1	200	3850	641	0	0	0
67	1.1	200	3956	748	0	0	0
68	1.1	200	4063	855	0	0	0
69	1.1	200	4170	962	0	0	0

At $t = 70$ min, the first interface has reached 4,277 ft, while the second interface is located at 1,069 ft. This means that the initial fluid is located in the range $5,000 > z > 4,277$. The first fluid is found in the range $4,277 > z > 1,069$, while the second appears in the range $1,069 > z > 0$.

70	1.2	200	4277	1069	0	0	0
71	1.2	200	4384	1176	0	0	0
72	1.2	200	4491	1283	0	0	0
73	1.2	200	4598	1390	0	0	0
74	1.2	200	4705	1497	0	0	0
75	1.2	200	4812	1604	0	0	0

At approximately $t = 76$ min, the first interface is located at 4,919 ft, while the second is found at 1,711 ft. Recall that the borehole length is assumed to be 5,000 ft. At $t = 77$ min, the first interface has been flushed out of the pipe and has flowed into the annulus. This is noted in remarks at the end of this table ("Fluid "1" enters annulus at: 77 min."). From the $t = 77$ line, the second interface is located at 1,818 ft. Thus, the first fluid is to be found in the the range $5,000 > z > 1,818$, while the second fluid is located in the range $1,818 > z > 0$.

76	1.3	200	4919	1711	0	0	0
77	1.3	200	0	1818	0	0	0
78	1.3	200	0	1925	0	0	0
79	1.3	200	0	2031	0	0	0
80	1.3	200	0	2138	0	0	0
81	1.4	200	0	2245	0	0	0
82	1.4	200	0	2352	0	0	0
83	1.4	200	0	2459	0	0	0
84	1.4	200	0	2566	0	0	0
85	1.4	200	0	2673	0	0	0
86	1.4	200	0	2780	0	0	0
87	1.5	200	0	2887	0	0	0
88	1.5	200	0	2994	0	0	0
89	1.5	200	0	3101	0	0	0
90	1.5	200	0	3208	0	0	0
91	1.5	200	0	3315	0	0	0
92	1.5	200	0	3422	0	0	0
93	1.5	200	0	3529	0	0	0
94	1.6	200	0	3636	0	0	0
95	1.6	200	0	3743	0	0	0
96	1.6	200	0	3850	0	0	0
97	1.6	200	0	3956	0	0	0
98	1.6	200	0	4063	0	0	0
99	1.6	200	0	4170	0	0	0

At $t = 100$ min, the second interface is located at 4,277 ft. This means that the first fluid is found in the range $5,000 > z > 4,277$, and the second is in the range $4,277 > z > 0$.

100	1.7	200	0	4277	0	0	0
101	1.7	200	0	4384	0	0	0
102	1.7	200	0	4491	0	0	0
103	1.7	200	0	4598	0	0	0
104	1.7	200	0	4705	0	0	0

105	1.8	200	0	4812	0	0	0
106	1.8	200	0	4919	0	0	0
107	1.8	200	0	0	0	0	0
108	1.8	200	0	0	0	0	0
109	1.8	200	0	0	0	0	0
110	1.8	200	0	0	0	0	0
111	1.9	200	0	0	0	0	0
112	1.9	200	0	0	0	0	0
113	1.9	200	0	0	0	0	0
114	1.9	200	0	0	0	0	0
115	1.9	200	0	0	0	0	0
116	1.9	200	0	0	0	0	0
117	2.0	200	0	0	0	0	0
118	2.0	200	0	0	0	0	0
119	2.0	200	0	0	0	0	0
120	2.0	300	0	0	0	0	0
121	2.0	300	0	0	160	0	0
122	2.0	300	0	0	320	0	0
123	2.0	300	0	0	481	0	0
124	2.1	300	0	0	641	0	0
125	2.1	300	0	0	802	0	0
126	2.1	300	0	0	962	0	0
127	2.1	300	0	0	1122	0	0
128	2.1	300	0	0	1283	0	0
129	2.2	300	0	0	1443	0	0

At $t = 130$ min, the third interface is located at 1,604 ft. This means that the second fluid is located in the range $5,000 > z > 1,604$, while the third fluid is found in the range $1,604 > z > 0$.

130	2.2	300	0	0	1604	0	0
131	2.2	300	0	0	1764	0	0
132	2.2	300	0	0	1925	0	0
133	2.2	300	0	0	2085	0	0
134	2.2	300	0	0	2245	0	0
135	2.2	300	0	0	2406	0	0
136	2.3	300	0	0	2566	0	0
137	2.3	300	0	0	2727	0	0
138	2.3	300	0	0	2887	0	0
139	2.3	300	0	0	3047	0	0
140	2.3	300	0	0	3208	0	0
141	2.3	300	0	0	3368	0	0
142	2.4	300	0	0	3529	0	0
143	2.4	300	0	0	3689	0	0
144	2.4	300	0	0	3850	0	0
145	2.4	300	0	0	4010	0	0
146	2.4	300	0	0	4170	0	0
147	2.5	300	0	0	4331	0	0
148	2.5	300	0	0	4491	0	0
149	2.5	300	0	0	4652	0	0

At $t = 150$ min, the third interface is located at 4,812 ft. Since the pipe length is 5,000 ft, it is about to be flushed out of the end of the pipe. In the next 30 minutes, approximately, there are no interfaces in the pipe. The "all 0" printout indicates that the only fluid in the pipe is the third fluid.

150	2.5	300	0	0	4812	0	0
151	2.5	300	0	0	4972	0	0
152	2.5	300	0	0	0	0	0
153	2.5	300	0	0	0	0	0
154	2.6	300	0	0	0	0	0
155	2.6	300	0	0	0	0	0
156	2.6	300	0	0	0	0	0
157	2.6	300	0	0	0	0	0
158	2.6	300	0	0	0	0	0
159	2.7	300	0	0	0	0	0

160	2.7	300	0	0	0	0	0
161	2.7	300	0	0	0	0	0
162	2.7	300	0	0	0	0	0
163	2.7	300	0	0	0	0	0
164	2.7	300	0	0	0	0	0
165	2.8	300	0	0	0	0	0
166	2.8	300	0	0	0	0	0
167	2.8	300	0	0	0	0	0
168	2.8	300	0	0	0	0	0
169	2.8	300	0	0	0	0	0
170	2.8	300	0	0	0	0	0
171	2.8	300	0	0	0	0	0
172	2.9	300	0	0	0	0	0
173	2.9	300	0	0	0	0	0
174	2.9	300	0	0	0	0	0
175	2.9	300	0	0	0	0	0
176	2.9	300	0	0	0	0	0
177	3.0	300	0	0	0	0	0
178	3.0	300	0	0	0	0	0
179	3.0	300	0	0	0	0	0
180	2.0	400	0	0	0	0	0

Now the fourth interface has entered the pipe. The third fluid is located in the range $5,000 > z > 213$, while the fourth fluid is found in the range $213 > z > 0$.

181	3.0	400	0	0	0	213	0
182	3.0	400	0	0	0	427	0
183	3.0	400	0	0	0	641	0
184	3.1	400	0	0	0	855	0
185	3.1	400	0	0	0	1069	0
186	3.1	400	0	0	0	1283	0
187	3.1	400	0	0	0	1497	0
188	3.1	400	0	0	0	1711	0
189	3.2	400	0	0	0	1925	0
190	3.2	400	0	0	0	2138	0
191	3.2	400	0	0	0	2352	0
192	3.2	400	0	0	0	2566	0
193	3.2	400	0	0	0	2780	0
194	3.2	400	0	0	0	2994	0
195	3.2	400	0	0	0	3208	0
196	3.3	400	0	0	0	3422	0
197	3.3	400	0	0	0	3636	0
198	3.3	400	0	0	0	3850	0
199	3.3	400	0	0	0	4063	0
200	3.3	400	0	0	0	4277	0

At $t = 201$ min, the fourth interface has migrated to 4,491 ft. The third fluid is located in the range $5,000 > z > 4,491$, while the fourth fluid occupies almost the entire length of the pipe in the range $4,491 > z > 0$. By $t = 204$ min, the fourth interface will have left the pipe and turned into the annulus. Then the fourth fluid completely occupies the pipe.

201	3.3	400	0	0	0	4491	0
202	3.4	400	0	0	0	4705	0
203	3.4	400	0	0	0	4919	0
204	3.4	400	0	0	0	0	0
205	3.4	400	0	0	0	0	0
206	3.4	400	0	0	0	0	0
207	3.5	400	0	0	0	0	0
208	3.5	400	0	0	0	0	0
209	3.5	400	0	0	0	0	0
210	3.5	400	0	0	0	0	0
211	3.5	400	0	0	0	0	0
212	3.5	400	0	0	0	0	0

213	3.5	400	0	0	0	0	0
214	3.6	400	0	0	0	0	0
215	3.6	400	0	0	0	0	0
216	3.6	400	0	0	0	0	0
217	3.6	400	0	0	0	0	0
218	3.6	400	0	0	0	0	0
219	3.7	400	0	0	0	0	0
220	3.7	400	0	0	0	0	0
221	3.7	400	0	0	0	0	0
222	3.7	400	0	0	0	0	0
223	3.7	400	0	0	0	0	0
224	3.7	400	0	0	0	0	0
225	3.8	400	0	0	0	0	0
226	3.8	400	0	0	0	0	0
227	3.8	400	0	0	0	0	0
228	3.8	400	0	0	0	0	0
229	3.8	400	0	0	0	0	0
230	3.8	400	0	0	0	0	0
231	3.8	400	0	0	0	0	0
232	3.9	400	0	0	0	0	0
233	3.9	400	0	0	0	0	0
234	3.9	400	0	0	0	0	0
235	3.9	400	0	0	0	0	0
236	3.9	400	0	0	0	0	0
237	4.0	400	0	0	0	0	0
238	4.0	400	0	0	0	0	0
239	4.0	400	0	0	0	0	0
240	4.0	500	0	0	0	0	0
241	4.0	500	0	0	0	0	267
242	4.0	500	0	0	0	0	534
243	4.1	500	0	0	0	0	802
244	4.1	500	0	0	0	0	1069
245	4.1	500	0	0	0	0	1336
246	4.1	500	0	0	0	0	1604
247	4.1	500	0	0	0	0	1871
248	4.1	500	0	0	0	0	2138
249	4.2	500	0	0	0	0	2406

At t = 250 min, the fifth interface is located at 2,673 ft. This means that the fourth fluid is found in the range $5,000 > z > 2,673$, while the fifth fluid is found in the range $2,673 > z > 0$.

250	4.2	500	0	0	0	0	2673
251	4.2	500	0	0	0	0	2940
252	4.2	500	0	0	0	0	3208
253	4.2	500	0	0	0	0	3475
254	4.2	500	0	0	0	0	3743
255	4.2	500	0	0	0	0	4010
256	4.3	500	0	0	0	0	4277
257	4.3	500	0	0	0	0	4545
258	4.3	500	0	0	0	0	4812

At t = 259 min, the fifth interface has left the pipe, and the fifth fluid now completely occupies it, as indicated by the "all 0" data that follow. Note that while, in Figure 9.7, we have allowed for a total of 600 minutes of simulation, the printout here terminates at t = 259 min because nothing of dynamical significance occurs beyond this time. (The only fluid in the pipe will be the fifth fluid, and this printout is eliminated for convenience.) The Fortran simulator used to calculate interfaces permits up to 10,000 minutes of rig-time modeling, or approximately one week of continuous pumping with six different fluids.

```
259   4.3  500    0            0   0   0   0
Borehole total length L, is:   5000 ft.
Fluid "1" enters annulus at:    77 min.
```

```
Fluid "2" enters annulus at:    107 min.
Fluid "3" enters annulus at:    152 min.
Fluid "4" enters annulus at:    204 min.
Fluid "5" enters annulus at:    259 min.
```

The summary above is printed for convenience and is obtained by interrogating the tabular pipe data. Now that we have completed the tracking of all five interfaces in the pipe, the software algorithm turns to interface tracking in the annulus. The middle diagram in Figure 9.2 used a "left-pointing, little z" coordinate system for pipe flow with the origin at the far right, but now, as shown in the bottom diagram of the figure, we use a "right-pointing, big Z" convention for annular flow with an origin at the far left.

Also note that the tabular interface headings for pipe flow took the form $z(1)$, $z(2)$, ..., $z(5)$. However, for annular flow we reverse the order of the tabulation: $Z(5)$, $Z(4)$, ..., $Z(1)$, as shown next. The reason for this is obvious. From the summary just given, the first interface does not enter the annulus until $t = 77$ min. Thus, prior to $t = 77$ min, only the initial fluid exists in the annulus. For this reason, the annular table that follows contains 0's everywhere until approximately $t = 78$ min. We will continue our discussion at the $t = 78$ min time entry.

ELAPSED	TIME	FLOW	Annular Fluid Interface (feet)				
Minutes	Hours	GPMs	Z(5)	Z(4)	Z(3)	Z(2)	Z(1)
0	0.0	100	0	0	0	0	0
1	0.0	100	0	0	0	0	0
2	0.0	100	0	0	0	0	0
3	0.1	100	0	0	0	0	0
4	0.1	100	0	0	0	0	0
5	0.1	100	0	0	0	0	0
6	0.1	100	0	0	0	0	0
7	0.1	100	0	0	0	0	0
8	0.1	100	0	0	0	0	0
9	0.2	100	0	0	0	0	0
10	0.2	100	0	0	0	0	0
11	0.2	100	0	0	0	0	0
12	0.2	100	0	0	0	0	0
13	0.2	100	0	0	0	0	0
14	0.2	100	0	0	0	0	0
15	0.2	100	0	0	0	0	0
16	0.3	100	0	0	0	0	0
17	0.3	100	0	0	0	0	0
18	0.3	100	0	0	0	0	0
19	0.3	100	0	0	0	0	0
20	0.3	100	0	0	0	0	0
21	0.3	100	0	0	0	0	0
22	0.4	100	0	0	0	0	0
23	0.4	100	0	0	0	0	0
24	0.4	100	0	0	0	0	0
25	0.4	100	0	0	0	0	0
26	0.4	100	0	0	0	0	0
27	0.4	100	0	0	0	0	0
28	0.5	100	0	0	0	0	0
29	0.5	100	0	0	0	0	0
30	0.5	100	0	0	0	0	0
31	0.5	100	0	0	0	0	0
32	0.5	100	0	0	0	0	0
33	0.6	100	0	0	0	0	0
34	0.6	100	0	0	0	0	0
35	0.6	100	0	0	0	0	0
36	0.6	100	0	0	0	0	0
37	0.6	100	0	0	0	0	0

38	0.6	100	0	0	0	0	0
39	0.6	100	0	0	0	0	0
40	0.7	100	0	0	0	0	0
41	0.7	100	0	0	0	0	0
42	0.7	100	0	0	0	0	0
43	0.7	100	0	0	0	0	0
44	0.7	100	0	0	0	0	0
45	0.8	100	0	0	0	0	0
46	0.8	100	0	0	0	0	0
47	0.8	100	0	0	0	0	0
48	0.8	100	0	0	0	0	0
49	0.8	100	0	0	0	0	0
50	0.8	100	0	0	0	0	0
51	0.9	100	0	0	0	0	0
52	0.9	100	0	0	0	0	0
53	0.9	100	0	0	0	0	0
54	0.9	100	0	0	0	0	0
55	0.9	100	0	0	0	0	0
56	0.9	100	0	0	0	0	0
57	0.9	100	0	0	0	0	0
58	1.0	100	0	0	0	0	0
59	1.0	100	0	0	0	0	0
60	1.0	200	0	0	0	0	0
61	1.0	200	0	0	0	0	0
62	1.0	200	0	0	0	0	0
63	1.0	200	0	0	0	0	0
64	1.1	200	0	0	0	0	0
65	1.1	200	0	0	0	0	0
66	1.1	200	0	0	0	0	0
67	1.1	200	0	0	0	0	0
68	1.1	200	0	0	0	0	0
69	1.1	200	0	0	0	0	0
70	1.2	200	0	0	0	0	0
71	1.2	200	0	0	0	0	0
72	1.2	200	0	0	0	0	0
73	1.2	200	0	0	0	0	0
74	1.2	200	0	0	0	0	0
75	1.2	200	0	0	0	0	0
76	1.3	200	0	0	0	0	0
77	1.3	200	0	0	0	0	0

At $t = 78$ min, we find that the first interface (under the $Z(1)$ heading) is located at $Z = 53$ ft. Thus, the first fluid is found in the range $0 < Z < 53$, while the initial fluid is found in the range $53 < Z < 5,000$ (again, "5,000" represents the surface).

78	1.3	200	0	0	0	0	53
79	1.3	200	0	0	0	0	106
80	1.3	200	0	0	0	0	160
81	1.4	200	0	0	0	0	213
82	1.4	200	0	0	0	0	267
83	1.4	200	0	0	0	0	320
84	1.4	200	0	0	0	0	374
85	1.4	200	0	0	0	0	427
86	1.4	200	0	0	0	0	481
87	1.5	200	0	0	0	0	534
88	1.5	200	0	0	0	0	588
89	1.5	200	0	0	0	0	641
90	1.5	200	0	0	0	0	695
91	1.5	200	0	0	0	0	748
92	1.5	200	0	0	0	0	802
93	1.5	200	0	0	0	0	855
94	1.6	200	0	0	0	0	909
95	1.6	200	0	0	0	0	962
96	1.6	200	0	0	0	0	1015

97	1.6	200	0	0	0	0	1069
98	1.6	200	0	0	0	0	1122
99	1.6	200	0	0	0	0	1176
100	1.7	200	0	0	0	0	1229
101	1.7	200	0	0	0	0	1283
102	1.7	200	0	0	0	0	1336
103	1.7	200	0	0	0	0	1390
104	1.7	200	0	0	0	0	1443
105	1.8	200	0	0	0	0	1497
106	1.8	200	0	0	0	0	1550
107	1.8	200	0	0	0	0	1604
108	1.8	200	0	0	0	53	1657
109	1.8	200	0	0	0	106	1711
110	1.8	200	0	0	0	160	1764
111	1.9	200	0	0	0	213	1818
112	1.9	200	0	0	0	267	1871
113	1.9	200	0	0	0	320	1925
114	1.9	200	0	0	0	374	1978
115	1.9	200	0	0	0	427	2031
116	1.9	200	0	0	0	481	2085
117	2.0	200	0	0	0	534	2138
118	2.0	200	0	0	0	588	2192
119	2.0	200	0	0	0	641	2245
120	2.0	300	0	0	0	695	2299
121	2.0	300	0	0	0	775	2379
122	2.0	300	0	0	0	855	2459
123	2.0	300	0	0	0	935	2539
124	2.1	300	0	0	0	1015	2620
125	2.1	300	0	0	0	1096	2700
126	2.1	300	0	0	0	1176	2780
127	2.1	300	0	0	0	1256	2860
128	2.1	300	0	0	0	1336	2940
129	2.2	300	0	0	0	1417	3021
130	2.2	300	0	0	0	1497	3101
131	2.2	300	0	0	0	1577	3181
132	2.2	300	0	0	0	1657	3261
133	2.2	300	0	0	0	1737	3342
134	2.2	300	0	0	0	1818	3422
135	2.2	300	0	0	0	1898	3502
136	2.3	300	0	0	0	1978	3582
137	2.3	300	0	0	0	2058	3662
138	2.3	300	0	0	0	2138	3743
139	2.3	300	0	0	0	2219	3823
140	2.3	300	0	0	0	2299	3903
141	2.3	300	0	0	0	2379	3983
142	2.4	300	0	0	0	2459	4063
143	2.4	300	0	0	0	2539	4144
144	2.4	300	0	0	0	2620	4224
145	2.4	300	0	0	0	2700	4304
146	2.4	300	0	0	0	2780	4384
147	2.5	300	0	0	0	2860	4464
148	2.5	300	0	0	0	2940	4545
149	2.5	300	0	0	0	3021	4625

At t = 150 min, the first interface is approaching the surface, since it is located at 4,705 ft (the surface location is 5,000 ft). The second, Z(2), interface is found at 3,101 ft. Thus, the initial fluid is found in the range $4{,}705 < Z < 5{,}000$, while the first fluid is in the range $3{,}101 < Z < 4{,}705$. The second fluid is located in the range $0 < Z < 3{,}101$.

150	2.5	300	0	0	0	3101	4705
151	2.5	300	0	0	0	3181	4785
152	2.5	300	0	0	0	3261	4865
153	2.5	300	0	0	80	3342	4946

Now the first interface has left the annulus and entered the mud tank at the surface. The second interface is located at 3,422 ft, while the third interface is found at 160 ft. Thus, the first fluid is found in the range $3,422 < Z < 5,000$, while the second is found in the range $160 < Z < 3,422$. The third fluid is located in the range $0 < Z < 160$.

154	2.6	300	0	0	160	3422	0
155	2.6	300	0	0	240	3502	0
156	2.6	300	0	0	320	3582	0
157	2.6	300	0	0	401	3662	0
158	2.6	300	0	0	481	3743	0
159	2.7	300	0	0	561	3823	0
160	2.7	300	0	0	641	3903	0
161	2.7	300	0	0	721	3983	0
162	2.7	300	0	0	802	4063	0
163	2.7	300	0	0	882	4144	0
164	2.7	300	0	0	962	4224	0
165	2.8	300	0	0	1042	4304	0
166	2.8	300	0	0	1122	4384	0
167	2.8	300	0	0	1203	4464	0
168	2.8	300	0	0	1283	4545	0
169	2.8	300	0	0	1363	4625	0

At $t = 170$ min, the second interface is located at 4,705 ft, while the third is found at 1,443 ft. Thus, the first fluid is located in the range $4,705 < Z < 5,000$, while the second is found in the range $1,443 < Z < 4,705$. The third fluid is found in the range $0 < Z < 1,443$. At approximately $t = 173$ min, the second interface leaves the annulus and completely disappears from the system. Then the second fluid is found in the range $1,684 < Z < 5,000$, while the third is located in the range $0 < Z < 1,684$.

170	2.8	300	0	0	1443	4705	0
171	2.8	300	0	0	1523	4785	0
172	2.9	300	0	0	1604	4865	0
173	2.9	300	0	0	1684	4946	0
174	2.9	300	0	0	1764	0	0
175	2.9	300	0	0	1844	0	0
176	2.9	300	0	0	1925	0	0
177	3.0	300	0	0	2005	0	0
178	3.0	300	0	0	2085	0	0
179	3.0	300	0	0	2165	0	0
180	3.0	400	0	0	2245	0	0
181	3.0	400	0	0	2352	0	0
182	3.0	400	0	0	2459	0	0
183	3.0	400	0	0	2566	0	0
184	3.1	400	0	0	2673	0	0
185	3.1	400	0	0	2780	0	0
186	3.1	400	0	0	2887	0	0
187	3.1	400	0	0	2994	0	0
188	3.1	400	0	0	3101	0	0
189	3.2	400	0	0	3208	0	0
190	3.2	400	0	0	3315	0	0
191	3.2	400	0	0	3422	0	0
192	3.2	400	0	0	3529	0	0
193	3.2	400	0	0	3636	0	0
194	3.2	400	0	0	3743	0	0
195	3.2	400	0	0	3850	0	0
196	3.3	400	0	0	3956	0	0
197	3.3	400	0	0	4063	0	0
198	3.3	400	0	0	4170	0	0
199	3.3	400	0	0	4277	0	0

At t = 200 min, the third interface is located at 4,384 ft. Thus, the second fluid is found in the range $4{,}384 < Z < 5{,}000$, while the first appears in the range $0 < Z < 4{,}384$ (recall that, at t = 173 min, the second interface has left the annulus). By now, the interpretation process for both pipe and annulus should be apparent. We turn finally to t = 296 min.

200	3.3	400	0	0	4384	0	0
201	3.3	400	0	0	4491	0	0
202	3.4	400	0	0	4598	0	0
203	3.4	400	0	0	4705	0	0
204	3.4	400	0	0	4812	0	0
205	3.4	400	0	106	4919	0	0
206	3.4	400	0	213	0	0	0
207	3.5	400	0	320	0	0	0
208	3.5	400	0	427	0	0	0
209	3.5	400	0	534	0	0	0
210	3.5	400	0	641	0	0	0
211	3.5	400	0	748	0	0	0
212	3.5	400	0	855	0	0	0
213	3.5	400	0	962	0	0	0
214	3.6	400	0	1069	0	0	0
215	3.6	400	0	1176	0	0	0
216	3.6	400	0	1283	0	0	0
217	3.6	400	0	1390	0	0	0
218	3.6	400	0	1497	0	0	0
219	3.7	400	0	1604	0	0	0
220	3.7	400	0	1711	0	0	0
221	3.7	400	0	1818	0	0	0
222	3.7	400	0	1925	0	0	0
223	3.7	400	0	2031	0	0	0
224	3.7	400	0	2138	0	0	0
225	3.8	400	0	2245	0	0	0
226	3.8	400	0	2352	0	0	0
227	3.8	400	0	2459	0	0	0
228	3.8	400	0	2566	0	0	0
229	3.8	400	0	2673	0	0	0
230	3.8	400	0	2780	0	0	0
231	3.8	400	0	2887	0	0	0
232	3.9	400	0	2994	0	0	0
233	3.9	400	0	3101	0	0	0
234	3.9	400	0	3208	0	0	0
235	3.9	400	0	3315	0	0	0
236	3.9	400	0	3422	0	0	0
237	4.0	400	0	3529	0	0	0
238	4.0	400	0	3636	0	0	0
239	4.0	400	0	3743	0	0	0
240	4.0	500	0	3850	0	0	0
241	4.0	500	0	3983	0	0	0
242	4.0	500	0	4117	0	0	0
243	4.1	500	0	4251	0	0	0
244	4.1	500	0	4384	0	0	0
245	4.1	500	0	4518	0	0	0
246	4.1	500	0	4652	0	0	0
247	4.1	500	0	4785	0	0	0
248	4.1	500	0	4919	0	0	0
249	4.2	500	0	0	0	0	0
250	4.2	500	0	0	0	0	0
251	4.2	500	0	0	0	0	0
252	4.2	500	0	0	0	0	0
253	4.2	500	0	0	0	0	0
254	4.2	500	0	0	0	0	0
255	4.2	500	0	0	0	0	0
256	4.3	500	0	0	0	0	0
257	4.3	500	0	0	0	0	0
258	4.3	500	0	0	0	0	0
259	4.3	500	0	0	0	0	0

260	4.3	500	133	0	0	0	0
261	4.3	500	267	0	0	0	0
262	4.4	500	401	0	0	0	0
263	4.4	500	534	0	0	0	0
264	4.4	500	668	0	0	0	0
265	4.4	500	802	0	0	0	0
266	4.4	500	935	0	0	0	0
267	4.4	500	1069	0	0	0	0
268	4.5	500	1203	0	0	0	0
269	4.5	500	1336	0	0	0	0
270	4.5	500	1470	0	0	0	0
271	4.5	500	1604	0	0	0	0
272	4.5	500	1737	0	0	0	0
273	4.6	500	1871	0	0	0	0
274	4.6	500	2005	0	0	0	0
275	4.6	500	2138	0	0	0	0
276	4.6	500	2272	0	0	0	0
277	4.6	500	2406	0	0	0	0
278	4.6	500	2539	0	0	0	0
279	4.7	500	2673	0	0	0	0
280	4.7	500	2807	0	0	0	0
281	4.7	500	2940	0	0	0	0
282	4.7	500	3074	0	0	0	0
283	4.7	500	3208	0	0	0	0
284	4.7	500	3342	0	0	0	0
285	4.8	500	3475	0	0	0	0
286	4.8	500	3609	0	0	0	0
287	4.8	500	3743	0	0	0	0
288	4.8	500	3876	0	0	0	0
289	4.8	500	4010	0	0	0	0
290	4.8	500	4144	0	0	0	0
291	4.8	500	4277	0	0	0	0
292	4.9	500	4411	0	0	0	0
293	4.9	500	4545	0	0	0	0
294	4.9	500	4678	0	0	0	0
295	4.9	500	4812	0	0	0	0

At $t = 296$ min, the $Z(5)$ interface is located at 4,946 ft, very close to the surface, which is located at 5,000 ft. Thus, the fifth fluid is found in the range $0 < Z < 4,946$, while the fourth fluid is found in the range $4,946 < Z < 5,000$. The computation of pressures in the annulus and in the pipe follows the general discussions given previously. For documentation purposes, we refer to both tables and their included explanations as Figure 9.8.

On real interfaces

In the calculations just given, we speak of interfaces as being located at "z = ..." or "Z = ..." In other words, "interfaces are flat." This description suffices from the macroscopic perspective. If we require details about the mixing zone between two contiguous fluids, we "zoom in" to perform boundary layer–type calculations using pressure gradient information obtained as discussed earlier. Typical mixing zones, shown in Figure 9.9, are clearly not planar in the detailed description.

296	4.9	500	4,946	0	0	0	0
297	4.9	500	0	0	0	0	0

FIGURE 9.8

Pipe and annular interface position table.

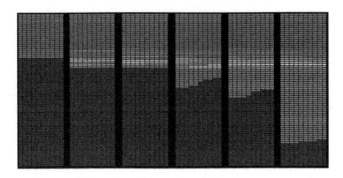

FIGURE 9.9

Propagating and diffusing front in time, constructed from movie frames for viscosity history using exaggerated diffusion.

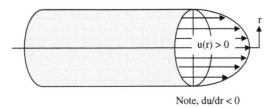

Note, du/dr < 0

FIGURE 9.10

Axisymmetric pipe flow.

DISCUSSION 9.3

Calculating Annular and Drillpipe Pressure Loss

Discussion 9.2 described our interface tracker, an important modeling tool that determines where our six fluids are at any instant in time. Once the length of a particular "fluid slug" is available, the volumetric pump rate Q at that instant is used to determine the pressure gradient applicable to the non-Newtonian fluid in question. The pressure loss associated with this slug is simply the product of length and pressure gradient. This idea was illustrated using both drillpipe and annular examples in the previous discussion.

For the sake of completeness, we now summarize key analytical results available for non-Newtonian pipe flows and recapitulate our new simulation capabilities for eccentric annular flows. Note that our pump schedule is transient, with Q's that vary in time; however, within a defined time interval, the Q in question is constant. In mathematics, this is known as a "piecewise-constant" specification. This approach makes it possible to use steady-state models within the framework of transient pumping.

Newtonian pipe flow model

Several exact, closed-form, analytical solutions are available in the literature for different types of rheologies of flow in circular pipe. We will review these results and offer key formulas without proof. Figure 9.10 illustrates straight, axisymmetric pipe flow, where the axial velocity, u(r) > 0, depends on the radial coordinate $r > 0$. With these conventions, the "shear rate," $du/dr < 0$, is negative; that is, $u(r)$ decreases as r increases. Very often, the notation $d\gamma/dt = -du/dr > 0$ is used. If the viscous shear stress, τ, and the shear rate are linearly related by

$$\tau = -\mu \, du/dr > 0 \tag{9.1a}$$

where μ is the viscosity, a constant or temperature-dependent quantity, then two simple relationships can be derived for pipe flow.

Let $\Delta p > 0$ be the (positive) pressure drop over a pipe of length L, and R be the inner radius of the pipe. Then the radial velocity distribution satisfies

$$u(r) = [\Delta p/(4\mu L)] \; (R^2 - r^2) > 0 \tag{9.1b}$$

Note that u is constrained by a "no-slip" velocity condition at $r = R$. If the product of "$u(r)$" and the infinitesimal ring area "$2\pi r \, dr$" is integrated over $(0, R)$, we obtain the volumetric flow rate expressed by

$$Q = \pi R^4 \Delta p/(8\mu L) > 0 \tag{9.1c}$$

which is the well-known Hagen-Poiseuille formula for flow in a pipe.

These solutions do not include unsteadiness or compressibility. The results are exact relationships derived from the Navier-Stokes equations, which govern viscous flows when the stress-strain relationships take the linear form in Equation 9.1a. We emphasize that the Navier-Stokes equations apply to Newtonian flows only and not to more general rheological models.

Note that viscous stress (and the wall value τ_w) can be calculated from Equation 9.1a, but the following formulas can also be used,

$$\tau(r) = r \, \Delta p/2L > 0 \tag{9.2a}$$

$$\tau_w = R \, \Delta p/2L > 0 \tag{9.2b}$$

Equations 9.2a and 9.2b apply generally to steady laminar flows in circular pipes and, significantly, whether the rheology is Newtonian or not. But they do not apply to ducts with other cross sections or to annular flows, even concentric ones, whatever the fluid.

Bingham plastic pipe flow

Bingham plastics satisfy a slightly modified constitutive relationship, usually in the form

$$\tau = \tau_0 - \mu \, du/dr \tag{9.3a}$$

where τ_0 represents the fluid yield stress. In other words, fluid motion will not initiate until stresses exceed yield; in a moving fluid, a "plug flow" moving as a solid body is always found below a "plug radius" defined by

$$R_p = 2\tau_0 L/\Delta p \tag{9.3b}$$

The "if-then" nature of this model renders it nonlinear, despite the (misleading) linear appearance in Equation 9.3a. Fortunately, simple solutions are known:

$$u(r) = (1/\mu)[\{\Delta p/(4L)\}(R^2 - r^2) - \tau_0(R - r)], R_p \le r \le R \tag{9.3c}$$

$$u(r) = (1/\mu)[\{\Delta p/(4L)\}(R^2 - R_p{}^2) - \tau_0(R - R_p)], 0 \le r \le R_p \tag{9.3d}$$

$$Q/(\pi R^3) = [\tau_w/(4\mu)][1 - 4/3(\tau_0/\tau_w) + 1/3(\tau_0/\tau_w)^4] \tag{9.3e}$$

Power law fluids in pipe flow

Power law fluids without yield stress satisfy Equation 9.4a and the rate solutions in Equations 9.4b and 9.4c.

$$\tau = K(-du/dr)^n \tag{9.4a}$$

$$u(r) = (\Delta p/2KL)^{1/n}[n/(n+1)](R^{(n+1)/n} - r^{(n+1)/n}) \tag{9.4b}$$

$$Q/(\pi R^3) = [R\Delta p/(2KL)]^{1/n}n/(3n+1) \tag{9.4c}$$

Herschel-Bulkley pipe flow model

The Herschel-Bulkley pipe flow model combines Power law with yield stress characteristics, with the result that

$$\tau = \tau_0 + K(-du/dr)^n \tag{9.5a}$$

$$\begin{aligned} u(r) = & K^{-1/n}(\Delta p/2L)^{-1}\{n/(n+1)\} \\ & \times [(R\Delta p/2L - \tau_0)^{(n+1)/n} - (r\Delta p/2L - \tau_0)^{(n+1)/n}], R_p \le r \le R \end{aligned} \tag{9.5b}$$

$$\begin{aligned} u(r) = & K^{-1/n}(\Delta p/2L)^{-1}\{n/(n+1)\} \\ & \times [(R\Delta p/2L - \tau_0)^{(n+1)/n} - (R_p\Delta p/2L - \tau_0)^{(n+1)/n}], 0 \le r \le R_p \end{aligned} \tag{9.5c}$$

$$\begin{aligned} Q/(\pi R^3) = & K^{-1/n}(R\Delta p/2L)^{-3}(R\Delta p/2L - \tau_0)^{(n+1)/n} \\ & \times [(R\Delta p/2L - \tau_0)^2n/(3n+1) + 2\tau_0(R\Delta p/2L - \tau_0)n/(2n+1) + \tau_0^2 n/(n+1)] \end{aligned} \tag{9.5d}$$

where the plug radius R_p is again defined by Equation 9.3b.

Ellis fluids in pipe flow

Ellis fluids satisfy a more complicated constitutive relationship, with the following known results:

$$\tau = -du/dr/(A + B\tau^{\alpha-1}) \tag{9.6a}$$

$$u(r) = A \,\Delta p(R^2 - r^2)/(4L) + B(\Delta p/2L)^\alpha(R^{\alpha+1} - r^{\alpha+1})/(\alpha+1) \tag{9.6b}$$

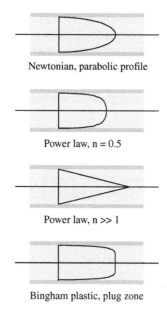

FIGURE 9.11

Typical non-Newtonian velocity profiles.

$$Q/(\pi R^3) = A\tau_w/4 + B\tau_w{}^\alpha/(\alpha + 3)$$
$$= A(R\Delta p/2L)/4 + B(R\Delta p/2L)^\alpha/(\alpha + 3) \tag{9.6c}$$

Dozens of additional rheological models appear in the literature, but the most common ones used in petroleum engineering are those given here. Typical qualitative features of the associated velocity profiles are shown in Figure 9.11.

Annular flow solutions

We next discuss annular flow solutions. As noted earlier in this book, annular flow solutions that are useful in petroleum engineering are lacking. The only known exact, closed-form, analytical solution is a classic one describing Newtonian flow in a concentric annulus. Let R be the outer radius and κR be the inner radius, so that $0 < \kappa < 1$. Then it can be shown that

$$u(r) = [R^2\Delta p/(4\mu L)]$$
$$\times \ [1 - (r/R)^2 + (1 - \kappa^2) \log_e (r/R)/\log_e(1/\kappa)] \tag{9.7a}$$

$$Q = [\pi R^4\Delta p/(8\mu L)][1 - \kappa^4 - (1 - \kappa^2)^2/\log_e(1/\kappa)] \tag{9.7b}$$

For non-Newtonian flows, even for concentric geometries, numerical procedures are required (see Fredrickson and Bird (1958); Bird, Stewart, and Lightfoot (1960); or Skelland (1967)).

Analytically based treatments for eccentric annuli formed from circles are available through bipolar coordinate formulations. These are ultimately numerical in nature and require significant amounts of algebra in their development. Because the methods are limited to circles and not generalizable to practical geometries with cuttings beds, washouts, and other borehole anomalies, they are not discussed here.

The mappings we have developed, we emphasize, can and have been extended to three-dimensional applications that allow changes of cross-sectional geometry along the borehole. Moreover, the effects of multiphase flow with diffusive mixing have been incorporated in the author's models. Recent publications describing these specialized efforts appear in Savery, Darby, and Chin (2007); Deawwanich, Liew, Nguyen, Savery, Tonmukayakul, and Chin (2008); Nguyen, Deawwanich, Tonmukayakul, Savery, and Chin (2008); Savery, Chin, and Babu Yerubandi (2008); and Savery, Tonmukayakul, Chin, Deawwanich, Liew, and Nguyen (2008).

We note that the algorithms developed in this book are faster and more stable than the models just referenced, particularly in handling spatial derivatives of apparent viscosity and the coupling of rotating flows to axial effects. We next review the eccentric annular flow capabilities with respect to their use in total pressure drop in Discussion 9.2.

Review of steady eccentric flow models

As noted, models do not presently exist for non-Newtonian yield stress fluids in arbitrary eccentric annuli, for either steady or transient flow, with or without pipe rotation, except for those developed in this book. Only software models that are fast and numerically stable are discussed and offered for general dissemination. We take this opportunity to summarize these methods now because Discussion 9.2 importantly described the roles played by our steady-state "building block" modules. From that discussion, we noted how the pressure profile in the drillpipe and borehole system (as a function of time) requires computations that look something like "$(100-80)$ $(\partial P/\partial z)_{pipe,0} + (80-75)(\partial P/\partial z)_{pipe,1} + (75-65)$ $(\partial P/\partial z)_{pipe,2} + (65-50)$ $(\partial P/\partial z)_{pipe,3} + (50-30)$ $(\partial P/\partial z)_{pipe,4} + (30-0)$ $(\partial P/\partial z)_{pipe,5} + \Delta + (100-0)$ $(\partial P/\partial z)_{annulus,0}$," where pipe flow equations were succinctly given previously and the annular pressure drops require our sophisticated computational modeling tools.

First, we emphasize the importance of our steady flow simulator, whose user interface is shown in Figure 9.12. This computes all flow properties for eccentric nonrotating annular flows (allowing washouts, cuttings beds, and other geometric anomalies), assuming general Herschel-Bulkley fluids, importantly, in the "volumetric flow rate specified" mode in which the required pressure gradients are automatically calculated without user intervention. Here the size and shape of all plug zones are calculated naturally using an extended Herschel-Bulkley model. The model includes borehole radius of curvature effects, should Figure 9.2 incorporate turns from vertical to horizontal.

Our steady two-dimensional simulator also includes analytical solutions for concentric annuli, as shown in Figure 9.13. These are "Newtonian, nonrotating, axial pipe motion"; "Herschel-Bulkley, no rotation or pipe movement"; and "Power law, rotating, no axial pipe movement." For eccentric flows, when detailed spatial plots for physical properties are not required, the fast mode shown in Figure 9.14 gives numerous pressure gradient results in one or two minutes of computing time. Our steady 2D eccentric solver assumes zero pipe rotation.

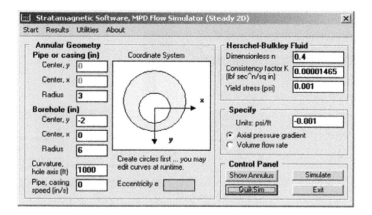

FIGURE 9.12

MPD Flow Simulator, Steady 2D.

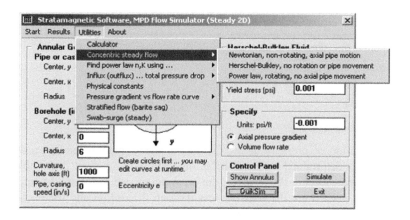

FIGURE 9.13

MPD Flow Simulator, Steady 2D utilities.

As noted elsewhere, the computation of steady flows with pipe rotation within the framework of a purely steady formulation is an unstable numerical process at the present time. This is not to say that steady flows with rotation cannot be computed. They can, as indicated in Figure 9.15, provided we treat the unsteady problem and carry out our computations for large times until steady conditions are reached. This often requires one minute or less for fluids with low specific gravity and sometimes as many as three minutes for heavy-weight muds or cements. Figure 9.15 shows how steady-state pressure gradients can be obtained for given flow rates. Once the target flow rate is given, the search for the required pressure gradient may take several intelligent guesses. We have summarized all of the methods we have devised to obtain pressure gradients when target flow rates are specified.

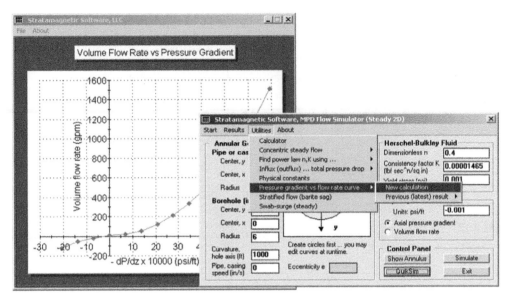

FIGURE 9.14

Rapid calculation of multiple flow solutions.

DISCUSSION 9.4

Herschel-Bulkley Pipe Flow Analysis

As noted, the calculation of pressure at the drillbit (in the formation) and of pressure along the borehole is completely determined by the distribution of pressure gradient in the hole and the value of pressure at the surface choke. If, however, the pressure needed at the mud pump to support the flow is required, also needed are the pressure loss through the drillbit and the pressure drop in the drillpipe. For nonrotating pipe flow, exact, closed-form circular pipe flow solutions for radial velocity distribution and total volumetric flow rate are available for Herschel-Bulkley fluids from Equations 9.5a through 9.5d. Thus, the same properties for the subsets including Newtonian, Power law, and Bingham plastic fluids are also available.

The general mathematical solution has been incorporated into two software programs for convenience. The first, shown in Figure 9.16(a), solves Equation 9.5d for pressure gradient when the flow rate is given. Note that this represents a nonlinear algebraic equation for the unknown. The example given here applies to a 10-cp Newtonian fluid. For the parameters shown, the required pressure gradient is about -0.001 psi/ft. In Figure 9.16(b), we introduce yield stress to this fluid, so that it now acts as a Bingham plastic. We expect that the pressure gradient should steepen because there is greater difficulty in moving the fluid. In fact, the pressure gradient is now about -0.015 psi/ft. Finally, in Figure 9.16(c) we change the fluid exponent from 1.0 to 0.8, so that the fluid is now of a Herschel-Bulkley type. In this case, the pressure gradient is obtained as -0.014 psi/ft. It is interesting how the presence of yield stress introduces large changes to pressure gradient over Newtonian flows.

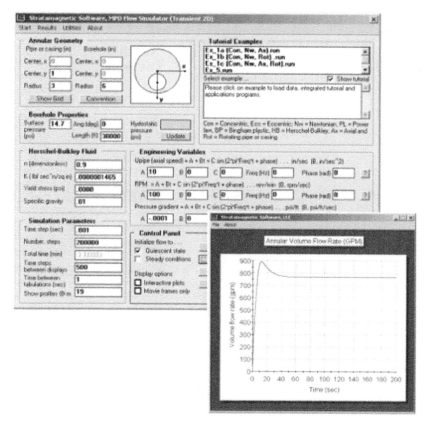

FIGURE 9.15

"Transient 2D" solver.

In Figure 9.17(a), we demonstrate our second use of Equations 9.5a through 9.5d—namely, computing total flow rate and radial velocity distribution for any Herschel-Bulkley fluid. Here a Newtonian fluid is assumed, and the classic paraboloidal velocity profile is obtained. In Figure 9.17(b), we illustrate this capability with a Herschel-Bulkley fluid. The graph clearly indicates the presence of a plug zone. The plug radius is also given in the output.

DISCUSSION 9.5

Transient, Three-Dimensional Eccentric Multiphase Flow Analysis for Nonrotating Newtonian Fluids

Here we introduce multiphase flow computations for a special limit of the general problem, one assuming Newtonian mixtures in concentric or eccentric annuli (with possible cross-sectional

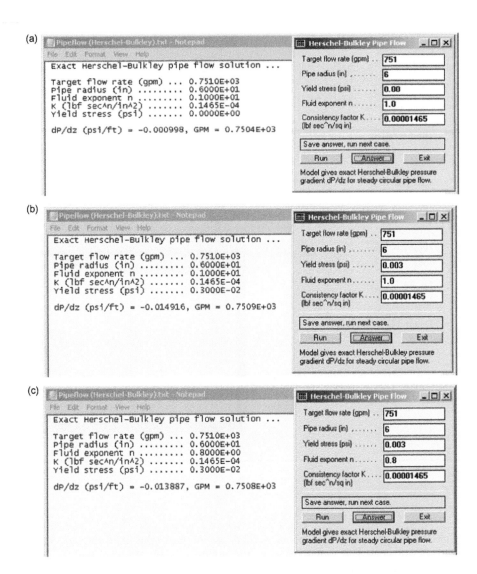

FIGURE 9.16

(a) Newtonian fluid, flow rate given. (b) Bingham plastic, flow rate given. (c) Herschel-Bulkley fluid, flow rate given.

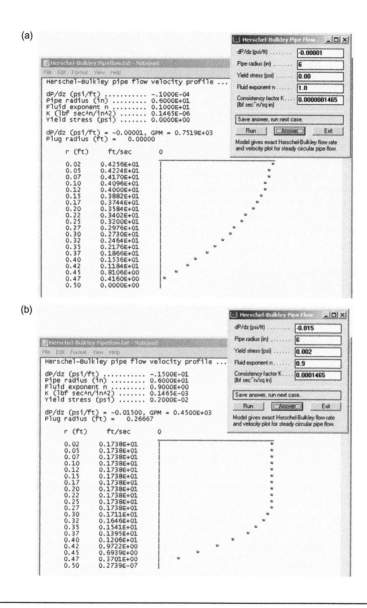

FIGURE 9.17

(a) Newtonian fluid, pressure gradient given. (b) Herschel-Bulkley fluid, pressure gradient fluid given.

changes in the axial direction)—however, without pipe or casing rotation. Later, we will remove our Newtonian, nonrotating flow restrictions and consider general non-Newtonian fluids in eccentric annuli with steady pipe rotation. Software for the present limit was developed because the solution process could be automated and Newtonian applications do exist. But our purposes are twofold: first to illustrate basic flow concepts and second to demonstrate that our formulation, solution, and software foundation for subsequent development are sound and correct.

DISCUSSION 9.5, EXAMPLE 9.1

We first show that our exact, steady concentric Newtonian flow solution and the transient numerical model under consideration are consistent in the concentric single-phase flow limit. This is intended to validate the software architecture, which is complicated and forms the basis for other models. The simulator for our exact solution is launched from the earlier "Steady 2D" menu in Figure 9.18, leading to the applications program in Figure 9.19. Note how the assumed parameters yield a flow rate of 947.1 gpm.

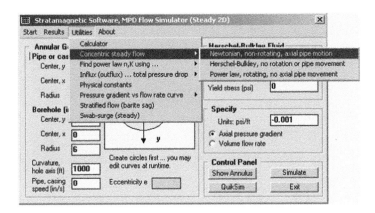

FIGURE 9.18

General "Steady 2D" menu.

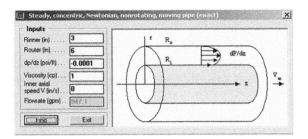

FIGURE 9.19

Exact two-dimensional Newtonian flow solution.

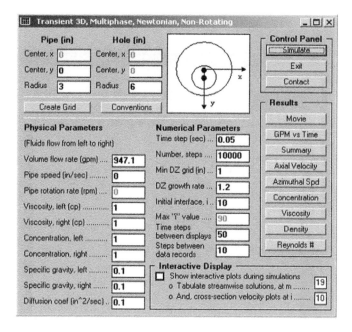

FIGURE 9.20

Consistent transient simulation parameters.

Next we launch the "Transient 3D, Multiphase, Newtonian, Non-Rotating" flow simulator in Figure 9.20. For multiphase problems, it is not meaningful to specify pressure gradients as in single-phase calculations; these gradients vary with space and time as local fluids mix, and it is impossible to state clearly what they are. One is therefore forced to specify total flow rate, at least approximately, and this specification must be used when dealing with multiphase applications. Our simulator operates in a "Specify flow rate" mode.

To be completely consistent with Figure 9.19, we assume a 1-cp viscosity for both "left" and "right" fluids, zero pipe speed, and identical geometries. We also assume identical small specific gravities; low mechanical inertias allow larger time steps and reduce integration times needed for convergence. Internal to the software, $C = 0$ means left properties, (i.e., μ_{left} and ρ_{left}), while $C = 1$ means right; since left and right properties are identical, the choice $C_{left} = C_{right} = 1$ ensures that $C = 1$ continuously throughout and the fluid is homogeneous.

Note that, in Figure 9.20, we have entered 947.1 as the target flow rate. Once numerical integrations begin, the imposed motion must overcome "nonuniformities" associated with the uniform (unsheared) flow used to initialize the calculation and, of course, the effects of inertia. After some time, the calculations converge. For example, Figure 9.21 gives a flow rate of 949.1 gpm for an error of 0.2 percent. For the 10,000 time steps shown, the computing time is about five minutes for this three-dimensional run. We have used the transient, three-dimensional, two-phase flow solver to reproduce an exact, steady, two-dimensional, single-phase flow result. In general, single-phase flows can be calculated this way, although this is obviously suboptimal. However, the example was designed to show that the numerical model is basically correct.

Transient Flow Subtleties

Again we remind the reader of certain difficulties encountered in transient flow modeling. In steady flow analysis, whether concentric or eccentric, computations for flow rate (when pressure gradients are given) are very rapid and vice versa. For linear Newtonian flows, these are especially fast. If $(\partial P/\partial z)_1$ corresponding to Q_1 is

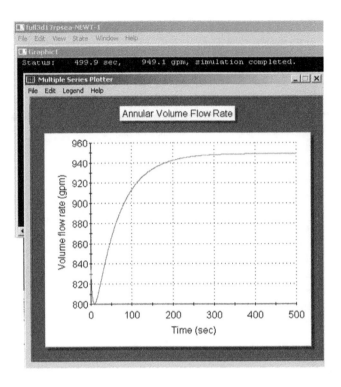

FIGURE 9.21

Example 1, smoothly convergent flow rate history.

known from just one eccentric or concentric calculation or experiment, then the identity $(\partial P/\partial z)_2/Q_2 = (\partial P/\partial z)_1/Q_1$ allows us to immediately obtain $(\partial P/\partial z)_2$ when Q_2 is given or Q_2 when $(\partial P/\partial z)_2$ is given. For non-Newtonian flows, the nonlinearity of the pressure gradient and flow rate relationship disallows this simple rescaling. However, the "Specify volumetric flow rate" option in Figure 9.18 does use a rapidly convergent half-step method to guess the pressure gradient corresponding to a target flow rate to within 1 percent accuracy.

In transient calculations, one can *in principle* specify total volumetric flow rate at each instant in time. However, to achieve the required solution, numerous attempts using different pressure gradients will have to be made at each time step. When this is repeated for the entire range of time integration, the computations needed are voluminous and require hours or overnight runs. This is particularly unacceptable if, perhaps during the calculations, instabilities are encountered; then all of the numerical effort expended will be wasted. Thus, we ask if there is an acceptable compromise: "Is there an approximate pressure gradient we can use in a constant flow rate process?"

For Newtonian nonrotating flows, the answer is yes. We recall from our theoretical discussion of steady single-phase flow that volumetric flow rate is directly proportional to the pressure gradient $\partial P/\partial z$ and inversely related to the viscosity μ. If "1" and "2" now denote two positions along the three-dimensional channel (at a fixed instant in time) without area changes, then constancy of flow rate implies that $(\partial P/\partial z)_1/\mu_1 = (\partial P/\partial z)_2/\mu_2$. Suppose that the volumetric flow rate at the (left) inlet and the starting viscosity are specified. Then the pressure gradient required for the eccentric Newtonian flow can be obtained from the "Steady 2D" solver in Figure 9.18. As the fluid at the inlet flows downstream, it mixes with the "right" fluid and local concentrations will change. The underlying viscosity will consequently change, in a manner consistent with an assumed mixing relationship (taken again as the Todd-Longstaff law). If the local viscosity is now μ_2, then the corresponding pressure gradient is $(\partial P/\partial z)_2 = (\partial P/\partial z)_1 \, \mu_2/\mu_1$, showing correctly, for instance, that an increase in viscosity will require an increase in pressure gradient.

This procedure has been programmed into the solver of Figure 9.20; there is no need to operate the simulator in Figure 9.18 because the procedure has been completely automated. Again, starting pressure gradients are obtained from inlet conditions and local values are obtained by concentration-dependent rescaling. This automation is only convenient for Newtonian mixtures where there is no pipe rotation. The "$(\partial P/\partial z)_1/\mu_1 = (\partial P/\partial z)_2/\mu_2$" law does not apply to eccentric problems with rotation, although it remains valid for concentric rotating flow because axial and azimuthal modes decouple. For more complicated problems, a more complete approach applies, with different degrees of complexity, depending on the nature of the underlying flow. The general problem will be considered in a separate discussion.

DISCUSSION 9.5, EXAMPLES 9.2 AND 9.3

For our second calculation, we repeat the above simulation except that we double the inlet-outlet viscosity and density ratios, as shown in Figure 9.22. Note that, in order to track two different phases, the concentrations at the inlet and outlet are set to 0 and 1, respectively. The calculation yields almost identical flow rates and flow rate history curves. Why? This occurs because, in Newtonian mixtures, the ratio of density to viscosity controls the dynamics and not either parameter alone; there is, however, an effect associated with the ratio of density to diffusion coefficient, which need not always be small. Thus, the effects of the doubling almost cancel. In our third simulation, we set our inlet-outlet viscosity and density ratios to 5 and 2, respectively. Figure 9.23 shows that the volumetric flow rate history changes somewhat, with the predominant effect being the time required to reach equilibrium.

A detailed description of the simulator appears in Discussion 9.6. The reader should study it, since many of its software features are shared by the more general solver introduced in Discussion 9.7, which deals with real two-phase flows in which the mixing of non-Newtonian fluids, in the presence of rotation, is addressed. Mixing is controlled by numerous factors: convection, diffusion, annular geometry, rheology, flow rate, and initial

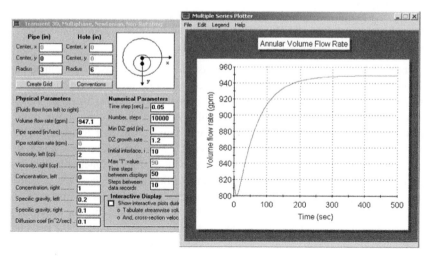

FIGURE 9.22

Example 2 calculation.

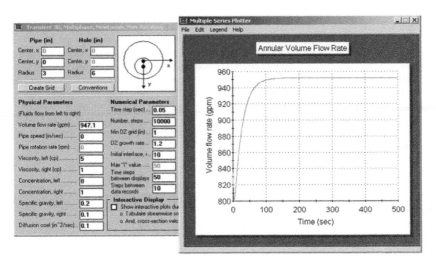

FIGURE 9.23

Example 3 calculation.

conditions. This complexity means that general conclusions are difficult to formulate and that each flow solution must be interpreted on a case-by-case basis. Predictions should be substantiated by laboratory experiment and field data whenever possible.

DISCUSSION 9.6

Transient, Three-Dimensional Eccentric Multiphase Analysis for Nonrotating Newtonian Fluids: Simulator Description

Here we describe in detail the operation of our "Transient 3D, Multiphase, Newtonian, Non-Rotating" flow simulator in Figure 9.24. Again, this stand-alone module was developed because the model could be rigorously formulated and fully automated; it is also, of course, useful as a planning tool in itself. We emphasize that the module applies to highly eccentric annuli and allows limited cross-sectional geometric modification along the borehole axis. Many of the user features described here are also incorporated in our more general multiphase solver for non-Newtonian rotating flow.

The upper left text boxes of Figure 9.24 host the annulus definition function common to all of our simulators, with "Create Grid" displaying the curvilinear grid chosen to host the eccentric annulus at run time; this feature provides needed error checking to ensure that circles do not cross over. Clicking "Create Grid," in this case, leads to Figure 9.25. The "Conventions" button provides explanations on azimuthal grid numbering conventions needed to select cross-section plots for run-time interactive displays and movies.

Our annular fluid flows from left to right, with the inlet at the left and the outlet at the right. Fluid properties are inputted in the lower left menu. We have selected default run inputs that will

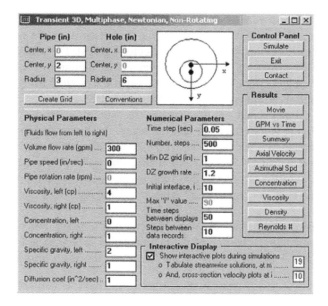

FIGURE 9.24

Basic user interface with default parameters.

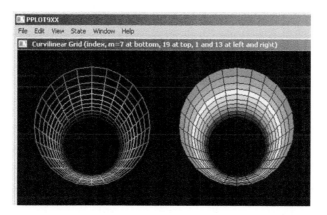

FIGURE 9.25

Curvilinear grid used in present example.

provide a good "fast start" user experience: Simply uncheck the "Interactive Display" box, click "Simulate," and allow the simulation to run to completion (this process requires less than one minute); finally, click "Movie." A movie showing computed results (e.g., see Figure 9.35 later) automatically launches, showing the evolution of the convection and mixing process. The inputs in Figure 9.24 show a heavier, more viscous fluid as the displacing fluid.

The diffusion coefficient used is unusually large, only to provide viewable results (such as those in Figure 9.35) because the graphical displays used at the present time are capable of providing 12 colors only (see this book's companion website for color figures). For actual use, diffusion coefficients available in the environmental or chemical engineering literature should be entered, or those obtained in laboratory studies. Detailed numbers are output for plotting using commercial software, and the manner in which these are accessed is described later.

Once the annular geometry and run-time inputs are entered, clicking "Simulate" leads to the status box shown in Figure 9.26, indicating that the main curvilinear grid for the eccentric annulus just input has been computed. Clicking "Yes" prompts the simulator to solve (using the grid just created) a steady, two-dimensional "Specify volumetric flow rate" problem for the inlet conditions and target low rate prescribed, a process that requires up to two to three seconds. When this is completed, the status box in Figure 9.27 appears. Clicking "Yes" leads to the query in Figure 9.28.

If this query is answered affirmatively, the submenu and message box in Figure 9.29 appears. This allows the user to redefine a portion of the main annulus, whose axial index "i" for the spatial coordinate z_i varies from 1 to 90. For the example shown, the main grid parameters repeated in

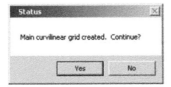

FIGURE 9.26

Curvilinear grid for main eccentric annulus created.

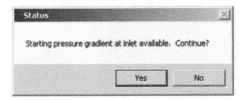

FIGURE 9.27

Starting pressure gradient computed using "Steady 2D" solver.

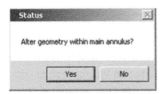

FIGURE 9.28

Option to alter annulus for limited axial extent.

Figure 9.28 are altered so that they are replaced by the concentric annulus in Figure 9.30. Clicking "Apply" leads to the display in Figure 9.31. For the present example, we repeat our steps but do not alter the main annulus; while the numerical engine is presently set up to correctly calculate the effects of this change, the graphical displays are still being developed at this time. (The annulus modification feature *is* usable except for this graphical limitation.)

We next explain the gridding system used. Axial z_i grid control is provided for in the central portion of the menu in Figure 9.24 (cross-sectional grid densities are hardcoded as suggested in Figures 9.25 and 9.31). The main grid is indexed from $i = 1$ at the inlet to $i = 90$ at the outlet, again with the flow moving from the left inlet to the right outlet. Initially, two fluids are permitted: the left with a concentration $C = 0$ and the right with $C = 1$. The initial (flat) interface is assumed at $i = i_{face}$ entered by the user. The finest z mesh length, or "Minimum DZ grid," is centered at this initial interface location and is defined by the user. The mesh amplification rate, or "DZ growth rate," is a number that equals or exceeds 1. A geometrically varying mesh is generated internally and used together with our curvilinear cross-sectional grid to provide three-dimensional simulation capabilities. If we had chosen to modify the main annular geometry, the cross-sectional metrics would have been automatically changed internally.

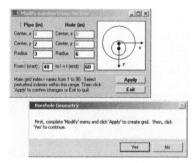

FIGURE 9.29

Perturbation annulus definition.

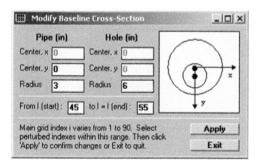

FIGURE 9.30

Concentric annulus defined, $45 < I < 55$.

While the gridding and display options presented here are somewhat awkward, we note that users with more computing resources have extended the algorithm and developed their own gridding and display capabilities. In one case, a fully three-dimensional grid was created that varied continuously in the z direction and could be updated in real time with borehole caliper measurements. Users interested in such capabilities should discuss their needs with the developers. Also note that, while we have discussed the initial condition for two phases, it is also straightforward to perform single-phase flow studies, if there is an interest in modeling single-phase flow in an annulus with internal cross-sectional changes as noted above.

In this case, the left and right concentrations can be set entirely to 0 or to 1, and solutions to the concentration equations will be entirely 0 or 1 (thus suppressing any internal variations to fluid properties). The time step shown in Figure 9.24 is large. Generally speaking, it needs to be much smaller to provide the needed physical resolution. These steps are constant during the simulation. The total time simulated is simply the product of "time step" and "number of steps."

Having made preliminary comments, we continue the simulation process. The final status box in Figure 9.32 appears. If interactive displays are desired, "Interactive Display" should be checked. Because our fully three-dimensional simulators at the present time do not allow true three-dimensional color displays, we offer the limited options available in the option box. First, we can display fluid properties in any single azimuthal "m = constant" plane (click "Conventions" for definitions). Second, we can give cross-sectional plots at any single "i = constant" location. Users with special requirements can contact the developers for source code access or other support. If the interactive

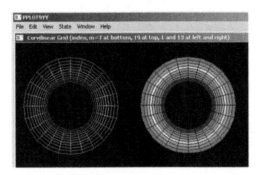

FIGURE 9.31

Concentric annulus redefinition.

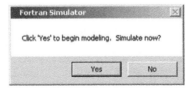

FIGURE 9.32

Simulation to commence.

display box is not checked, a simple status box showing time and "percent complete" appears on screen.

Upon run termination, all results are written to text output files and movie displays for the time evolution of axial velocity, and concentration-dependent viscosity and fluid density are available. If interactive displays are required at periodic user-defined intervals, multiple screens appear, the first being that shown in Figure 9.33. All three diagrams have flow moving downward. The left diagram, here for "m = 19," gives the axial velocity. Blue represents low (zero) speeds at the pipe and annular surfaces, while the uniform red display indicates a high uniform velocity in the annular space. The middle and right diagrams show displacement of one fluid by a second, starting near "i = 10." These are accompanied by velocity plots in Figure 9.34. Closing these windows allows the simulation to continue.

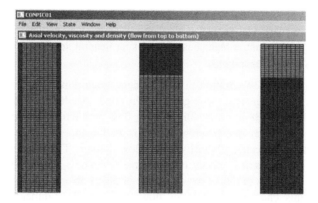

FIGURE 9.33

Axial velocity, viscosity, and density at m = 19.

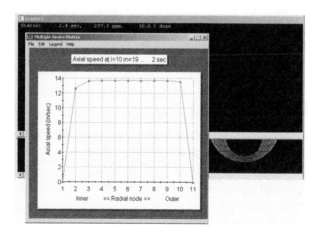

FIGURE 9.34

Velocity graph with cross-section plot in background.

Plots like those in Figure 9.33 are automatically generated internally (whether or not interactive displays are selected) and are assembled to create movies available for user access by clicking "Movie." Example frames are shown in Figure 9.35. The complete output menu is shown in Figure 9.36. The buttons labeled "Axial Velocity," "Azimuthal Velocity," "Concentration," "Viscosity," "Density," and "Reynolds Number" provide spreadsheet-style numerical output for the respective quantities. (Azimuthal velocities are identically zero for the present Newtonian flow simulator, but generally need not be.) Figure 9.37 shows numerical output in the case of concentrations, for azimuthal location $m = 19$, where the axial index "i" varies from 1 to 90 and the radial-like index varies from 1 at the pipe surface to 11 at the annular wall.

DISCUSSION 9.7

Transient, Three-Dimensional Eccentric Multiphase Analysis for General Rotating Non-Newtonian Fluids: Simulator Description

Here we give a brief qualitative description of the transient, three-dimensional, multiphase flow model considered in this book. Again, general rheologies are permitted, together with highly eccentric borehole annular cross sections. Figure 9.38 illustrates the rotating flow problem considered here, but for simplicity displays only two contiguous non-Newtonian fluids. At the top, we have an initial condition in which a flat fluid interface is located arbitrarily in the flow domain. The situation shown at the bottom is a diffused interface, not necessarily planar or uniform in thickness, encountered at a later instant in time. Our objective, of course, is to model the dynamics of this problem.

The remainder of this chapter demonstrates how transient, three-dimensional, multiphase flow fields can be obtained computationally. Figures 9.39(a) and 9.39(b) illustrate, for instance, "movies" (with time increasing downward) in which a purely eccentric annulus that does not vary axially is considered, followed by a mixed geometry having concentric and eccentric sections. These movies can be accessed from the "Start" menu for the "Transient 3D multiphase" solver shown in Figure 1.33.

Note that the plots in Figures 9.39(a) and 9.39(b) were created using Tecplot 360™ software (see *www.tecplot.com*).

DISCUSSION 9.8

Transient, Three-Dimensional Eccentric Multiphase Analysis for General Rotating Non-Newtonian Fluids with Axial Pipe Movement: Validation Runs for Completely Stationary Pipe

Here we will start with simple examples and graduate to more complicated ones, demonstrating first that the three-dimensional algorithm is correct.

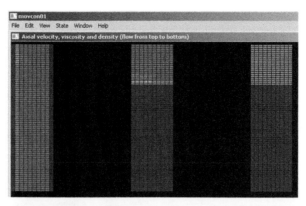

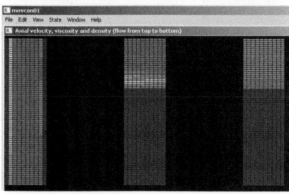

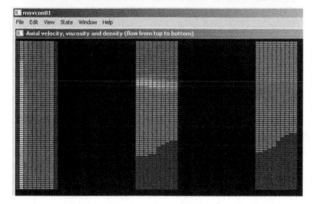

FIGURE 9.35

Movie frames at different times showing mixing.

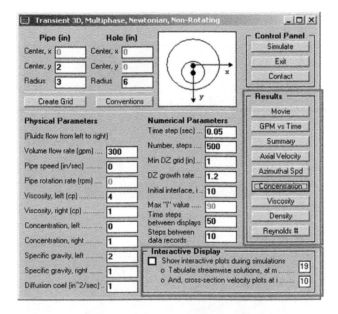

FIGURE 9.36

Output menu.

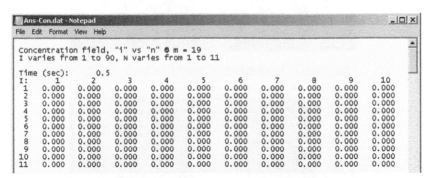

FIGURE 9.37

Tabulated numerical output (for concentration shown).

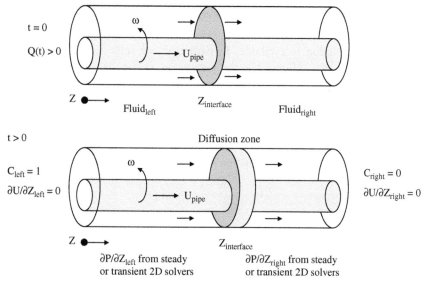

$t = 0$

$Q(t) > 0$

ω

U_{pipe}

Z

Fluid$_{left}$

$Z_{interface}$

Fluid$_{right}$

$t > 0$

Diffusion zone

$C_{left} = 1$

$\partial U/\partial Z_{left} = 0$

ω

U_{pipe}

$C_{right} = 0$

$\partial U/\partial Z_{right} = 0$

Z

$Z_{interface}$

$\partial P/\partial Z_{left}$ from steady
or transient 2D solvers

$\partial P/\partial Z_{right}$ from steady
or transient 2D solvers

FIGURE 9.38

Mathematical problem formulation.

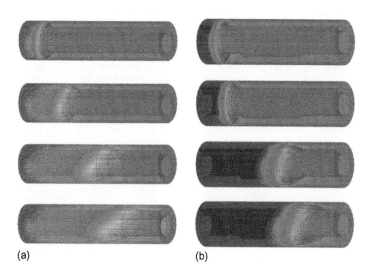

(a)   (b)

FIGURE 9.39

(a) Purely eccentric annulus. (b) Mixed concentric-eccentric flow.

Validation 1: Concentric single-phase Newtonian flow

In this example, we wish to demonstrate that our transient, three-dimensional simulator is correct in a limit where an exact solution is available. In particular, we refer to the concentric Newtonian flow solver in Figure 9.43 that follows. For the parameters shown, the exact volumetric flow rate given in the bottom shaded box is 947.1 gpm. We ask, "Can we solve a transient, three-dimensional problem for a long annulus with the same cross section and obtain a 947.1 gpm flow rate in the steady asymptotic limit?"

To answer this question, we select the simulation option indicated in Figure 9.40. This launches two screens: the main module in Figure 9.41 and the pump schedule and fluid properties menu in Figure 9.42. In Figure 9.42, we have populated both inlet and outlet boxes with Newtonian fluid

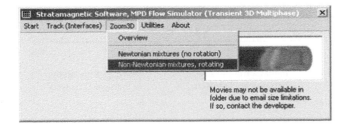

FIGURE 9.40

General "Transient 3D Multiphase" menu.

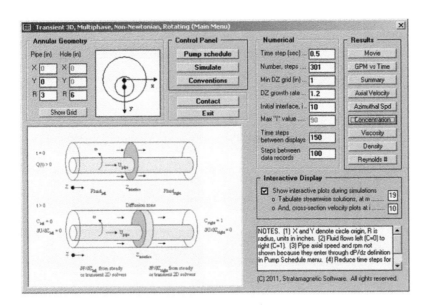

FIGURE 9.41

Main simulation menu.

parameters consistent with Figure 9.43 and assumed a pressure gradient of -0.0001 psi/ft everywhere. A low value of specific gravity is used to minimize mechanical inertia, so that convergence to steady state can be accelerated (larger values will yield the same answers, but they require greater computing). It is important to click "Save" next.

In the simulator of Figure 9.41, we have entered the foregoing concentric geometry and assumed suitable computational parameters, noting in particular a somewhat large time step size of 0.5 sec. Clicking "Simulate" leads to a picture of the assumed annulus and grid shown in Figure 9.44,

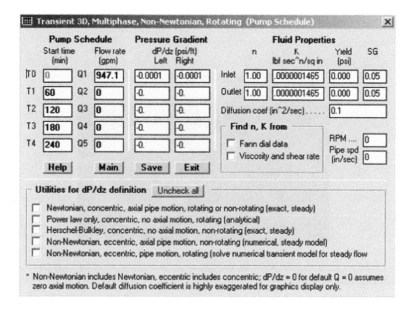

FIGURE 9.42

Pump schedule and fluid properties definition.

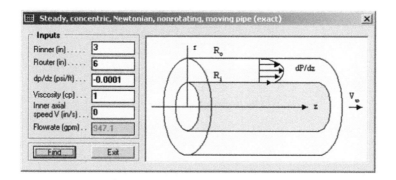

FIGURE 9.43

Exact concentric Newtonian solution.

provided for error checking, and the setup menus in Figure 9.45. Intermediate results (as requested in Figure 9.41) are displayed in Figure 9.46. Similar results appear at periodic intervals in simulations, and we will not duplicate them.

On completion of the simulation, the volumetric flow rate versus time history is given as shown in Figure 9.47, and a final value of 927.6 gpm is calculated. This is to be compared with the exact value of 947.1 gpm, and it is seen that the two simulators are consistent to within an acceptable 2 percent error. Of course, solving a steady, two-dimensional problem with an unsteady, three-dimensional solver is not an efficient use of computing resources. Our only purpose here is in validating the three-dimensional code logic, which, as we have explained, involves a great deal of

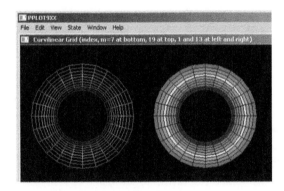

FIGURE 9.44

Geometry displayed for error checking.

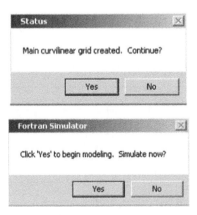

FIGURE 9.45

Setup menu status.

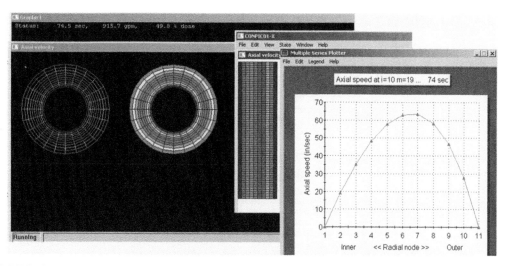

FIGURE 9.46

Intermediate axial velocity displays requested by user, cross-section color plot, and line plot at given azimuthal station.

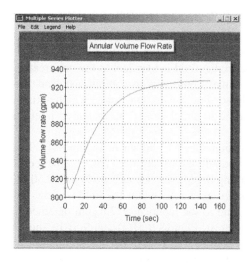

FIGURE 9.47

Volumetric flow rate history at end of simulation.

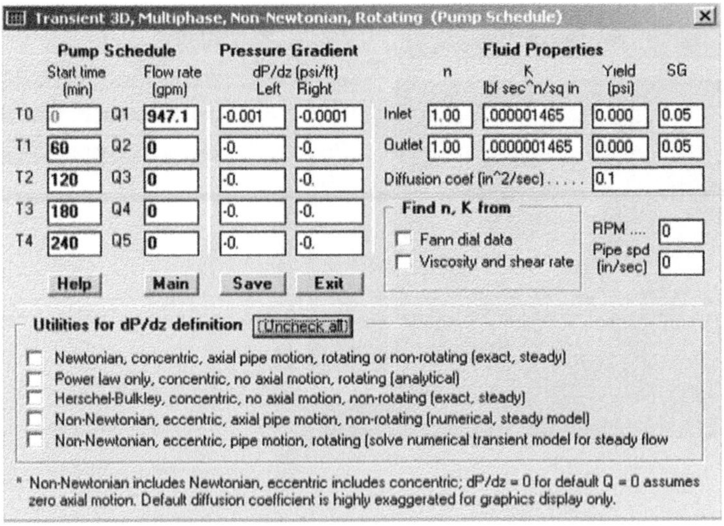

FIGURE 9.48

Viscosity and pressure gradient increased tenfold.

subtlety. The ultimate purpose is adaptation of the software platform to handle problems that are truly transient and three-dimensional—namely, those that involve convection and diffusive mixing.

Validation 2: Concentric two-phase Newtonian flow

In this example, we extend our discussion of Validation 1 and ask how the previous setup can be modified to handle the displacement of a thin fluid by a thicker one. For illustrative purposes, let us assume that the displaced (right) fluid is identical to the one treated in the earlier example, while the displacing fluid is ten times more viscous. While we can certainly use the calculator in Figure 9.43, we need not do so. For Newtonian fluids, which satisfy linear pressure gradient and flow rate relationships, we need to simply enter the increased inlet viscosity and pressure gradient as indicated in Figure 9.48.

The corresponding simulation menu is shown in Figure 9.49. Use of the 0.5-sec time step in Validation 1 will lead to computational instabilities. Thus, a smaller 0.05 sec is taken for this example; nonetheless, total computing time is just seconds. In Figure 9.50, it is significant that the specification of discontinuous pressure gradients within the field of flow (where the interface is moving) leads to stable computations and to the identical 927.6 gpm obtained earlier. However, the intermediate results are of greater interest. Figures 9.51(a) and 9.51(b) show results at time steps 150 and 300, respectively, while in Figure 9.51(c) we reran the simulation to 2,000 time steps (requiring about one minute of computing). These plots show the velocity and viscosity mixing thickness.

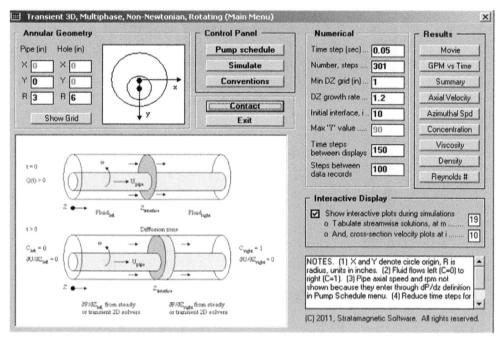

FIGURE 9.49

Simulation menu, with reduced time step. Note displays selected at azimuthal station m = 19 and axial location i = 10.

Validation 3: Concentric single-phase Herschel-Bulkley flow

Here we repeat the example of Validation 1, except that instead of a Newtonian fluid we consider a Herschel-Bulkley fluid with nonvanishing yield stress. Checking the "Herschel-Bulkley" box in the "Pump Schedule" menu automatically launches our exact solver for concentric, nonrotating, Herschel-Bulkley flow. In fact, we run this solver with the inputs and results shown in Figure 9.52, noting in particular the 471.9 gpm computed for this problem. The corresponding transient, three-dimensional calculation is performed in Figure 9.53, in which a steady flow rate of about 487 gpm is shown, for a modest 3 percent error.

Again, we have obtained an exact solution using our three-dimensional computational logic and demonstrated its correctness. We do emphasize that in all of our yield stress work, our "extended Herschel-Bulkley" model is *not* the same as the "conventional Herschel-Bulkley" offered in the literature, since a smooth (but rapid) transition from sheared to plug flows is allowed. Thus, *agreement will not always be found and discrepancies can be significant for "small n" flows*. This, we emphasize, is to be expected.

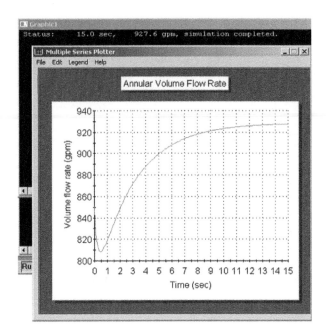

FIGURE 9.50

Volumetric flow rate history.

Validation 4: Concentric two-phase Herschel-Bulkley flow

Here we repeat Validation 2 except that extended Herschel-Bulkley flow is considered. In fact, we will displace water with the Herschel-Bulkley fluid analyzed previously. In Figures 9.54(a) and 9.54(b), we run our exact, two-dimensional, concentric model for Newtonian flow in the annulus shown to give a flow rate of 471.5 gpm. As evident from the line plot, the transient, three-dimensional solver leads to the same flow rate as required (actually, it is 462 gpm, for an error of less than 2 percent).

In Figure 9.55, we have set up the problem so that our thick Herschel-Bulkley fluid is displacing water at the 472-gpm flow rate. (An actual rate of 486 gpm is successfully obtained, for an error of less than 3 percent, again noting that our extended Herschel-Bulkley model is not the conventional one.) Of interest is the mixing result obtained at the end of the calculations, shown in Figure 9.56. Computation times in both three-dimensional runs are less than one minute.

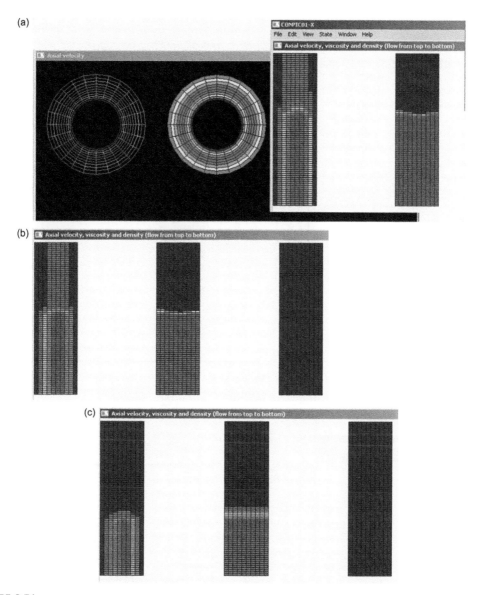

FIGURE 9.51

(a) Result at 150 time steps. (b) Result at 300 time steps. (c) Result at 2,000 time steps (requiring one minute).

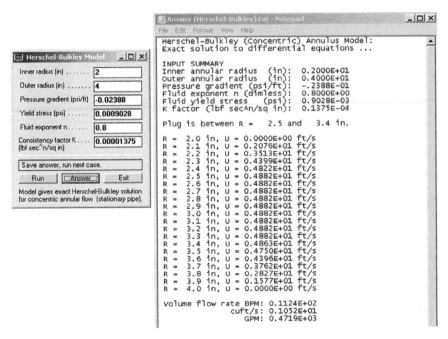

FIGURE 9.52

Exact, steady, two-dimensional Herschel-Bulkley solution.

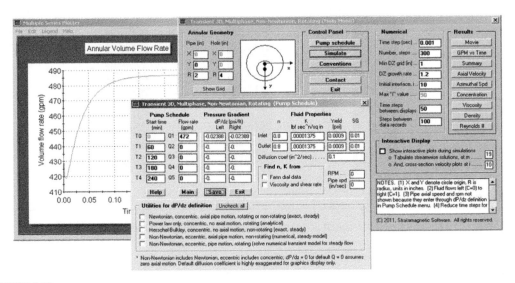

FIGURE 9.53

Transient, three-dimensional flow.

(a)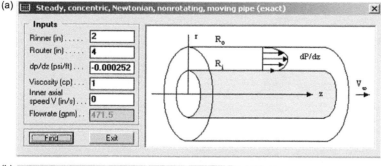

(b)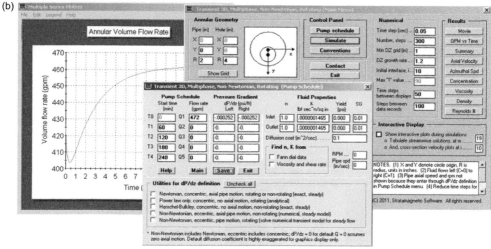

FIGURE 9.54

Newtonian flow validation (exact steady solution).

Validation 5: Eccentric single-phase and multiphase non-Newtonian flow

Here we consider the challenging problem dealing with transient displacement and convective-diffusive mixing of Herschel-Bulkley fluids in highly eccentric three-dimensional annuli. We also address some subtleties of the formulations employed in this book and deal with practical simulation ideas. These discussions are given to promote well-considered, and not blind, use of our simulation models. We will discuss the issues as they arise in the simulations.

First we examine the eccentric annulus defined in Figure 9.57. Here the "Steady 2D" simulator is operated in "Volumetric flow rate specified" mode with a target flow rate of 500 gpm. The result of the iterative calculation gives a pressure gradient of -0.002881 psi/ft. The computed velocity field and curvilinear grid used are shown in Figure 9.58.

Next we run our transient, three-dimensional non-Newtonian flow simulator with the -0.002881 psi/ft gradient specified throughout, as shown in Figure 9.59, in order to replicate

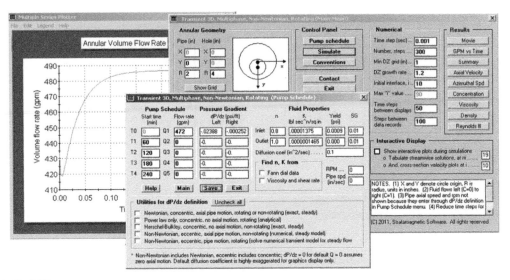

FIGURE 9.55

Displacement calculation setup.

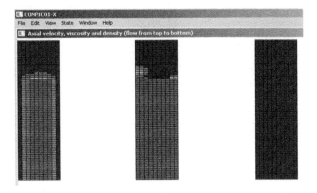

FIGURE 9.56

Mixing solutions for axial velocity, viscosity, and density.

results consistent with Figure 9.58. However, the line graph shows an asymptotic flow rate of 430 gpm and not the 500 gpm assumed previously. What happened? What is the simulation doing? Are there errors in the formulation?

Fortunately, the result is not incorrect. In order to understand the boxed entries, it is important to understand the underlying algorithm. The "500" in Figure 9.59 is not, in any sense, a boundary condition: It is only used to provide starting velocities to initialize the time integration; its effect dampens out with time, and in fact one could have taken "1234" and the steady flow rate computed would be the same. The driving terms of dynamical significance insofar as the differential

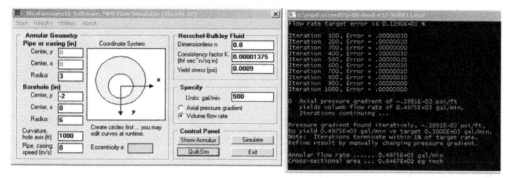

FIGURE 9.57

"Steady 2D" menu calculation with target 500-gpm flow rate.

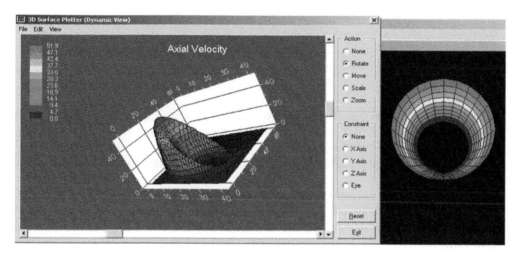

FIGURE 9.58

Computed velocity field and curvilinear grid used.

equations are concerned are the applied pressure gradients. One might then ask why the "−0.002881 psi/ft" did not lead to 500 gpm.

The reason lies in the formulations used. In "Steady 2D," the computations are exact in the sense that the variable apparent viscosity function $N(x, y, z)$, and all of its spatial derivatives are included. In "Transient 3D" this is not the case. While $N(x, y, z)$ itself is included, its derivatives are not; this approximation is consistent with the use of Landau's ad hoc concentration model. The approach is not unlike the use of significant digits in data interpretation (e.g., there is no reason to keep three-decimal-place accuracy if some effects are only known to two places). The problem does not arise in Newtonian fluids, as we have shown by example, since derivatives of the constant viscosity vanish identically.

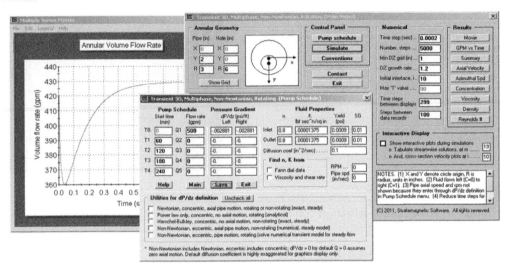

FIGURE 9.59

The "wrong" answer (subject to explanation given).

All of this does not mean that simulations are not possible. Knowing now how the code is structured, we simply ignore the "500" in Figure 9.60 and, through trial and error, determine the pressure gradient that will yield "500" in the final line graph for flow rate. For the present example, the author obtained the " −0.00335 psi/ft" shown after four tries, requiring about five minutes of desktop effort.

Now let us turn to our second fluid, which we assume for simplicity as Newtonian. For the same eccentric annulus, running the exact "Steady 2D" solver in "Volumetric flow rate specified" mode with a target 500-gpm flow rate leads to a pressure gradient of −0.00003281 psi/ft, as shown in Figure 9.61.

As suggested earlier, there is no problem replicating this result using the transient, three-dimensional solver for Newtonian fluids. As shown in Figure 9.62, a flow rate of 490 gpm is computed, which differs from 500 gpm by only 2 percent. (The "minus" signs in the pressure gradient boxes do not appear because they have scrolled to the left, but they *are* entered.) In summary thus far, we have obtained the pressure gradients for two different fluids in the same eccentric annulus needed to achieve a flow rate of 500 gpm. In order to model both the displacement of the second fluid by the first and the convection and diffusive mixing process, we combine the pressure gradients and fluid properties as shown in Figure 9.63.

For the two-fluid system assumed in Figure 9.63, we have input strongly discontinuous axial pressure gradients that differ by two orders of magnitude. This difference is needed because the two fluids have contrasting rheological properties. Moreover, the discontinuous pressure gradients are applied to the fluid system while it is moving and diffusing. The plots in Figure 9.64 give sectional properties at the azimuthal index "m = 19" (for the wide side of the annulus) with time

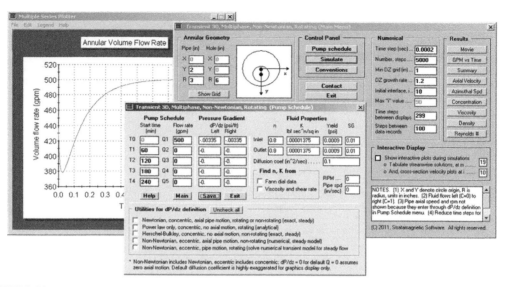

FIGURE 9.60

Hand calculation result for target 500 gpm.

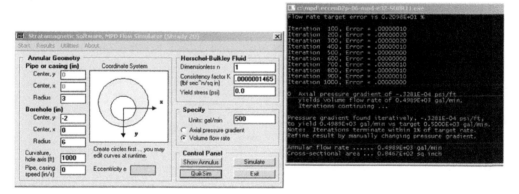

FIGURE 9.61

"Steady 2D" menu calculation for second fluid.

increasing as the figures progress downward. (For a color version of this and other figures, see this book's companion website.) The flow is moving from top to bottom. The left axial velocity plot correctly shows a uniformly lower "blue speed" for the non-Newtonian fluid, while the displaced Newtonian fluid is more colorful, with blues, yellows, oranges, and reds being indicative of the parabolic shape we expect. The viscosity plot, in fact, clearly shows how the mixing interface moves downward with time and widens.

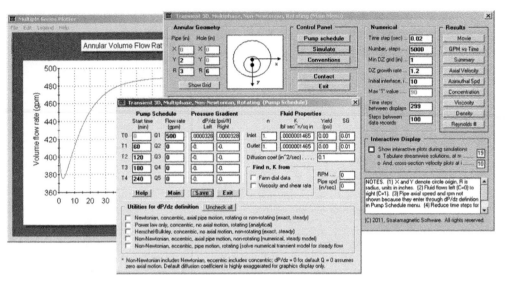

FIGURE 9.62

Newtonian flow model setup.

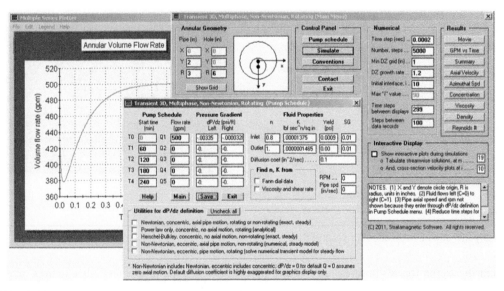

FIGURE 9.63

Two-fluid displacement and mixing flow setup.

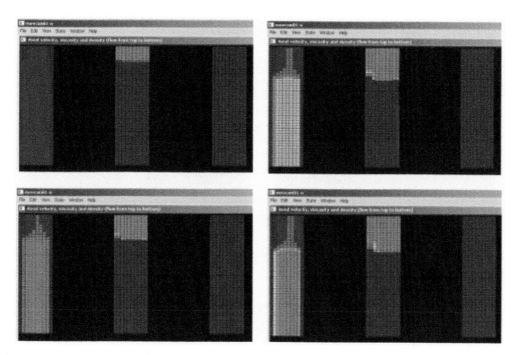

FIGURE 9.64

"Wide-side" axial velocity and fluid mixing.

DISCUSSION 9.9

Transient, Three-Dimensional Concentric Multiphase Analysis For Rotating Power Law Fluids without Axial Pipe Movement

In the present calculation, we demonstrate how the foregoing procedures apply when the host pressure solver is the host model for Power law fluids in concentric annuli. Figures 9.65(a) and 9.65(b) show two calculations for pressure gradient with identical volumetric flow rates and rotational speeds. The differences between the two are the fluid properties. The pressure gradients shown at the bottoms of the respective text output screens differ by a factor of 10.

Calculated results are shown in Figures 9.65(c) and 9.65(d). Here it is important to note that the input 100 gpm in the software screens of Figures 9.65(a) and 9.65(b) are not replicated in the line graph shown, although the "84" is not significantly different. The reason for this discrepancy lies in the nature of the simulator in Figures 9.65(a,b). By referring to the mathematics in Example 5.6, where simplifications to boundary condition implementation were made to enable closed-form analytical solutions that can be rapidly evaluated by computer.

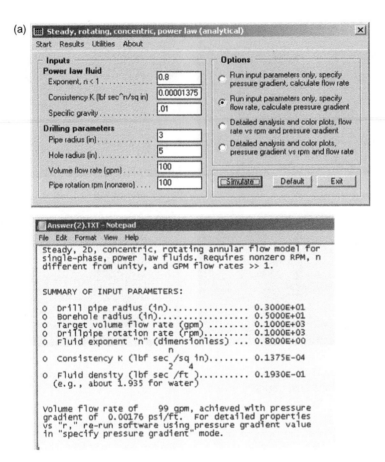

FIGURE 9.65a

Pressure gradient for thin Power law fluid.

DISCUSSION 9.10

Transient, Three-Dimensional Eccentric Multiphase Analysis for General Rotating Non-Newtonian Fluids with Axial Pipe Movement: Validation Runs for Constant-Rate Rotation and Translation

In this example, we consider a very complicated annular flow problem typical of those encountered in field operations. We study the highly eccentric annulus in Figure 9.66(b). The pipe or casing is moving in the direction of flow at 10 in./sec and simultaneously rotating at 100 rpm. The total volumetric flow rate is 100 gpm.

A Herschel-Bulkley fluid—again, one with nonzero yield stress—is entering at the inlet and displacing a 10-cp Newtonian fluid that is partially present in the annulus. The fluid system is initially quiescent. We wish to calculate the time-dependent axial and azimuthal velocities and apparent

(b)

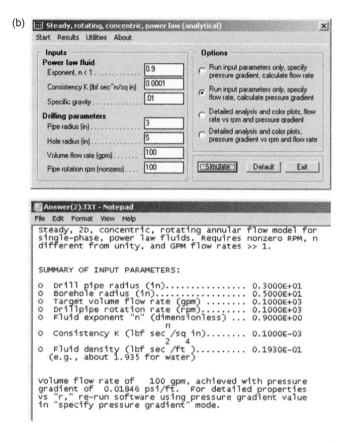

FIGURE 9.65b

Pressure gradient calculation for thick Power law fluid.

viscosity fields along with the position and mixing zone history associated with the fluid interface undergoing transient movement. Following the strategy developed in this chapter, we first calculate the (very different) pressure gradients present near the inlet and outlet and flow in single-phase manner. This is accomplished using our exact, transient two-dimensional solvers, a process that requires only seconds. Then both pressure gradients are used in the combined problem addressing convective and diffusive mixing to solve the questions posed in this paragraph. The two-dimensional solutions are fast, taking only seconds in computing time. We summarize our calculations.

Steady, rotating, non-Newtonian single-phase eccentric flow solution

We once again remind the reader that solutions for rotating eccentric flow problems using purely steady flow formulations are presently numerically unstable for parameters of drilling and cementing interest. However, solutions can be found by solving the transient problem asymptotically for

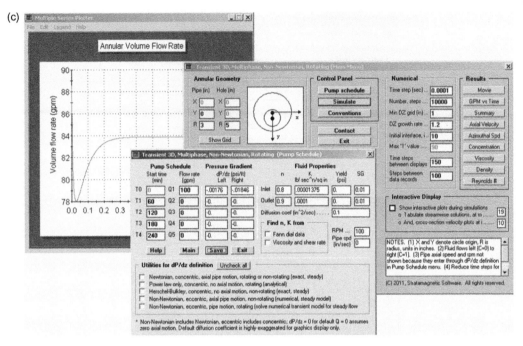

FIGURE 9.65c

Mixing calculation setup and results.

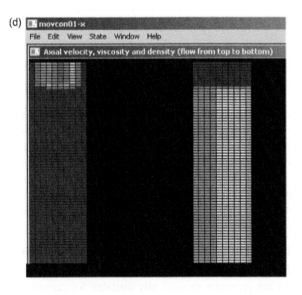

FIGURE 9.65d

Diffusion solutions in a problem with 100-rpm rotation.

(a)
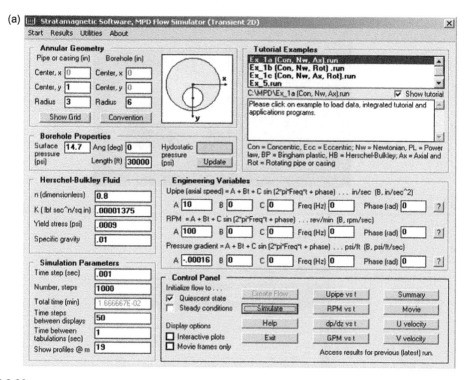

FIGURE 9.66a

Non-Newtonian, single-phase flow setup.

large times. This is possible using the "Transient 2D" simulator developed in this book. As noted in prior discussions, it is not possible, for mathematical reasons, to specify volumetric flow rate and obtain pressure gradient in a single pass. But because the two-dimensional solver is extremely fast, requiring only seconds or up to a minute per computation, we can determine pressure gradient by trial and error, entering various test values and "hand-converging" the solutions for the targeted 100 gpm. For the problem at hand, the author was able to complete the entire example in about 15 minutes of desk time. The input assumptions are shown in Figure 9.66(a). For the targeted flow rate of 100 gpm, the required axial pressure gradient is −0.00016 psi/ft, as indicated in Figures 9.66 (a) and 9.66(c).

Steady, rotating, Newtonian single-phase eccentric flow solution

For our Newtonian fluid (zero yield stress) with a 10-cp viscosity, the unsteady formulation in Figure 9.66(d) leads to a pressure gradient of −0.000026 psi/ft for the 100-gpm target. (The minus sign has scrolled to the left.) Entries hidden by the graph are all zero as in Figure 9.66(a). The axial velocity field is shown in Figure 9.66(e), with high (red) velocities at the pipe because the velocity in the pipe exceeds those in the annulus. There is no symmetry about the vertical line passing

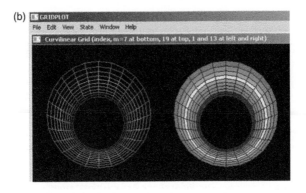

FIGURE 9.66b

Eccentric annulus.

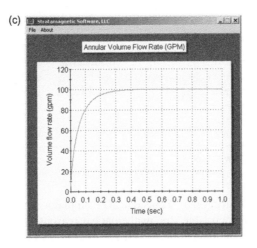

FIGURE 9.66c

Volumetric flow rate history for non-Newtonian fluid.

through the center because rotation destroys the symmetry. The azimuthal picture is similar to this one because the rotational speeds are highest at the pipe and vanish at the annular wall.

Mixing problem

Now we solve the problem for the combined fluids using the "Zoom3D" solver shown at the top of Figure 9.66(f). The target flow rate of 100 gpm is achieved with a 2 percent error. For the parameters indicated, about one minute of computing time is required. To create the color profiles shown in Figure 9.67, the 10,000-step run selected requires about 10 minutes. In the screen

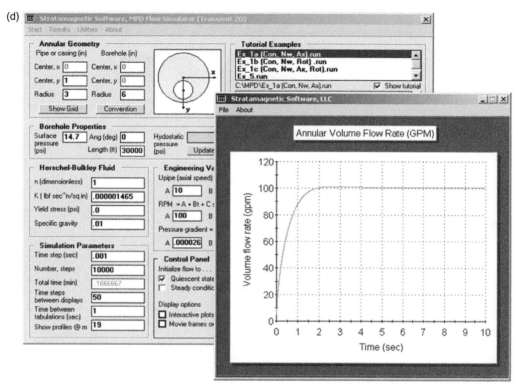

FIGURE 9.66d

Newtonian flow formulation and solution.

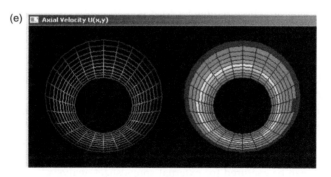

FIGURE 9.66e

Axial velocity profile in rotating flow.

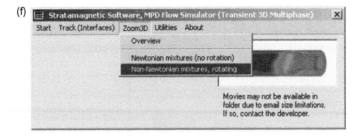

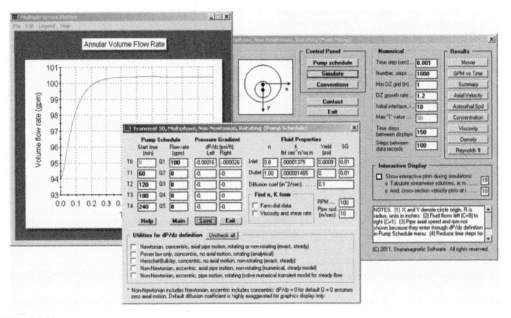

FIGURE 9.66f

Transient, three-dimensional, two-phase mixture formulation.

captures given, time increases downward from frame to frame; each snapshot displays the "m = 19" azimuthal solution selected in Figure 9.66(f), which shows axial velocity and apparent viscosity in the "streamwise-radial" plane.

The initial position index of the interface is 10 out of a maximum 90 grids in the direction of flow. In these snapshots, the flow moves downward, and the interface is seen progressing downward. As expected, diffusion causes this interface to widen with time. Clicking the right-side buttons in Figure 9.66f leads to numerical output captured in text files, as shown in Figures 9.68, 9.69, 9.70, and 9.71, which can be captured for external spreadsheet analysis.

Note that the very low Reynolds numbers in Figure 9.71 indicate fluid stability on a single-phase-flow basis. The interface in Figure 9.67 is seen to widen gradually as it convects downward. Our analysis does not include computations for interfacial stability; an extremely difficult problem

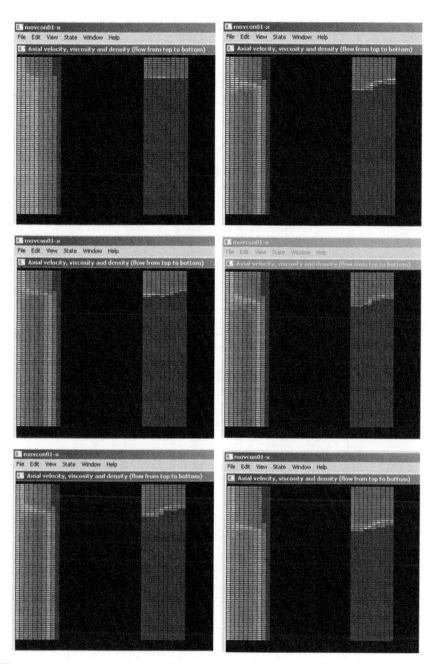

FIGURE 9.67

Axial velocity (*left* of each frame); apparent viscosity (*right* of each frame). The flow moves downward in each frame; time increases downward from frame to frame.

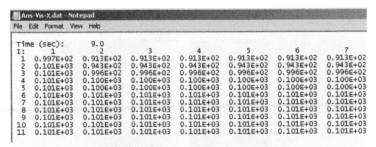

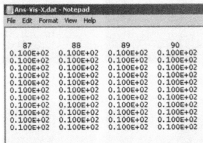

FIGURE 9.68

Apparent viscosity for "constant m" or azimuthal angle.

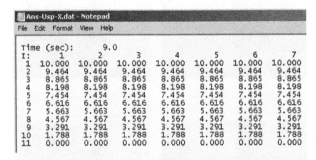

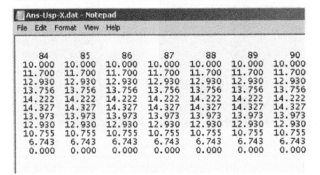

FIGURE 9.69

Axial velocity solution.

```
Ans-Vth-X.dat - Notepad
File  Edit  Format  View  Help

Time (sec):      9.0
I:      1           2           3           4           5           6           7
 1   0.314E+02   0.314E+02   0.314E+02   0.314E+02   0.314E+02   0.314E+02   0.314E+02
 2   0.277E+02   0.277E+02   0.277E+02   0.277E+02   0.277E+02   0.277E+02   0.277E+02
 3   0.242E+02   0.242E+02   0.242E+02   0.242E+02   0.242E+02   0.242E+02   0.242E+02
 4   0.208E+02   0.208E+02   0.208E+02   0.208E+02   0.208E+02   0.208E+02   0.208E+02
 5   0.176E+02   0.176E+02   0.176E+02   0.176E+02   0.176E+02   0.176E+02   0.176E+02
 6   0.145E+02   0.145E+02   0.145E+02   0.145E+02   0.145E+02   0.145E+02   0.145E+02
 7   0.114E+02   0.114E+02   0.114E+02   0.114E+02   0.114E+02   0.114E+02   0.114E+02
 8   0.851E+01   0.851E+01   0.851E+01   0.851E+01   0.851E+01   0.851E+01   0.851E+01
 9   0.564E+01   0.564E+01   0.564E+01   0.564E+01   0.564E+01   0.564E+01   0.564E+01
10   0.281E+01   0.281E+01   0.281E+01   0.281E+01   0.281E+01   0.281E+01   0.281E+01
11   0.000E+00   0.000E+00   0.000E+00   0.000E+00   0.000E+00   0.000E+00   0.000E+00
```

FIGURE 9.70

Azimuthal velocity solution.

```
Ans-Rey-X.dat - Notepad
File  Edit  Format  View  Help

Time (sec):      9.0
I:      1           2           3           4           5           6           7
 1   0.192E+01   0.209E+01   0.209E+01   0.209E+01   0.209E+01   0.209E+01   0.209E+01
 2   0.180E+01   0.192E+01   0.192E+01   0.192E+01   0.192E+01   0.192E+01   0.192E+01
 3   0.168E+01   0.170E+01   0.170E+01   0.170E+01   0.170E+01   0.170E+01   0.170E+01
 4   0.156E+01   0.156E+01   0.156E+01   0.156E+01   0.156E+01   0.156E+01   0.156E+01
 5   0.141E+01   0.142E+01   0.142E+01   0.142E+01   0.142E+01   0.142E+01   0.142E+01
 6   0.125E+01   0.126E+01   0.126E+01   0.126E+01   0.126E+01   0.126E+01   0.126E+01
 7   0.107E+01   0.107E+01   0.107E+01   0.107E+01   0.107E+01   0.107E+01   0.107E+01
 8   0.866E+00   0.864E+00   0.864E+00   0.864E+00   0.864E+00   0.864E+00   0.864E+00
 9   0.624E+00   0.622E+00   0.622E+00   0.622E+00   0.622E+00   0.622E+00   0.622E+00
10   0.339E+00   0.338E+00   0.338E+00   0.338E+00   0.338E+00   0.338E+00   0.338E+00
11   0.000E+00   0.000E+00   0.000E+00   0.000E+00   0.000E+00   0.000E+00   0.000E+00
```

FIGURE 9.71

Reynolds number solution.

is formulated and solved rigorously. Finally, we emphasize that large diffusion coefficients were assumed only for visualization purposes so that fluid movement could be seen using our somewhat crude 12-color plotter. Also, very small specific gravities were taken in order for our transient results to approach steady conditions quickly. In general, smaller time steps will be required for higher fluid densities and rotational rates.

Closing Remarks

10

In this book, and particularly in Chapter 9, the objectives of our flow simulation efforts, focusing on the complete system for drilling and cementing shown in Figure 10.1, were brought to closure. In broad terms, the technical objectives are easily expressed:

- Finding pressures everywhere
- Allowing a general pump schedule for non-Newtonian fluids
- Supporting pipe and casing that may be rotating or moving axially in any transient combination
- Permitting real-world rheologies that may lead to all-important plug flows associated with yield stress fluids

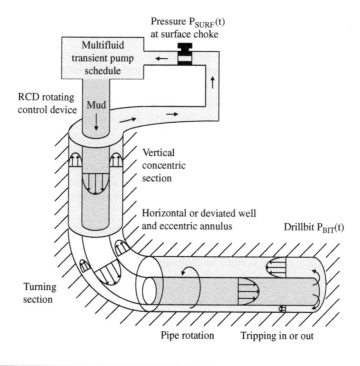

FIGURE 10.1

General system for drilling and cementing.

Managed Pressure Drilling: Modeling, Strategy and Planning
© 2012 Elsevier Inc. All rights reserved.

- If need be, determining interfacial mixing details for applications such as cementing in which diffusion and miscible mixing can be significant.

Operational objectives included "managed pressure drilling," where pressures along the borehole and especially at the drillbit are required as functions of time, and cementing and completions, in which mixing details are needed to assess potential problems with zonal isolation.

The overall problem strategy devised was simple conceptually: develop fundamental "building blocks" that, in themselves, represent useful simulation tools but, when assembled, address the transient, three-dimensional multiphase problem in a general manner. First we developed a "Steady 2D" capability that permits exact numerical modeling of single-phase, non-Newtonian flow in general eccentric annular cross sections.

This required us to use boundary-conforming curvilinear grids to model the annulus, refining the ideas first reported by the author in *Borehole Flow Modeling* (1992) and *Computational Rheology* (2001). Also, stable and highly accurate methods were designed to handle apparent viscosity, particularly its spatial derivatives, and "extended Herschel-Bulkley" constitutive relationships were employed to reach across and into plug zones for fluids with yield stress, allowing complete determination of plug zone size, shape, and location for accurate pressure drop analysis. In addition, new modeling capabilities included axial pipe movement, centrifugal effects due to borehole axis curvature, and, finally, pressure gradient versus flow rate specification, all handled without approximation.

The fast and robust "Steady 2D" building block simulator was augmented with integrated graphics that automatically display field properties such as axial velocity, apparent viscosity, shear rate, and viscous stress—quantities that have proven useful for engineering correlation in specific applications. Substantial effort was expended to design a simulation interface that was easy to use, requiring no mathematics or computational expertise and enabling fast, accurate solutions the first time and every time. The methods provide new ways to accurately study swab-surge, hole cleaning, and pressure drop analysis.

Our second building block is encapsulated in the "Transient 2D" simulator. Aside from its obvious transient applications—for example, fully unsteady axial reciprocation, pipe rotation, pumping rate taken in any combination, and helical cuttings transport—this method provides the first generally available means to study the effects of pipe rotation in eccentric annuli with non-Newtonian flow. Importantly, this capability implies more than academic interest. We demonstrated the role of eccentricity in pressure control while rotating. For instance, when pressure gradient is fixed, shear thinning in concentric flows leads to increased flow rate; new convective terms that appear in eccentric problems usually lead to decreased throughput.

The foregoing results, demonstrated in numerous calculations and consistent with laboratory and field observations for concentric and eccentric applications, indicate that drillpipe rotation can be used for real-time pressure control in managed pressure drilling. Presently, three means are typically employed:

1. Change in pump rate to effect dynamic friction, a method that may be dangerous, since strong transients are involved.
2. Altering mud rheology and weight, a slow process that may not be responsive to danger indicators.
3. Direct choke control to adjust background pressures.

Drillpipe rotation now provides a fourth means. It is an easy-to-implement procedure that can be modeled conveniently and whose slower transients are unlikely to induce fracturing or rapid influx or outflux.

In addition to the "Steady 2D" and "Transient 2D" building blocks, we developed many "utilities" addressing flows in more idealized geometries for rapid and accurate pressure analysis. These include, among others,

- Analytical solutions for rotating Power law flows in concentric annuli
- Exact Herschel-Bulkley concentric solutions for flows with stationary boundaries
- Newtonian solutions for fluid flow past concentric annuli with axial pipe movement
- Recirculating flow in the presence of barite sag

Taken together, a broad collection of simulators was developed, and all were assembled for multiple applications.

Importantly, in Chapters 8 and 9 we demonstrated how the foregoing building blocks can be used to model the fully transient flow of multiple non-Newtonian fluids down the drillpipe following a general pump schedule. We studied pressures both along the borehole and at the drillbit as a function of time, and showed how, when required, interfacial details related to diffusion and flow convection can be obtained with fully coupled momentum and species concentration models in a reasonable amount of computing time.

Validation questions invariably arise with simulator use. Engineers ask, "How have computed results been validated?" As far as math is concerned, our simpler utility models were used to validate more complicated models. In fact, in areas of common overlap, models developed under different general assumptions and solved with contrasting numerical methods often differed by no more than 2 to 3 percent without "fudging." Solutions for simple geometries were used to check curvilinear coordinate approaches, two-dimensional solutions validated three-dimensional approaches, and steady curvilinear grid solutions were shown to be in agreement with transient solutions in the limit of large time.

Our efforts at validation did not end with mathematics; computational consistency was merely the beginning. Starting with *Borehole Flow Modeling* (1992) and *Computational Rheology* (2001), we analyzed in detail cuttings transport data obtained in laboratory flow loops, spotting fluid results in jarring applications, vortex formation in flows with barite sag, long boreholes with bends, and the like, with careful observation being the only final arbiter. A major accomplishment of the present research is accurate modeling of pipe or casing rotation in eccentric geometries. Consistency with field observation (see, for example, Figure 2.2) provided a high degree of credibility.

Our multiphase efforts were also validated through experiment. For example, lab setups and results, duplicated in Figures 10.2 and 10.3, were reported in detail in Deawwanich, Liew, Nguyen, Savery, Tonmukayakul, and Chin (2008); Nguyen, Deawwanich, Tonmukayakul, Savery, and Chin (2008); and Savery, Tonmukayakul, Chin, Deawwanich, Liew, and Nguyen (2008). The math model in these papers applied to general transient, multiphase three-dimensional flow.

Some more recent work expands on the earlier method by providing a fast means to calculate borehole pressures using our "Steady 2D" and "Transient 2D" building blocks, the latter providing the first mathematically rigorous approach dealing with rotating pipe flows. In addition, newer "zoom" capabilities for interfacial mixing stably and accurately model the suddenly changing pressure gradients acting on contiguous fluids without using cruder approximations.

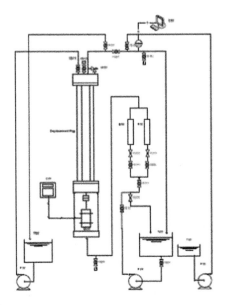 

FIGURE 10.2

Multiphase flow visualization experimental setup.

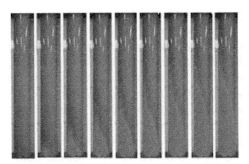 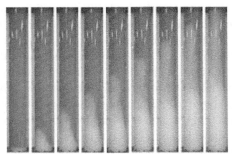

FIGURE 10.3

Typical eccentric flow mixing (time increases left to right).

Needless to say, our modeling efforts will not end here. In any research, more questions are raised than are answered, and our research is no exception. By reporting our work in complete mathematical and numerical detail and making the simulators available for wide dissemination, we hope that the experiences and comments of users will help us accelerate our progress in addressing a very challenging and interesting technical and operational problem.

Cumulative References

Anderson, A.G., 1961. Sedimentation. In: Streeter, V.L. (Ed.), Handbook of Fluid Mechanics. McGraw-Hill, New York.

Batchelor, G.K., 1970. Fluid Dynamics, Cambridge University Press, Cambridge.

Becker, T.E., Azar, J.J., Okrajni, S.S., October 8–11, 1989. Correlations of Mud Rheological Properties with Cuttings Transport Performance in Directional Drilling, Paper No. 19535, SPE 64th Annual Technical Conference and Exhibition, Society of Petroleum Engineers, San Antonio.

Becker, T.E., Morgan, R.G., Chin, W.C., Griffith, J.E., March 22–25, 2003. Improved Rheology Model and Hydraulics Analysis for Tomorrow's Wellbore Fluid Applications, Paper 82415, SPE Production and Operations Symposium, Oklahoma City.

Bird, R.B., Armstrong, R.C., Hassager, O., 1987. Dynamics of Polymeric Liquids, Volume 1: Fluid Mechanics, John Wiley & Sons, New York.

Bird, R.B., Stewart, W.E., Lightfoot, E.N., 2002. Transport Phenomena, second ed. John Wiley & Sons, New York.

Chhabra, R.P., Richardson, J.F., 1999. Non-Newtonian Flow in the Process Industries, Butterworth-Heinemann, Oxford.

Chin, W.C., 1992. Borehole Flow Modeling in Horizontal, Deviated and Vertical Wells, Gulf Publishing, Houston.

Chin, W.C., 2001. Computational Rheology for Pipeline and Annular Flow, Elsevier Science, London.

Chin, W.C., 2002. Quantitative Methods in Reservoir Engineering, Elsevier Science, London.

Chin, W.C., Zhuang, X., June, 2010. Exact Non-Newtonian Flow Analysis of Yield Stress Fluids in Highly Eccentric Borehole Annuli with Pipe or Casing Translation and Rotation, Paper 131234-PP, CPS/SPE International Oil & Gas Conference and Exhibition, Beijing.

Chin, W.C., Zhuang, X., April 12–14, 2011(a). Effect of Rotation on Flowrate and Pressure Gradient in Eccentric Holes, Paper AADE-11-NTCE-45, AADE National Technical Conference and Exhibition, Houston.

Chin, W.C., Zhuang, X., April 12–14, 2011(b). Advances in Swab-Surge Modeling for Managed Pressure Drilling, Paper AADE-11-NTCE-46, AADE National Technical Conference and Exhibition, Houston.

Chin, W.C., Zhuang, X., April 12–14, 2011(c). Transient, Multiphase, Three-Dimensional Pumping Models for Cementing and Drilling, Paper AADE-11-NTCE-72, AADE National Technical Conference and Exhibition, Houston.

Chin, W.C., Zhuang, X., April 12–14, 2011(d). Comprehensive Annular Flow Models for Drilling and Completions, Paper AADE-11-NTCE-73, AADE National Technical Conference and Exhibition, Houston.

Chin, W.C., Zhuang, X., May 2011(e). Advances in Swab-Surge Modeling for Managed Pressure Drilling, Paper OTC-21115-PP, Offshore Technology Conference, Houston.

Cole, J.D., 1968. Perturbation Methods in Applied Mathematics, Blaisdell Publishing, Waltham, MA.

Deawwanich, T., Liew, J.C., Nguyen, Q.D., Savery, M., Tonmukayakul, N., Chin, W.C., September 28–October 1, 2008. Displacement of Viscoplastic Fluids in Eccentric Annuli: Numerical Simulation and Experimental Validation," paper presented at the Chemeca Conference, Newcastle, Australia.

Dodge, D.W., Metzner, A.B., 1959. Turbulent flow of non-Newtonian systems. AICE Journal 5(2), 189–204.

Escudier, M.P., Gouldson, I.W., Oliveira, P.J., Pinho, F.T., 2000. Effects of inner cylinder rotation on laminar flow of a Newtonian fluid through an eccentric annulus. Int. J. Heat and Fluid Flow 21, 92–103.

Fang, P., Manglik, R.M., 2002. The influence of inner cylinder rotation on laminar axial flows in eccentric annuli of drilling bore wells. Int. J. Transp. Phenomena 4, 257–274.

Fraser, L.J., February–March 1990. Field Application of the All-Oil Drilling Fluid Concept, Paper No. 19955, IADC/SPE Drilling Conference, Houston.

Fraser, L.J., March 1990. Green canyon drilling benefits from all oil mud. Oil and Gas J. 33–39.

Fraser, L.J., November, 1990. Effective ways to clean and stabilize high-angle hole. Petroleum Engineer Int., 30–35.

Fredrickson, A.G., Bird, R.B., 1958. Flow of non-Newtonian fluids in annuli. Ind. Eng. Chem. 50, 347–352.

Halliday, W.S., and Clapper, D.K., February 28–March 3, 1989. Toxicity and Performance Testing of Non-Oil Spotting Fluid for Differentially Stuck Pipe, Paper No. 18684, SPE/IADC Drilling Conference, New Orleans.

Hussain, Q.E., Sharif, M.A.R., 2000. Numerical modeling of helical flow of viscoplastic fluids in eccentric annuli. AIChE Journal 46 (10), 1937–1946.

Ibraheem, S.O., Adewumi, M.A., Savidge, J.L., 1998. Numerical simulation of hydrate transport in natural gas pipeline. J. Energy Resources Technol., Trans. ASME 120, 20–26.

Kapfer, W.H., 1973. Flow of sludges and slurries. In: King, R. (Ed.), Piping Handbook. McGraw-Hill, New York.

Landau, L.D., Lifschitz, E.M., 1959. Fluid Mechanics, Pergamon Press, London.

Milne-Thomson, L.M., 1968. Theoretical Hydrodynamics, fifth ed. McMillan Company, New York.

Morrison, F.A., 2001. Understanding Rheology, Oxford University Press, Oxford.

Nayfeh, A., 1973. Perturbation Methods, John Wiley & Sons, New York.

Nguyen, Q.D., Deawwanich, T., Tonmukayakul, N., Savery, M.R., Chin, W.C., August 2008. Flow Visualization and Numerical Simulation of Viscoplastic Fluid Displacements in Eccentric Annuli, paper presented at the XVth International Congress on Rheology (ICR), Society of Rheology 80th Annual Meeting, Monterey, CA.

Outmans, H.D., 1958. Mechanics of differential pressure sticking of drill collars. Petroleum Trans., AIME 213, 265–274.

Press, W.H., Teukolsky, S.A., Vetterling, W.T., Flannery, B.P., 1992. Numerical Recipes, Cambridge University Press, Cambridge, UK.

Quigley, M.S., Dzialowski, A.K., Zamora, M., February 27–March 2, 1990. A Full-Scale Wellbore Friction Simulator, Paper No. 19958, IADC/SPE Drilling Conference, Houston.

Savery, M., Chin, W.C., Babu Yerubandi, K., April 2008. Modeling Cement Placement Using a New Three-Dimensional Flow Simulator, Paper AADE-08-DF-HO-08, 2008 AADE Fluids Conference and Exhibition, American Association of Drilling Engineers, Houston.

Savery, M., Darbe, R., Chin, W.C., April 30–May 3, 2007. Modeling Fluid Interfaces During Cementing Using a Three-Dimensional Mud Displacement Simulator, OTC Paper 18513, Offshore Technology Conference, Houston.

Savery, M., Tonmukayakul, P., Chin, W.C., Deawwanich, T., Liew, J., Q. D. Nguyen, August 2008. Laminar Displacement of Viscoplastic Fluids in Eccentric Annuli–Numerical Simulation and Experimental Validations, paper presented at the XXII International Congress of Theoretical and Applied Mechanics (ICTAM), Adelaide.

Savins, J.G., 1958. Generalized Newtonian (pseudoplastic) flow in stationary pipes and annuli. Petroleum Trans., AIME 213, 325–332.

Savins, J.G., Wallick, G.C., 1966. Viscosity profiles, discharge rates, pressures, and torques for a rheologically complex fluid in a helical flow. AIChE Journal 12(2), 357–363.

Schlichting, H., 1968. Boundary Layer Theory, McGraw-Hill, New York.

Seeberger, M.H., Matlock, R.W., Hanson, P.M., February 28–March 3, 1989. Oil Muds in Large Diameter, Highly Deviated Wells: Solving the Cuttings Removal Problem, Paper No. 18635, SPE/IADC Drilling Conference, New Orleans.

Skelland, A.H.P., 1967. Non-Newtonian Flow and Heat Transfer, John Wiley & Sons, New York.

Slattery, J.C., 1981. Momentum, Energy, and Mass Transfer in Continua, Robert E, Krieger Publishing, New York.

Souza Mendes, P.R., Braga, A.M.B., Azevedo, L.F.A., Correa. K.S., September 1999. Resistive force of wax deposits during pigging operations. J. Energy Resources Technol., Trans. ASME 121, 167–171.

Souza Mendes, P.R., Dutra, E.S.S., 2004. A viscosity function for viscoplastic liquids. Annu. Trans. Nordic Rheology Soc. vol. 12.

Streeter, V.L., 1961. Handbook of Fluid Dynamics, McGraw-Hill, New York.

Tamamidis, P., Assanis, D.N., 1991. Generation of orthogonal grids with control of spacing. J. Comput. Phys. 94, 437–453.

Thompson, J.F., 1984. Grid Generation Techniques in Computational Fluid Dynamics. AIAA Journal 22(11), 1505–1523.

Thompson, J.F., Warsi, Z.U.A., Mastin, C.W., 1985. Numerical Grid Generation, Elsevier Science, New York,

Turner, J.S., 1973. Buoyancy Effects in Fluids, Cambridge University Press, London.

van Dyke, M., 1964. Perturbation Methods in Fluid Mechanics, Academic Press, New York.

Walton, I.C., Bittleston, S.H., 1991. The axial flow of a bingham plastic in a narrow eccentric annulus. J. Fluid Mech. 222, 39–60.

Yih, C.S., 1960. Exact solutions for steady two-dimensional flow of a stratified fluid. J. Fluid Mech. 9, 161–174.

Yih, C.S., 1969. Fluid Mechanics, McGraw-Hill, New York.

Yih, C.S., 1980. Stratified Flows, Academic Press, New York.

Index

Printed and bound by CPI Group (UK) Ltd, Croydon, CR0 4YY

03/10/2024

01040301-0001